FUNDAMENTALS OF ASTRONOMY

FUNDAMENTALS OF ASTRONOMY

C BARBIERI
University of Padua, Italy

Taylor & Francis
Taylor & Francis Group
New York   London

Taylor & Francis is an imprint of the
Taylor & Francis Group, an informa business

Published in 2007 by
CRC Press
Taylor & Francis Group
6000 Broken Sound Parkway NW, Suite 300
Boca Raton, FL 33487-2742

International Standard Book Number-10: 0-7503-0886-9 (Softcover)
International Standard Book Number-13: 978-0-7503-0886-1 (Softcover)
Library of Congress Card Number 2006003157

Library of Congress Cataloging-in-Publication Data

Barbieri, Cesare, 1942-
     Fundamentals of astronomy / Cesare Barbieri.
          p. cm.
     Includes bibliographical references and index.
     ISBN-13: 978-0-7503-0886-1 (alk. paper)
     ISBN-10: 0-7503-0886-9 (alk. paper)
     1. Astronomy--Textbooks. 2. Astrophysics--Textbooks. I. Title.

QB43.3.B37 2006
520--dc22   2006003157

Taylor & Francis Group
is the Academic Division of Informa plc.

**Visit the Taylor & Francis Web site at
http://www.taylorandfrancis.com**

**and the CRC Press Web site at
http://www.crcpress.com**

To Gianna

Preface

Fundamentals of Astronomy originates from the introductory course to astronomy and astrophysics given at the University of Padova. The book provides a broad overview of several classical topics, from spherical astronomy to celestial mechanics, closing with two chapters giving elements of astronomical photometry and spectroscopy. Explanatory notes and exercises have been inserted at the end of many chapters.

The selection of the arguments has been guided by my conviction that classical astronomy deserves a basic role in the education of a serious student. Therefore, spherical astronomy is presented in great detail. To some students, this material can appear obsolete and unduly formal, and therefore, discouraging. However, the possibility of performing exacting measurements from the ground and from space has revived practical interest in astrometry. Very long baseline interferometry at radio wavelength, excellent meridian circles such as the Carlsberg, the recent Hipparcos satellite, the ambitious projects of the near future, such as Gaia, require fresh treatment of the very fundamentals of coordinates and time measurements. Yet, while broad in coverage, this book does not aim to reach the level of theoretical knowledge or the numerical accuracy required by these future applications. In particular, general relativity will be mentioned only occasionally. However, I hope that this introduction, with the indication of its limits when appropriate, will stimulate young researchers to go beyond the present treatment, in order to bring personal contributions to this developing field.

Obviously, some important arguments have been left out. The decision not even to attempt to cover in a single book everything in astronomy and astrophysics, is undoubtedly facilitated by the availability in the astronomical literature of excellent textbooks at all levels. In the Bibliography I have recommended a mixture of old and new books, selected by personal taste, again with the intention to stimulate further reading.

During the writing of the Italian version of my lectures, a great revolution happened in the way science is made available; the Internet gives easy access to both well-established and developing knowledge. Many research and educational institutions provide excellent websites where the reader can follow the unfolding of experiments and observations, almost in real time. Navigation of the Internet takes the reader through an incredible number of sites, some of which are very good, but many are very bad, and even deceiving. Therefore, I have indicated some of the sites I find reliable and valuable to science and education. I have not attempted in any way to present a complete list. The student will quickly develop the capability to judge what is most useful.

It must be underlined that during the final editing of the text, the *Astronomical Almanac* carried out a revision of its content, which will take effect starting in 2006. Some of these changes have been briefly mentioned.

Acknowledgments

I wish to thank all the students who used the Italian edition, discovered many errors, and asked for better clarity. My hope is that the present version will provide a useful textbook for students in other countries.

My warmest gratitude goes to Michael Mendillo, of Boston University, for his kind hospitality in the Center for Space Physics, where most of the writing took place.

Author

Cesare Barbieri is professor of astronomy at the University di Padova. He directed the Astronomical Observatory of Padova, and the construction of the 3.5 m Telescopio Nazionale Galileo (Canary Islands). He held several positions in the European Space Agency and European Southern Observatory advisory committees, with an active role in several space missions (GIOTTO to Comet Halley, the Faint Object Camera on the Hubble Space Telescope, the imaging system OSIRIS on board ROSETTA).

His main research themes are: astrometry, photometry and discovery of quasi-stellar objects; telescopes and their instrumentation for ground and space; spectrophotometry and discovery of asteroids; the exosphere of Mercury. More recently, he became interested in quantum optics applied to astronomy. He has organized several international conferences and exhibits. He has received the NASA Group Award for the FOC/HST, and the Gold Medal of Italian Ministry for Public Education.

Contents

Chapter 10 The Astronomical Times

Chapter 11 The Terrestrial Atmosphere

Chapter 12 The Two-Body Problem

1

Spherical Astronomy

The name spherical astronomy indicates the systematic and formal representation of the apparent positions and motions of the heavenly bodies such as they appear projected on the celestial sphere. The name astrometry indicates the several methods of measurements and reductions. Dynamical astronomy studies those positions and motions in the Cartesian space, and investigates the responsible forces; the name celestial mechanics is also used for this discipline.

In this chapter, we begin the study of spherical astronomy, using essentially only geometric concepts, and without considering the physical bases of the phenomena. Historically, this is the way astronomy began several millennia ago, leading to the geocentric conception of the world that found its maximum expression in the Ptolemaic theory, which survived until Copernicus (15th century). This geocentric vision is fully adequate for the present purposes. In subsequent chapters, when we shall investigate the dynamics of planets, asteroids, comets, or finally the motions of the stars with respect to the Sun and to the entire Milky Way, we will be led to the operational definition of reference frames that are as inertial as possible, in order to correctly apply the Newtonian dynamical laws. Occasionally, we will discuss the inadequacies of the Newtonian theory and introduce some concepts of general relativity.

The few notions of plane and spherical trigonometry needed in the rest of the book are given in the following sections.

1.1 Elements of Plane Trigonometry

In theoretical expressions, such as trigonometric functions, plane angles must be expressed in radians, but units such as sexagesimal degrees (symbol $°$, some authors would rather say *nonagesimal* degrees), sexagesimal minutes of arc (denoted by the symbol $'$, $1° = 60'$) and seconds of arc (denoted by the symbol $''$, $1' = 60''$) are employed in practical applications. Notice that in geodetic applications, centesimal degrees (400 to the circle) are sometimes employed, but we will avoid this system.

Given that $\pi/2$ radians correspond to $90°$, one radian is:

$$1 \text{ radian} \approx 57°.2957795 \approx 3437'.74677 \approx 206264''.806$$

Inversely:

$$1'' \approx 0.000004848 \text{ radians}, \quad 1' \approx 0.000290888 \text{ radians},$$

$$1° \approx 0.017453292 \text{ radians}$$

Therefore for instance:

$$\text{angle } \theta \text{ (radians)} = \theta''/206264.806, \quad \theta'' = 206264.806\theta \text{ (radians)}$$

With a repeated application:

$$\theta = \frac{\theta''}{206264.806} \text{ radians}, \quad \theta^n = \frac{(\theta'')^n}{(206264.806)^n},$$

$$(\theta'')^n = (\theta'')\left(\frac{(\theta'')}{206264.806}\right)^{n-1} = (\theta'')\theta^{n-1}$$

The constant $R'' = 206264.806$, often written as $R'' = 1/\sin 1''$, will appear frequently throughout the book.

The reason for the difference between theoretical and practical units is the following: by definition, the radian is the arc equal to the radius, but the circumference is not a rational multiple of the arc. However, in order to perform measurements with graduated circles, or digital encoders, rational units are needed.

The ($° \, ' \, ''$) system is not the only possible one; the other system used in astronomy is that of hours (^{h}), minutes (^{m}) and seconds (^{s}); although the names are equal, those units represent angles, not times! Given that the circumference is divided in 24^h, the conversion factors are:

$$24^h = 360° = 2\pi \text{ radians}, \quad 1^h = 15° \approx 0.26179935 \text{ radians}$$

$$4^m = 1° \approx 0.017453292 \text{ radians}, \quad 1^s = 15'' \approx 0.000072722 \text{ radians}$$

Some useful formulae of plane trigonometry are the following:

$$\sin(\alpha \pm \beta) = \sin\alpha\cos\beta \pm \cos\alpha\sin\beta \tag{1.1}$$

$$\cos(\alpha \pm \beta) = \cos\alpha\cos\beta \mp \sin\alpha\sin\beta \tag{1.2}$$

$$\tan(\alpha \pm \beta) = \frac{\tan\alpha \pm \tan\beta}{1 \mp \tan\alpha\tan\beta} \tag{1.3}$$

which, in particular, give the duplication formulae when $\alpha = \beta$.

Frequent use will be made of the series expansions:

$$\sin\theta = \frac{e^{i\theta} - e^{-i\theta}}{2i} = \theta - \frac{1}{3!}\theta^3 + \frac{1}{5!}\theta^5 - \cdots + (-1)^k\frac{\theta^{2k+1}}{(2k+1)!} + \cdots \quad (\text{all } \theta's) \tag{1.4}$$

$$\cos\theta = \frac{e^{i\theta}+e^{-i\theta}}{2} = 1 - \frac{1}{2!}\theta^2 + \frac{1}{4!}\theta^4 - \cdots + (-1)^k\frac{\theta^{2k}}{(2k)!} + \cdots \quad \text{(all } \theta's) \quad (1.5)$$

$$\tan\theta = \frac{1}{i}\frac{e^{2i\theta}-1}{e^{2i\theta}+1} = \theta + \frac{1}{3}\theta^3 + \frac{2}{15}\theta^5 + \cdots (-1)^{n-1}\frac{2^{2n}(2^{2n}-1)B_{2n}\theta^{2n-1}}{(2n)!}$$
$$+ \cdots (-\pi/2 < \theta < \pi/2) \quad (1.6)$$

$$e^{\sin\theta} = 1 + \theta + \frac{1}{2}\theta^2 - \frac{1}{6}\theta^4 + \cdots$$

(where i is the imaginary unit, and B_{2n} are Bernoulli's numbers), together with their inverse expressions:

$$\theta = \sin\theta + \frac{1}{6}\sin^3\theta + \frac{3}{40}\sin^5\theta + \frac{5}{112}\sin^7\theta + \cdots \quad (1.7)$$

$$\theta = \tan\theta - \frac{1}{3}\tan^3\theta + \frac{1}{5}\tan^5\theta - \frac{1}{7}\tan^7\theta + \cdots \quad (1.8)$$

These formulae show that the first differences between the functions $\sin\theta$ and $\tan\theta$ and the arc θ, are quantities of third degree (and of opposite sign), while the function $\cos\theta$ differs from 1 by a term of second degree. When the angles are very small it is therefore legitimate to put arc θ equal to $\sin\theta$ or $\tan\theta$, and with less precision $\cos\theta$ equal to 1 (note, it is very inefficient to calculate the angle from its cosine for small angles).

We have already said that when approximating the arc with its sin (or with the tan), the function restitutes the value of the arc in radians. To obtain it in seconds of arc, we must multiply the value in radians by $206264''.806$. For instance, if $\sin a = 0.00141$, then $a = 0.00141 \times 206264''.806 = 290''.83$.

Let us see an example where practical and theoretical units are mixed together: suppose we wish to determine the maximum value of θ for which $\tan\theta - \theta \leq 1''$. We obtain:

$$\tan\theta - \theta \approx \frac{\theta^3}{3} \leq \frac{1}{206264.806}, \quad \theta \leq \sqrt[3]{3/206264.806} \approx 0.02441 \text{ radians,}$$

$$\theta'' = R''\theta \leq 5034''.9 \approx 1°24'$$

Another example is the calculation of the Taylor expansion of a function $y = f(x)$ using small finite increments $\Delta\theta$:

$$\Delta f = \frac{df}{d\theta}\Delta\theta + \frac{1}{2!}\frac{d^2 f}{d\theta^2}\Delta^2\theta + \cdots$$

Suppose we have chosen $\Delta\theta = 50'' = 0.00024$ radians. The square of the increment is $(\Delta\theta)^2 = 0.00024 \times 50'' = 0''.012$ (notice, arcsec, *not* arcsec square!). Ignoring the second and higher terms thus produces an error of $\pm 0''.01$ on the function (provided the amplitude of the derivatives does not exceed unity); should a precision of $0''.1$ on the function be sufficient, we could raise the increment on the variable to $150''$, and to $500''$ if an error of $\pm 1''.0$ is unimportant. These considerations are also applicable when deriving the value of the angles from expansions of trigonometric formulae; for instance, near $0°$ the error on the angle θ can be much greater than the error on $\cos\theta$, while near $90°$ the error on the angle can be much greater than the error on $\sin\theta$. As shown in the exercises, it is more precise to derive θ from its tangent instead than from its sine or its cosine.

1.2   Some Properties of Plane Triangles

Let us now consider a plane triangle, having vertices A, B, C, sides a, b, c and angles α, β, γ (the general convention is to name a the side opposite to angle α, and so on). It is well known that $\alpha + \beta + \gamma = \pi$, so that the knowledge of two angles is sufficient to determine the third one, and in general the knowledge of three elements is sufficient to find the other three, with a notable exception: given the three angles, the three sides are not unequivocally determined (while given the three sides, the three angles are unequivocally determined). Useful relations are:

$$a^2 = b^2 + c^2 - 2bc\cos\alpha \tag{1.9}$$

$$\frac{\sin\alpha}{a} = \frac{\sin\beta}{b} = \frac{\sin\gamma}{c} = \frac{1}{2R}, \quad \frac{a-b}{a+b} = \frac{\tan\frac{1}{2}(\alpha-\beta)}{\tan\frac{1}{2}(\alpha+\beta)},$$

$$\sin\frac{\alpha}{2} = \sqrt{\frac{(s-b)(s-c)}{bc}}, \quad \cos\frac{\alpha}{2} = \sqrt{\frac{s(s-a)}{bc}}, \quad \tan\frac{\alpha}{2} = \sqrt{\frac{(s-b)(s-c)}{s(s-a)}} \tag{1.10}$$

(known as Brigg's formulae),

$$K = sr = \frac{1}{2}bc\sin\alpha = c^2\frac{\sin\alpha\sin\beta}{2\sin\gamma} = \frac{abc}{2R}$$

where $a + b + c = 2s$ is the perimeter, R the radius of the circumscribed circle, r the radius of the inscribed one, K the area.

It is useful in the context of plane triangles to recall the binomial series expansion:

$$(1-y)^{-1/2} = 1 + \frac{1}{2}y + \frac{1\times3}{2\times4}y^2 + \frac{1\times3\times5}{2\times4\times6}y^3 + \cdots \quad (|y|<1) \tag{1.11}$$

Consider then Equation 1.9 in the case $b/c \ll 1$, and write:

$$a^2 = c^2\left[1 + \frac{b^2 - 2bc\cos\alpha}{c^2}\right], \quad b\cos\alpha = x, \quad \frac{b}{c} \approx 0$$

After simple passages, taking into account that a, b, and c are all positive quantities, we derive:

$$\frac{1}{a} \approx \frac{1}{c} \frac{1}{\sqrt{1 - \frac{2x}{c}}} = \frac{1}{c}\left[1 + \frac{x}{c} + \frac{(3x^2 - b^2)}{2c^2} + \frac{(5x^3 - 3xb^2)}{2c^3} + \cdots\right] \quad (1.12)$$

a relation which will be used in several occasions.

1.3 Elements of Spherical Trigonometry

Spherical trigonometry finds its origins in the Greek world, with Hipparchus of Nicea (circa 180 B.C.), and then with Claudius Ptolomeus (second century A.D.). In the 19th century, Gauss put many of the relations used in the following paragraphs in a systematic form. In the 3-dimensional Cartesian space, the sphere is the locus of points having the same distance R from a given center O. Its surface is finite but unlimited. Each plane passing through the center defines a great circle with radius R on the sphere; one of these planes is assumed to be the equatorial plane. A straight line passing through O and perpendicular to this plane intersects the sphere on two points that are the poles of the plane. Let us take for the moment a radius $R = 1$, so that the area on the surface can be expressed in steradians (sr), or for practical applications in square degrees. The area of the entire sphere is then:

$$4\pi \, \mathrm{sr} = 4\pi(360/2\pi)^2 = 129600/\pi \approx 41252.96125 \text{ sq deg}$$

Let us now consider a plane parallel to the equator; its intersection with the sphere is a small circle, which we call a parallel. Draw two great circles passing through the pole P and making an angle λ among them (see Figure 1.1), circles that we call meridians; let us call H, H', K, K' the intersections of the two meridians with the parallel and the equator, and let z be the angle POH = POK. Given that the length of arc $H'K'$ is equal to the angle λ, the length of arc HK is $\lambda \sin z$. Later, we shall show that this length is larger than the distance measured along the great circle passing for H and K. Also from Figure 1.1, we see that three great circles divide the sphere in eight portions: a spherical triangle is that part whose sides are all smaller than π, namely that part that is all contained in a hemisphere.

The sum of the angles at the vertices of the spherical triangle is always larger than π; indeed, a spherical triangle can have all three angles = $\pi/2$. Following the widespread convention, we indicate with capital italics letters A, B, C the angles at the vertices A, B, C, and with a, b, c the length of the opposing arcs, all six quantities being expressed in circular units, and satisfying the limitations:

$$0 < a + b + c < 2\pi, \quad \pi < A + B + C < 3\pi$$

The difference $\sigma = A + B + C - \pi$ is called spherical excess.

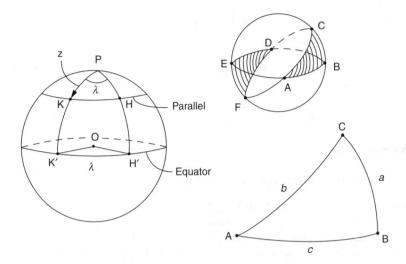

FIGURE 1.1
A sphere with great circles. Three intersecting great circles define spherical triangles, such as PH'K', ABC, DEF. Instead, PHK is not a spherical triangle.

The principal relations among the six elements A, B, C, a, b, and c are usually indicated as Gauss's groups. For a complete list we refer, for instance, to the book of Danjon (1980) and to the chapter *Computational Techniques* in the *Explanatory Supplement to the Astronomical Almanac* by Duncombe (see Seidelman et al., 1992, Bibliography). The chapter by Duncombe gives definitions and formulae valid for triangles with sides larger than π.

It is instructive to derive at least one of those formulae. Following Smart (1965), refer to Figure 1.2: from the center O we project the vertices B and C on the plane tangent to the sphere through A, obtaining three plane triangles, AB'C', OAB', OAC'. On OAB' we have:

$$OA = 1, \text{ angle } OAB' = \pi/2, \text{ angle } B'OA = c, \ AB' = \tan c, \ OB' \cos c = 1$$

In the same manner we find $AC' = \tan b$, $OC' \cos b = 1$.

In the plane triangle B'AC', angle B'AC' is, by definition, the spherical angle A, hence:

$$B'C'^2 = AB'^2 + AC'2 - 2AB'AC' \cos A$$

In the plane triangle OB'C':

$$B'C'^2 = OB'^2 + OC'^2 - 2OB'OC' \cos a$$

Substituting the lengths by the corresponding trigonometric expressions (e.g., $OB' = 1/\cos c$, etc.), and recalling that $1 + \tan^2 = 1/\cos^2$, after simple manipulation, we obtain the final result:

$$\cos a = \cos b \cos c + \sin b \sin c \cos A$$

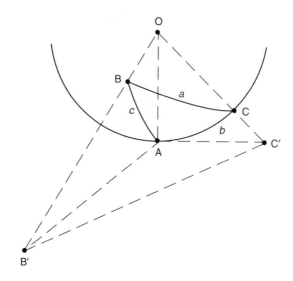

FIGURE 1.2
Demonstration of the first formula of the first Gauss's group. (Adapted from Smart, W.M., 1965, *Text Book on Spherical Astronomy*, Cambridge University Press, Cambridge.)

which is the first relation of the first group of Gauss. The three relations of the first such group are:

$$\cos a = \cos b \cos c + \sin b \sin c \cos A \tag{1.13}$$

$$\sin a \cos B = \cos b \sin c - \sin b \cos c \cos A \tag{1.14}$$

$$\frac{\sin A}{\sin a} = \frac{\sin B}{\sin b} = \frac{\sin C}{\sin c} \tag{1.15}$$

Another relation, and a variant of the first formula we shall apply are:

$$\cos a \cos C = \sin a \cot b - \sin C \cot B,$$

$$\cos A = -\cos B \cos C + \sin B \sin C \cos a$$

where "cot" means "1/tan" (not the inverse tan!).

Analogous to Brigg's formulae are the following ($2s = a + b + c$):

$$\sin\frac{A}{2} = \sqrt{\frac{\sin(s-b)\sin(s-c)}{\sin b \sin c}}, \quad \cos\frac{A}{2} = \sqrt{\frac{\sin s \sin(s-a)}{\sin b \sin c}},$$

$$\tan\frac{A}{2} = \sqrt{\frac{\sin(s-b)\sin(s-c)}{\sin s \sin(s-a)}} \tag{1.16}$$

Therefore, in a spherical triangle, the three sides are unequivocally determined by the three angles, a property not present in plane trigonometry.

When the sides of the spherical triangle are very small in comparison with the radius R of the sphere, one can resort to the approximation of plane trigonometry, by considering the plane triangle on the plane tangent to the sphere. However, a better approximation can be obtained by using the spherical excess σ. It can be shown that, in general:

$$\tan\frac{a}{2} = \sqrt{\frac{\sin\dfrac{\sigma}{2}\sin\left(A - \dfrac{\sigma}{2}\right)}{\sin\left(B - \dfrac{\sigma}{2}\right)\sin\left(C - \dfrac{\sigma}{2}\right)}} \qquad (1.17)$$

similarly for b and c. Legendre showed that a very small spherical triangle is equivalent to a plane triangle having the same sides a, b, c and angles:

$$\alpha = A - \sigma/3, \quad \beta = B - \sigma/3, \quad \gamma = C - \sigma/3$$

The spherical excess σ also determines the area of the spherical triangle:

$$K = \sigma\,\mathrm{sr}$$

1.4   Cartesian and Polar Coordinates

Let us now consider a Cartesian rectangular reference frame passing through O, $Oxyz$, as in Figure 1.3. Operatively, there will be cases in which it

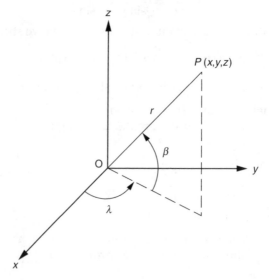

FIGURE 1.3
Cartesian and polar spherical coordinates.

is convenient to define the vertical Oz with appropriate devices (e.g., a plumb line); then the horizontal plane Oxy is derived as the perpendicular to Oz. In other cases, the horizontal plane can be better measured (e.g., with a mercury bath), and then the vertical is derived. Given a point P, its position is specified by the three Cartesian coordinates (x, y, z); alternatively, by the distance r to O and the two angles (λ, β), namely by the vector $\mathbf{r}(r, \lambda, \beta)$. The transformation laws between the two systems are:

$$\begin{cases} x = r \cos \beta \cos \lambda \\ y = r \cos \beta \sin \lambda, \\ z = r \sin \beta \end{cases} \quad \begin{cases} \lambda = \arctan \dfrac{y}{x} \\ \beta = \arcsin \dfrac{z}{r} \\ r = \sqrt{x^2 + y^2 + z^2} \end{cases} \tag{1.18}$$

The differentials, useful to evaluate both small variations and the influence of the measurement errors, are:

$$\begin{cases} dx = \dfrac{x}{r} dr - z \cos \lambda \, d\beta - y \, d\lambda \\ dy = \dfrac{y}{r} dr - z \sin \lambda \, d\beta + x \, d\lambda, \\ dz = \dfrac{z}{r} dr + r \cos \beta \, d\beta \end{cases}$$

$$\begin{cases} dr = \cos \beta \cos \lambda \, dx + \cos \beta \sin \lambda \, dy + \sin \beta \, dz \\ r \, d\beta = -\sin \beta \cos \lambda \, dx - \sin \beta \sin \lambda \, dy + \cos \beta \, dz \\ r \cos \beta \, d\lambda = -\sin \lambda \, dx + \cos \lambda \, dy \end{cases}$$

Usually, the latitude β is given in degrees minutes and seconds of arc ($^\circ\,'\,''$) between $(0^\circ, \pm 90^\circ)$, positive to the North of the fundamental plane and negative to the South. Instead of β, it is often useful to give its complement to 90°, namely the polar distance $z = 90^\circ - \beta$, $0^\circ \leq z \leq 180^\circ$. There is more freedom in the choice of units for the longitude λ: degrees, or radians, but often hours, minutes and seconds ($^{\rm h}, {}^{\rm m}, {}^{\rm s}$) are employed. The direction of the longitude can also vary according to the application, sometimes it increases in the conventional way of trigonometry (counterclockwise, or direct or prograde), in other instances the opposite verse is adopted (clockwise or retrograde).

In astronomy, the direct measurement of the distance r to the generic celestial body P is usually impossible, except for a few particular cases (solar system bodies, nearby stars). For the vast majority of stars and galaxies we must resort to indirect distance indicators, which are subject to a variety of systematic errors. Therefore, the astronomer cannot directly obtain (x, y, z), while he can measure with utmost accuracy the two angles (λ, β). The consequence of the immense distance of stars and galaxies is that we see all celestial bodies as sources of light (point-like in the case of stars) projected on a 2-dimensional celestial sphere with the observer at its center. The radius of the sphere is arbitrary, we can consider it of infinite extent (when we look at it from the inside) or of unit radius (when we look at it from the outside).

On such a sphere, we can measure the relative angular distance between two points even before having constructed a true reference system; therefore, we can immediately utilize points and arcs on the sphere. The Moon and the Sun provide a rough comparison for visual observations, having both an apparent diameter of approximately 0.5°. With the purely mechanical devices available until the 16th century, the human eye can reach a precision of one or two arcminutes (remember that the vision provides impressions associated with arcs, not with angles; for this reason for instance, the Moon at the horizon appears much larger than at the Zenith). With optical devices, the precision can reach one thousandth of an arcsec.

Much more delicate is the problem of defining and maintaining an absolute reference system, coherent on the whole celestial sphere, in order to work with angles and directions from the center. The European astrometry satellite Hipparcos (1997), whose data became available in 1997, has lead to substantial improvements with respect to previous star catalogues, because of two crucial advantages, namely the absence of atmospheric refraction and of the horizon. The general problem of astronomical reference catalogues will be examined in detail in the following chapters.

1.5   Terrestrial Latitude and Longitude on the Spherical Earth

As an example of the previous considerations, let us assume that the Earth is a sphere of unit radius; let us draw from its center O a line, coinciding with the rotation axis, that intersects the terrestrial poles in the points P and Q. The plane perpendicular to axis PQ through O will define a circle named terrestrial equator on the surface. By extending the rotation axis and the equatorial plane to intersect the celestial sphere, we define the celestial poles and the celestial equator. The real Earth's figure is slightly more complex, but irrespective of its true shape the celestial poles and equator are univocally defined by the diurnal rotation axis, which passes through the barycenter but does not necessarily coincide with the axis of figure, as we will see in detail in later chapters. However, let us ignore these complications for now and imagine regarding this from outside the unitary terrestrial sphere. Consider a particular place J on the surface (see Figure 1.4), and draw the great semi-circle through J passing through P and Q: this semicircle will be the terrestrial meridian of J, intersecting the equator in J'. Now draw the small circle parallel to the equator through J, namely the terrestrial parallel passing through J. Notice that through any given two points on the sphere, there is only one great circle joining them (unless the points are diametrically opposed),

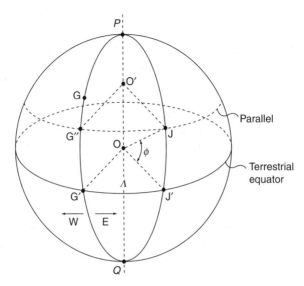

FIGURE 1.4
The terrestrial unit sphere. The generic point J is determined by the ordered pair of angles
$\Lambda = $ J'OG', $\phi = $ J'OJ.

but infinite small circles. The angular distance between them is therefore univocally defined by the arc of the great circle (we show in the exercises that this is the minimum distance; great circle arcs therefore are the geodesics on the sphere, equivalent to straight lines in Euclidean geometry).

Let us now arbitrarily assume a second point G as the origin of the terrestrial coordinates; more precisely, the intersection G' of the meridian of G on the equator will be the zero point of the longitudes. At present, this arbitrary origin G coincides with the point where the meridian circle of the Royal Astronomical Observatory in Greenwich (England) was located in the second half of the 19th century. The decision to give Greenwich this special status was reached only in 1887, after considerable debate. For instance, the great geographer Mercator, in the 16th century, had assumed the island of Hierro in the Canaries as zero point.

Now, arc(J'J) or angle J'OJ is the terrestrial latitude ϕ, arc(J'G') or angle(J'OG') is the terrestrial longitude Λ of J. In this manner, any point on the Earth has associated the ordered pair (Λ,ϕ), except the two poles P and Q for which Λ is undetermined. Instead of arc(J'J), its complementary arc(PJ) = $90° - \phi$ (named co-latitude) is often used.

Traditionally, Λ is given in (^{h}, ^{m}, ^{s}) East or West of Greenwich, while ϕ is expressed in ($°$ $'$ $''$) North or South of the equator, $-90° \le \phi \le +90°$. As already remarked, care must be exercised for Λ, which can be found expressed in circular units and increasing toward East ($-180° \le \Lambda \le 180°$, West longitudes being regarded as negative), but also toward West.

In *The Astronomical Ephemeris* the longitudes West of Greenwich are negative, for instance the Observatory of La Palma in the Canaries has $\Lambda \approx -17°$.

It is clear that (Λ, ϕ) are angular coordinates on a unitary sphere. If linear distances are needed, one must take into account that the radius of the Earth is approximately $a_\oplus = 6380$ km, so that for instance arc $(J'J) = a_\oplus\phi$ (ϕ in radians) gives the distance in kilometer from the equator to J along the meridian, while arc $(G''J) = a_\oplus\Lambda \cos \phi$ (Λ in radians) gives the distance in kilometer from the fundamental meridian to J along the parallel of latitude. Another way of giving linear distances is by using the nautical mile, namely that arc on the sphere corresponding to $1'$; according to the definition of 1929, a nautical mile equals 1852 m (1.15 statute miles). Because the Earth is not really a sphere, this length of the nautical mile corresponds to that of $1'$ on the meridian at 45° latitude. Other useful numbers to remember are the following: 1° corresponds to approximately 111.13 km (68 miles), the circumference at the equator is approximately 40 million km long (25 million miles). The original definition of the meter was indeed in terms of the 10-millionth part of the length of arc from pole to equator, and this definition was the basis for relating the measurements of distance made in the laboratory to those made on heavenly bodies by means of the diurnal parallax.

See Chapter 2 for a more precise discussion and values on the ellipsoidal Earth.

For everyday life, particular importance is attached to the 23 meridians at 15° (1^h) and multiples from Greenwich, which define the time zones. However, by international convention the borders of the time zones do not exactly coincide with the meridians, rather they follow the borders of the states. In Italy for instance the mean time of central Europe (TMEC) is adopted, in the contiguous U.S.A. there are four time zones (Eastern, Central, Mountain, Pacific, plus Alaska and Hawaii), and so on. The meridian opposite to Greenwich is the so-called date line: on the West side of the meridian, the date is 1 day more than on the East side. A traveler crossing that line going Eastward has to decrease his calendar date by 1 day, and *vice versa* going Westward. Probably, the first practical awareness of this fact (already supposed by medieval philosophers and geographers) was reached by members of Magellan's expedition in the 16th century, by discovering that counting actual days gave a result differing by one full day with respect to calendar dates.

It is often convenient to employ Cartesian coordinates, by directing the *x*-axis toward a given point X in the equatorial plane (in this particular case, $X = G'$), the *y*-axis toward a point Y 90° counterclockwise on the equatorial plane, and the *z*-axis toward Z (in this case, the north pole P). Mathematical convenience will not conceal the fact that the geocentric Cartesian system is not directly accessible to observation. Attention must also be made not to confuse the Cartesian distance between two points (necessary, for example, in interferometric applications) with the distance along the great circle joining them.

1.6 Elements of Vector Calculus

Vector calculus finds a most powerful application in the case of Solar System objects, because their distance can be accurately specified, but it is useful also in handling generic problems related to the celestial sphere.

Let $\mathbf{i}$, $\mathbf{j}$, $\mathbf{k}$ be the unit vectors along (x, y, z); the position of a point J on the unitary sphere can be represented by the unit vector r_J:

$$\mathbf{r}_J(x, y, z) = \mathbf{r}_J(\Lambda, \phi) = x\mathbf{i} + y\mathbf{j} + z\mathbf{k}$$

where $x = \cos xJ$, $y = \cos yJ$, $z = \cos zJ$, are the direction cosines of the line OJ. The transformations between Cartesian and polar spherical coordinates are:

$$x = \cos \Lambda \cos \phi, \quad y = \sin \Lambda \cos \phi, \quad z = \sin \phi$$

We might also write:

$$\hat{\mathbf{r}}_J = (x, y, z) = \begin{pmatrix} \cos \Lambda \cos \phi \\ \sin \Lambda \cos \phi \\ \sin \phi \end{pmatrix} \tag{1.19}$$

The general problem of coordinate transformations will be examined in Chapter 3.

It is useful to recall some basic notions of vector algebra. The sum of two vectors $\mathbf{a}$, $\mathbf{b}$ is a third vector $\mathbf{c}$ lying in the plane defined by the first two:

$$\mathbf{c} = \mathbf{a} + \mathbf{b} = (a_x + b_x)\mathbf{i} + (a_y + b_y)\mathbf{j} + (a_z + b_z)\mathbf{k}, \quad \mathbf{a} - \mathbf{b} = \mathbf{a} + (-\mathbf{b})$$

where $\mathbf{i}$, $\mathbf{j}$, $\mathbf{k}$ are the unit vectors along, respectively, x, y, and z. The sum has the commutative and associative properties:

$$\mathbf{c} = \mathbf{a} + \mathbf{b} = \mathbf{b} + \mathbf{a}, \quad \mathbf{a} + (\mathbf{b} + \mathbf{c}) = (\mathbf{a} + \mathbf{b}) + \mathbf{c}$$

The scalar (or inner) product c between two vectors whose directions make an angle θ ($0° \leq \theta \leq 180°$) is the number:

$$c = \mathbf{a} \cdot \mathbf{b} = \mathbf{b} \cdot \mathbf{a} = ab \cos \theta = a_x b_x + a_y b_y + a_z b_z$$

The vector product $\mathbf{c}$ between $\mathbf{a}$ and $\mathbf{b}$ is a third vector $\mathbf{c}$ perpendicular to the plane defined by the first two:

$$\mathbf{c} = \mathbf{a} \times \mathbf{b} = (a_y b_z - a_z b_y, \ a_z b_x - a_x b_z, \ a_x b_y - a_y b_x) = ab \sin \theta \, \mathbf{u}$$

being $\mathbf{u}$ the unit vector parallel to $\mathbf{c}$. Notice that the vector product has the distributive property $\mathbf{a} \times (\mathbf{b} + \mathbf{c}) = \mathbf{a} \times \mathbf{b} + \mathbf{a} \times \mathbf{c}$, but not the commutative property, because $\mathbf{c} = \mathbf{a} \times \mathbf{b} = -\mathbf{b} \times \mathbf{a}$. In particular:

$$\mathbf{i} \times \mathbf{j} = \mathbf{k}, \quad \mathbf{j} \times \mathbf{k} = \mathbf{i}, \quad \mathbf{k} \times \mathbf{i} = \mathbf{j}$$

and therefore the vector product can be written in matrix form:

$$\mathbf{c} = \mathbf{a} \times \mathbf{b} = \begin{vmatrix} \mathbf{i} & \mathbf{j} & \mathbf{k} \\ a_x & a_y & a_z \\ b_x & b_y & b_z \end{vmatrix}$$

The double mixed product

$$[\mathbf{a}, \mathbf{b}, \mathbf{c}] = \mathbf{a} \cdot (\mathbf{b} \times \mathbf{c}) = \mathbf{b} \cdot (\mathbf{c} \times \mathbf{a}) = \mathbf{c} \cdot (\mathbf{a} \times \mathbf{b}) = \begin{vmatrix} a_x & a_y & a_z \\ b_x & b_y & b_z \\ c_x & c_y & c_z \end{vmatrix}$$

is a number that can be interpreted as the volume of the parallepiped of sides **a, b, c**.

The triple vector product is defined by:

$$\mathbf{a} \times (\mathbf{b} \times \mathbf{c}) = (\mathbf{a} \cdot \mathbf{c})\mathbf{b} - (\mathbf{a} \cdot \mathbf{b})\mathbf{c}, \quad (\mathbf{a} \times \mathbf{b}) \times \mathbf{c} = (\mathbf{a} \cdot \mathbf{c})\mathbf{b} - (\mathbf{b} \cdot \mathbf{c})\mathbf{a} \neq \mathbf{a} \times (\mathbf{b} \times \mathbf{c})$$

Sometimes, the notation $\mathbf{a} \wedge \mathbf{b}$ instead of $\mathbf{a} \times \mathbf{b}$ is used. The vector product can be applied in particular to define the moment of a force **F**: consider a point O in space, and a force **F** whose direction does not intercept O. Draw from O a vector **r** to any point along the direction of **F**. The moment of **F** with respect to O is the vector **K**:

$$\mathbf{K} = \mathbf{r} \times \mathbf{F} = (rF \sin \theta)\mathbf{k} \tag{1.20}$$

θ being the angle between the two vectors, perpendicular therefore to the plane defined by **r** and **F**; notice that **K** is independent of where **r** terminates on the direction of **F**.

The derivative and the integral of a vector are, respectively:

$$\frac{d\mathbf{a}}{dt} = \frac{da_x}{dt}\mathbf{i} + \frac{da_j}{dt}\mathbf{j} + \frac{da_z}{dt}\mathbf{k}, \quad \frac{d}{dt}(\mathbf{a} \cdot \mathbf{b}) = \mathbf{b} \cdot \frac{d\mathbf{a}}{dt} + \mathbf{a} \cdot \frac{d\mathbf{b}}{dt},$$

$$\frac{d}{dt}(\mathbf{a} \times \mathbf{b}) = \frac{d\mathbf{a}}{dt} \times \mathbf{b} + \mathbf{a} \times \frac{d\mathbf{b}}{dt}$$

$$\int \mathbf{a}(t)dt = \mathbf{i} \int a_x(t)dt + \mathbf{j} \int a_y(t)dt + \mathbf{k} \int a_x(t)dt$$

and, if a vector **b** exists such that $\mathbf{a} = d\mathbf{b}/dt$, then:

$$\int \mathbf{a}(t)dt = \mathbf{b}(t) + \mathbf{c} \quad (\mathbf{c} = \text{constant vector}), \quad \int_{t_1}^{t_2} \mathbf{a}(t)dt = \mathbf{b}(t_2) - \mathbf{b}(t_1)$$

In Dynamics and in Celestial Mechanics it is usual to denote the first and second derivatives of a vector with the notations:

$$\frac{d\mathbf{a}}{dt} = \dot{\mathbf{a}}, \quad \frac{d^2\mathbf{a}}{dt^2} = \frac{d\dot{\mathbf{a}}}{dt} = \ddot{\mathbf{a}}$$

It is important to remember that not all entities specified by a modulus and a direction are *bona fide* vectors. The composition law and the commutative property are indeed essential to the definition; in particular, finite angular rotations of a rigid body are not vectors, they do not obey the commutative property! However, infinitesimal rotations, and therefore also angular velocities, do obey all vector laws. In the following chapters, we will encounter vectors **r** referred to a coordinate system itself rotating with angular velocity ω with respect to an inertial system. The total derivative of **r** relative to inertial axes will be:

$$\left(\frac{d\mathbf{r}}{dt}\right)_{\text{inertial}} = \left(\frac{d\mathbf{r}}{dt}\right)_{\text{relative}} + \boldsymbol{\omega} \times \mathbf{r} \qquad (1.21)$$

Notes

1. Experience tells that all too often the students forget to check the quadrant of the angle. Remember, between 0 and π, the inverse cosine is single valued, the inverse sine is not!

2. In the following chapter a spherical triangle will be determined, for instance by the Sun moving along the ecliptic (vertex A), by its projection on the celestial equator (vertex B), and by the intersection between the two planes (vertex C). With the passage of time, the two sides AC and BC of the triangle will exceed π, and therefore the figure ABC will cease to be a spherical triangle as defined here. However, it will be easy to reason on the supplementary triangle CAB. See also the already quoted article by Duncombe in the *Explanatory Supplement*.

3. Inverting series. Any converging series of the type:

$$\theta = \varphi + a\varphi^2 + b\varphi^3 + c\varphi^4 + d\varphi^5 + \cdots$$

    can be inverted as:

$$\varphi = \theta + \alpha\theta^2 + \beta\theta^3 + \gamma\theta^4 + \delta\theta^5 + \cdots$$

    where:

$$a = \alpha, \quad \beta = b - 2a^2, \quad \gamma = c + 5a^3 - 5ab,$$
$$\delta = d - 14a^2 - 3b^2 + 18a^2b - 4ac$$

4. Precision of trigonometric functions. Suppose your calculator gives the value of the trigonometric functions f with n decimal digits. The difference between the true value of the function and that given by the calculator can be at most 1 on the last digit, namely $\Delta f \leq 1.10^{-n}$, which can be regarded small enough to be considered as the

differential of f in such a point. This is because:

$$d \operatorname{sen} \theta = \cos \theta \, d\theta, \quad d \cos \theta = -\operatorname{sen} \vartheta \, d\theta, \quad d \tan \theta = \frac{1}{\cos^2 \vartheta} d\theta$$

the errors $\varepsilon_i = d\theta_i$ on the angles due to the errors on the functions will be:

$$\varepsilon_1 = \left| \frac{1}{\cos \theta} \right| 10^{-n}, \quad \varepsilon_2 = \left| \frac{1}{\operatorname{sen} \theta} \right| 10^{-n}, \quad \varepsilon_3 = \left| \cos^2 \theta \right| 10^{-n}$$

While the third expression deriving from the tangent never exceeds 10^{-n} for any angle, the first two can become very large if the angle approaches $\pi/2$ or 0, respectively. For instance, suppose that your calculator gives eight decimal digits (perhaps it is prudential to assume as exact the one before the last visualized digit), and you wish ε_3 in arcsec. We have to multiply the previous expression by $R'' = 206264.8$:

$$\varepsilon_3 \leq 2.06 \times 10^5 \times 10^{-8} \approx 2 \times 10^{-3} \text{ arcsec}$$

for all θ's. Should we have used ε_1 in the proximity of 90°, e.g., at 89.5°, the precision would have deteriorated by $1/\cos \theta \approx 115$ times.

As a general rule, in order to have an error less than 30″, four digits are needed, six for a precision better than 0″.5, and eight for 0″.005. Should astrometry become more precise than a millionth of an arcsec (as in future space missions such as the European GAIA), the task of the computer will become truly formidable.

5. Precision of series developments. Let us recall the expressions:

$$\sin a = a - \frac{1}{3!}a^3 + \frac{1}{5!}a^5 + \cdots \approx a\left(1 - \frac{1}{6}a^2\right),$$

$$\tan a = a + \frac{1}{3}a^3 - \frac{2}{15}a^3 + \cdots \approx a\left(1 + \frac{1}{3}a^2\right)$$

When arc a is small, the error made by confusing sin with arc is smaller than the first ignored term, namely the term in a^3; in the case of the tangent the error is slightly larger than twice that made with the sin, and with opposite sign (the term in a^5 exceeds 0″.01 only above 5°). Let us calculate the maximum value of a before the error exceeds a wanted amount, as given by Table 1.1. For instance, the Moon has an average diurnal parallax of 3422″.70, so that if we retain only the first two terms:

$$\sin 3422''.70 \approx 3422''.70 \left[1 - \frac{1}{6}\left(\frac{3422.70}{206264.8} \right)^2 \right]$$

$$\approx 3422''.70[1 - 0.000046] = 3422''.54$$

TABLE 1.1

Admitted Error and Maximum Angles

$a - \sin a$	a_{max}	$a - \tan a$	a_{max}	$1 - \cos a$	a_{max}
$0''.001$	$0°10'34''.3$	$0''.001$	$0°08'23''.5$	10^{-6}	$0°04'51''.7$
$0''.010$	$0°22'46''.7$	$0''.010$	$0°18'04''.7$	10^{-5}	$0°15'22''.5$
$0''.100$	$0°49'04''.4$	$0''.100$	$0°38'57''.0$	10^{-4}	$0°48'37''.0$
$1''.000$	$1°45'43''.7$	$1''.000$	$1°23'54''.5$	10^{-3}	$2°33'43''.2$

Namely, by considering only the second terms the sin differs from the parallax by $0''.16$, a quantity that can or cannot be considered negligible, according to the application.

Exercises

1. Find the elements and the area of the spherical triangles having the following vertices:

    $(\Lambda, \phi) = (0^h, 0°), \quad (\Lambda, \phi) = (4^m, 0°), \quad (\Lambda, \phi) = (4^m, 1°)$

    $(\Lambda, \phi) = (0^h, 0°), \quad (\Lambda, \phi) = (1^h, 0°), \quad (\Lambda, \phi) = (1^h, 45°)$

    $(\Lambda, \phi) = (0^h, 0°), \quad (\Lambda, \phi) = (6^h, 0°), \quad \text{the North pole}$

    $(\Lambda, \phi) = (0^h, 0°), \quad (\Lambda, \phi) = (18^h, 0°), \quad \text{the North pole}$

    $(\Lambda, \phi) = (0^h, 0°), \quad (\Lambda, \phi) = (21^h, 0°), \quad \text{the North pole}$

2. Calculate the distance between two Earthly sites A and B both north of the equator, and the coordinates of northernmost point on the great circle joining them, in the hypothesis of spherical Earth of radius $a_\oplus = 6380$ km.

    Let the two places in the Northern hemisphere have the geographic coordinates as shown in Table 1.2: We wish to determine the following:

    - The length of the great circle arc passing through A, B and of the joining chord.

    - P being the North pole, and PAB the corresponding spherical triangle, the amplitude of the angles $A = PAB$, $B = PBA$.

    - That $0 < \sum(\text{arcs}) < 2\pi$, $\pi < (\sum \text{angles}) < 3/2\pi$.

    Recall the first formula of the 1st Gauss's group (Equation 1.13), which we write here as:

    $$\cos p = \cos a \cos b + \sin a \sin b \cos P$$

    $$= \sin \phi_A \sin \phi_B + \cos \phi_A \cos \phi_B \cos \Delta\Lambda \qquad (1.22)$$

TABLE 1.2

Geographic Coordinates of the Two Sites

	Longitude Λ	Latitude ϕ
A	133°.65 E	+24°.3
B	125°.40 W	+36°.8

where: arc $AB = p$ unknown, arc $PA = b = 90° - \phi_A = 65°.7$, arc $PB = a = 90° - \phi_B = 53°.2$, angle $P = \Lambda_A - \Lambda_B = 100°.95$ (A is East, B is West of Greenwich). From formula 1.22 we derive:

$$p = 83°.80676 = 83°48'24'' = 5028'.4$$

By definition, the nautical mile is the arc corresponding to $1'$, so that 5028.4 is also the distance in nautical miles, equivalent to 9328 km, between A and B. The length of the chord, useful in measurement of the diurnal parallax or in interferometric applications, can be derived from:

$$L = 2a_\oplus \sin \frac{p}{2} = 8521.5 \text{ km}$$

The amplitude of the other two angles can be derived from the tangent formula 1.16, where:

$$2s = a + b + p = 202°.2, \quad s = 101°.1.$$

Whence $A = 52°.26$; by the same procedure, $B = 64°.17$, so that: $b + c + p = 202°.70$, $A + B + P = 217°.38$.

To determine C, the northernmost point on circle AB, we notice that the meridian for C will be perpendicular in C to the circle. Therefore, in the spherical triangle PCA (or PCB), the angle in C will be 90°. All other elements being known, we immediately obtaint arc PC (the co-latitude) and the angle APC, namely the longitude:

$$\Lambda_C = 164°.38 \text{ W}, \quad \phi_C = 43°.88 \text{ N}$$

3. Calculate the distance between the astronomical observatories of Padova (Italy) and La Silla (Chile), and the longitude where the great circle between the two sites cuts the equator (ignore the elevation above the sea level).

   The coordinates of the two observatories are:

$$\Lambda_{Pd} = 11°52'.3 \text{ E}, \quad \phi_{Pd} = +45°24',$$
$$\Lambda_{LS} = 70°43'.8 \text{ W}, \quad \phi_{LS} = -20°14'$$

For the first question, consider the spherical triangle between Padova, the North Pole P and La Silla, and let p be the angle at the pole between the meridians of the two places. This angle coincides with the difference in longitude between Padova and La Silla, with due regard to the first being East and the second West of Greenwich. Two sides of the triangles are given by the colatitudes. Then, as in the previous exercise, Equation 1.13 easily provides the third side, namely the wanted angular distance of 105°.5, corresponding to 6336 nautical miles, or to 11,733 km.

For the second question, if E is the intersection point between the arc through Padova, La Silla, and the equator, notice that the latitude of E is zero, therefore the spherical triangle drawn through P, E, and La Silla has one side of 90°. The angle opposite to this side is easily found, and after a few calculations we finally obtain $\Lambda_E = 44°$ W.

4. Two observers on the Earth surface, A and B, separated by a distance of 50 ± 0.1 km, observe a satellite C. The value of the angle CAB, as measured by A, is (87.5 ± 0.01) deg; the value of the angle CBA, as measured by B, is (88.5 ± 0.01) deg. Determine the other elements of the triangle ABC, and discuss the relative influence of the errors of the result.

2

Astronomical Reference Systems

The measurement of the positions on the celestial sphere was among the earliest operations performed on heavenly bodies. On several aspects, the procedure is similar to that performed on the Earth by the geographer, however, there are peculiarities that have led to the definition of reference systems unique to astronomy. Some of these systems have originated from observations carried out over several millennia from the surface of the Earth; an astronomer in the extra-terrestrial space, or living on another planet, might have found 'solutions more appropriate to his particular location.

The solid Earth prevents seeing the entire celestial sphere, the line of sight is limited above a great circle which we call the horizon, and which is more readily visible from a high mountain, or on the open sea. Due to the diurnal direct rotation of the Earth, the celestial sphere appears to rotate from East to West around an ideal axis, which cuts the sphere at two points called the celestial poles, North P and South Q (a generic observer only sees one of them). Due to the great distances of all celestial bodies, the axis passing for any observer on the Earth and parallel to the rotation axis defines exactly the same two points P and Q. Therefore, in the course of the day, each star will appear to describe exactly the same minor circle, irrespective of the position of the observer. In the same manner, the plane passing through any observer and perpendicular to the rotation axis defines the same celestial equator. Only for very few objects, and with difficult measurements, are we able to detect the daily rotation of the observer itself.

The enormous distances to the stars have another important consequence, that their relative motions are small, and remained undetected until the 17th century; for this reason, the term "fixed stars" was used, and is encountered even today. On the contrary, the Sun, the Moon, the planets, and the occasional comets, display an appreciable motion with respect to this fixed canopy of stars; the name planet itself has the meaning of "wandering body" in Greek. The motions of the Sun and the Moon were used to define units of time, as we will see in detail in later chapters. While the diurnal rotation is at the basis of the concept of the day, the movement of the Sun through the constellations defines the solar year, and that of the Moon defines the lunar month. In particular, the yearly motion of the Sun takes place along a great circle inclined by approximately 23°27′ to the celestial equator. This circle is named the ecliptic, and the angle ε between equator and ecliptic is called obliquity of the ecliptic. Horizon, equator, and ecliptic are three great circles on the celestial sphere, which can be used to define different astronomical reference systems. The common principle at the basis of each system has been already expounded:

- A plane through the observer is selected as a fundamental plane, to define a fundamental great circle on the sphere, and a particular point G′ on this fundamental circle is chosen as origin.
- A great circle is drawn through the pole of the fundamental plane and any star X, intersecting the fundamental plane in X′.
- Arc G′X′ is the angular abscissa λ, arc X′X the angular ordinate β of X, so that any star is univocally determined by the ordered pair (λ, β), with the obvious exceptions of the two poles, for which λ is undefined.

2.1   The Alt-Azimuth System

The Alt-Azimuth (or horizontal) system is based on the horizontal plane and its perpendicular, namely the vertical. This system can be realized, at least in principle, by very simple devices sensitive to gravity, such as a plumb-line or the free surface of a liquid. The points where the vertical cuts the celestial sphere are called, respectively, Zenith Z (above), and Nadir Z′ (below, unobservable from the Earth's surface). The plane passing through the observer and perpendicular to the vertical cuts the celestial sphere in the astronomical horizon. Consider now the celestial pole P, and draw the great circle through P and Z; this circle is called the meridian of the observer, and obviously it must contain also the Nadir Z′ and the other pole Q. The meridian cuts the horizon on two points, the so-called true North N (on the same side of P with respect to Z) and true South S. Any other great circle passing through Z and Z′, and thereby containing the vertical, is called a vertical circle; in particular that at 90° to the meridian is named "first vertical"; it defines on the horizon the points of true East E and true West W. Those four points on the horizon are referred to as cardinal points.

In order to determine the two angular coordinates of a star X, let us draw the vertical circle through Z and X (see Figure 2.1) that intersects the horizon in X′. Two arcs are therefore defined:

1. The Azimuth *A*, namely the arc SX′ counted clockwise from the South through the West, and usually measured in (° ′ ″), $0° \leq A \leq 360°$. This convention of origin and direction has been adopted here in order to coincide with that of the Hour Angle that will be

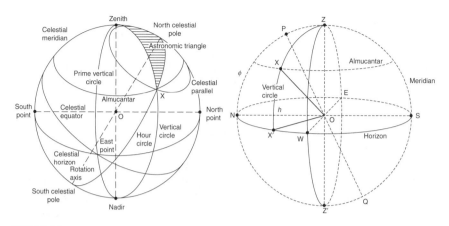

FIGURE 2.1
On the left, the local celestial sphere and the astronomic triangle as seen from the East. On the right the Alt-Az coordinates of star X with Azimuth origin from the South (as seen from the West). Angle SOX' is the Azimuth A, angle X'OX the altitude h (its complement ZOX is said Zenith distance z of X). The altitude of the celestial pole P (namely the angle NOP, or the arc NP) is the astronomical latitude ϕ of the observer.

described in the following paragraph. Several authors, and many telescope pointing algorithms, prefer to count A from the North through the East; others use a counterclockwise direction, and still others have $-180° \leq A \leq 180°$.

2. The altitude (or elevation) h, namely the arc X'X counted from the horizon toward the star, and measured in $(° ' '')$, $-90° \leq h \leq +90°$. Alternatively, the Zenith distance $z = 90° - h$ might be preferred, with $0° \leq z \leq 180°$. A star below the horizon ($h < 0°$, $z > 90°$) cannot be seen by the terrestrial observer. With an Arabic term, the parallels of elevation are called *almucantar* (or *almucantarat*). The height of the visible pole P above the horizon, namely the arc NP, defines the astronomical latitude of the site.

Therefore, each site has associated with it a system of fixed cardinal points and great circles, with respect to which the celestial sphere is in continuous rotation. In other words, the pair (A, h) attached to any star X depends on the location of the observer, and changes with the passage of time. Therefore, it is necessary to discuss more precisely the relationship between the geographic and astronomic coordinates on one hand, and the influence of time on the other. We stated in Chapter 1 that the Earth is only approximately a sphere; indeed, already in the 18th century the accuracy of the measurements conclusively showed that the shape is better approximated by an ellipsoid of revolution (called also a spheroid) having the polar axis slightly shorter than the equatorial one. Some authors call geoid the equilibrium surface which coincides with the surface of the free water in the oceans; this ellipsoid is usually below the geoid on the land and above it on the oceans, but the difference never exceeds 100 m. In the following, we will usually ignore the distinction, using rather loosely the terms spheroid, ellipsoid, or geoid, and the further deviations of the real Earth from it (the Earth is indeed remarkably smooth, the difference in elevation between the highest and the lowest point is less than 20 km, namely less than 1/300). The consequences of this slight ellipticity are, however, far reaching (see Figure 2.2):

1. The geodetic vertical is mathematically defined as the normal to the ellipsoid. It does not encounter the center O of the Earth, it intersects the rotation axis in a latitude-dependent point O′.

2. The astronomic vertical is physically defined by the overall gravitational field in any given place. This "true" vertical does not encounter the center O, and indeed it does not necessarily intersect the rotation axis. This small deviation, which can be accurately mapped, is caused by the presence of nearby masses or voids; these so-called anomalies of the gravity rarely exceed 10″, although in peculiar places they can reach 50″. Accurate measurements, such as those made at radiofrequency with the VLBI technique, clearly reveal these two slightly different Zeniths (see Chapter 6).

3. The convenience of using geocentric Cartesian coordinates requires the consideration of the direction from the center of the Earth to any particular place on the surface, namely of the geocentric vertical. It will be shown that the latitude-dependent difference between geocentric and astronomic vertical can reach 12′ at mid-latitudes.

4. The axis of symmetry of the ellipsoid does not necessarily coincide with the instantaneous rotation axis, as was theoretically shown by Euler in the 18th century and observed in the 19th century

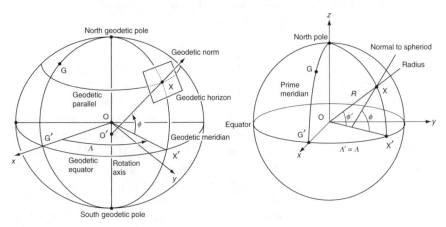

FIGURE 2.2
Geodetic, astronomic and geocentric coordinates of a point X on the surface of the ellipsoid.

(see Chapter 6; we shall see that the astronomical latitude and longitude of any site are not really constant, varying with amplitudes and time scales that can be measured).

In most of the following considerations, we shall ignore the small differences between geodetic and astronomic vertical, using an ellipsoidal model of the Earth with two equal equatorial axes and a slightly shorter polar axis (namely an oblate spheroid with rotational symmetry). Therefore, the terrestrial meridians are ellipses, for which several mathematical expressions can be given. Many such models have been adopted in the course of the last centuries, for instance the one named WGS84, having equatorial radius $a_\oplus = 6378.137$ km and polar axis $c = 6356.752$ km, finds a widespread utilization for navigation. Important parameters of each spheroidal model are the flattening f and the eccentricity e, namely the quantities:

$$f = \frac{a_\oplus - c}{a_\oplus} \approx 0.00335 \approx \frac{1}{298.25}, \quad 1 - f = \frac{c}{a_\oplus},$$

$$e^2 = \frac{a_\oplus^2 - c^2}{a_\oplus^2} \approx 0.006694, \quad e \approx 0.0818 \tag{2.1}$$

The two parameters e, f are connected by the following relationships:

$$e^2 = f(2 - f), \quad (1 - f)^2 = 1 - e^2, \quad c^2 = a_\oplus^2(1 - e^2) = a_\oplus^2(1 - f)^2 \tag{2.2}$$

With reference to Figure 2.2, let (r, z) be the coordinates of a point X on the ellipse (namely on the mathematical surface), distant $R = a_\oplus \rho$ from the center O:

$$r = R \cos \phi', \quad z = R \sin \phi', \quad \tan \phi' = \frac{z}{r}, \quad 0 \le r \le a_\oplus, \quad 0 \le z \le c,$$

$$\left(\frac{r}{a_\oplus}\right)^2 + \left(\frac{z}{c}\right)^2 = 1$$

or equivalently:

$$(1 - f)^2 r^2 + z^2 = a_\oplus^2(1 - f)^2, \quad (1 - e^2)r^2 + z^2 = a_\oplus^2(1 - e^2)$$

The geodetic latitude ϕ is defined by the angle between the direction of the semi-major axis and the normal to the tangent to the ellipse for X. The angular coefficient m of the tangent is given by:

$$m = -\frac{dz}{dr} = \pm \left(\frac{c}{a_\oplus}\right)^2 \frac{r}{z}$$

The angular coefficient m' of the normal is immediately found by $mm' = -1$, so that:

$$m' = \tan \phi = \mp \left(\frac{a_\oplus}{c}\right)^2 \frac{z}{r}$$

On its turn, the geocentric latitude is simply:

$$\tan\phi' = \frac{z}{r} = \left(\frac{c}{a_\oplus}\right)^2 \tan\phi = (1-f)^2 \tan\phi = (1-e^2)\tan\phi \qquad (2.3)$$

Numerically:

$$\tan\phi' \approx 0.99330552\,\tan\phi, \quad \tan\phi \approx 1.00673960\,\tan\phi' \qquad (2.4)$$

Given that the latitude is constrained between $\pm 90°$, no quadrant ambiguity can arise. The module of the geodetic latitude is therefore always greater or equal to the geocentric one.

From the trigonometric formulae seen in Chapter 1, we also derive that:

$$\tan(\phi' - \phi) = \frac{\tan\phi' - \tan\phi}{1 + \tan\phi'\tan\phi} = \frac{q\sin 2\phi}{1 - q\cos 2\phi}, \quad q = -\frac{e^2}{2 - e^2}$$

In our case, the difference between the two angles is always a small quantity, so we can apply the following series development, demonstrated by Lagrange using the complex expressions of the trigonometric functions:

$$\phi' = \phi - (f + \frac{1}{2}f^2)\sin 2\phi + \frac{1}{2}f^2(1+f)\sin 4\phi + \cdots$$

$$\approx \varphi - 692''.737\sin 2\phi + 1''.163\sin 4\phi \qquad (2.5)$$

where we have made use of the relations (2.2). The difference $|\phi' - \phi|$ reaches its maximum value of $11'.54$ at latitude $45°$. The results given by Equation 2.5 do not differ by those given by Equation 2.4 by more than $0''.1$.

It can be useful to remember that this result helps to solve the transcendent equation of frequent utilization:

$$\tan x = m\tan y \quad (m > 0) \qquad (2.6)$$

Putting $\dfrac{1-m}{1+m} = q$ ($|q| < 1$) the solution is:

$$x = y - q\sin 2y + \frac{q^2}{2}\sin 4y - \frac{q^3}{3}\sin 6y + \frac{q^4}{4}\sin 8y + \cdots \qquad (2.7)$$

In the same manner, R is found as function of ϕ or ϕ':

$$r = \frac{a_\oplus \cos\phi}{\sqrt{1 - e^2\sin^2\phi}} = R\cos\phi', \quad z = \frac{a_\oplus(1 - e^2)\sin\phi}{\sqrt{1 - e^2\sin^2\phi}} = R\sin\phi'$$

$$R = \sqrt{r^2 + z^2} = a_\oplus\sqrt{\frac{1 - e^2(2 - e^2)\sin^2\phi}{1 - e^2\sin^2\phi}} \qquad (2.8)$$

Or else, by dropping the terms in e^4 and higher, and using (2.2):

$$R = a_\oplus\left[\left(1 - \frac{1}{2}f + \frac{5}{16}f^2\right) + \frac{1}{2}f\cos 2\phi - \frac{5}{16}f^2\cos 4\phi + \cdots\right]$$

$$\approx 6367.45 + 10.69\cos 2\phi - 0.02\cos 4\phi + \cdots \text{(km)} \qquad (2.9)$$

At 28° latitude, R is approximately 4.9 km shorter than $a_{\oplus}$, and 10.6 km shorter at 45°.

We can also express the radius of curvature k (in the same units of $a_{\oplus}$, say in km) in $X(r, z)$:

$$\frac{dz}{dr} = -\frac{1}{\tan\phi}, \quad \frac{d^2z}{dr^2} = \frac{d\tan\phi}{dr}\frac{1}{\tan^2\phi},$$

$$k = \frac{\left(1+(\frac{dz}{dr})^2\right)^{3/2}}{\left|\frac{d^2z}{dr^2}\right|} = a_{\oplus}\frac{1-e^2}{(1-e^2\sin^2\phi)^{3/2}}$$

Therefore, the length K (say in km) of the 1° arc of meridian passing for X is:

$$K = 2\pi\frac{k}{360} = a_{\oplus}\frac{\pi}{180}\frac{1-e^2}{(1-e^2\sin^2\phi)^{3/2}}$$

By measuring K at several places along the meridian (ideally from pole to pole) the overall curvature of the surface is found, and from the comparison of its values for the different places the quantities $a_{\oplus}$, c, e, f can be derived. At the equator $K \approx 110.6$ km, at the pole $K \approx 111.7$ km.

Notice that we do not need a third dimension to determine the radius of curvature of the surface. Of course, observations of the Earth from the outer space are of fundamental value to determine its true shape.

Let us now consider a point X at an altitude H (measured along the geodetic vertical) above the surface of the Earth, such as the top of a mountain, or a low-flying satellite such as the Hubble Space Telescope, which is orbiting at approximately $H = 550$ km above the ground. Because:

$$dr = H\cos\phi, \quad dz = H\sin\phi$$

we also have:

$$dR = H\cos(\phi-\phi'), \quad d\phi' = \frac{H}{R}\sin(\phi-\phi')$$

From the geodetic coordinates longitude Λ, latitude ϕ, and height H we can derive the geocentric Cartesian coordinates (x, y, z) by taking into account H and the deviation of the vertical. An approximate procedure is the following: calculate the geocentric latitude ϕ' by means of Equation 2.5, and the radius R by means of Equation 2.9. Add H to R (no sensible error here from ignoring the deviation of the vertical), and then calculate:

$$x = (R(\phi) + H)\cos\phi'\cos\Lambda, \quad y = (R(\phi) + H)\cos\phi'\sin\Lambda,$$

$$z = (R(\phi) + H)\sin\phi'$$

with due account to the quadrant of Λ.

The *Astronomical Almanac* provides a table of reference ellipsoids from the time of Airy (*circa* 1830) to the present. Furthermore, it uses a slightly different formalism to calculate the geocentric Cartesian coordinates (x, y, z) of the point X, by means of two auxiliary functions C, S:

$$C = \frac{1}{\sqrt{\cos^2\phi + (1-f)^2\sin^2\phi}}, \quad S = (1-f)^2 C$$

which give:

$$\begin{cases} x = a_\oplus\rho\cos\phi'\cos\lambda = (a_\oplus C + H)\cos\phi\cos\lambda \\ y = a_\oplus\rho\cos\phi'\sin\lambda = (a_\oplus C + H)\cos\phi\sin\lambda \\ z = a_\oplus\rho\sin\phi' = (a_\oplus S + H)\sin\phi \end{cases} \quad (2.10)$$

The inverse transformation to derive (λ, ϕ, H) from (x, y, z) is usually performed by successive iterations.

As an example of the previous considerations, Table 2.1 shows the geocentric, geodetic, and astronomical positions of the intersections of the principle axes of the Copernicus Telescope at Cima Ekar (Asiago, Italy) and of the Telescopio Nazionale Galileo TNG in La Palma (Canary Islands, Spain). Notice that the geodetic altitude H does not coincide with the so-called height above sea level (a.s.l., given in parentheses), being determined by a best fitting procedure with a reference ellipsoidal surface that, for example in the Canaries, is approximately 50 m below the average sea level. Table 2.1 also shows the appreciable gravity anomaly at the two sites.

To derive these coordinates, the Global Positioning System (GPS, a network of satellites whose positions have been referred to that of WGS 84), and a set of stars from the fundamental catalog FK5 were used.

TABLE 2.1

Coordinates of the Copernicus and Galileo Telescopes in the WGS84 Reference Ellipsoid

Telescope	182 cm Copernicus	352 cm Galileo (TNG)
Cartesian geocentric		
x	$+4360893.8$	$+5327423.3$
y	$+892690.4$	-1719592.5
z	$+4554619.0$	$+3051176.2$
Geodetic		
Λ	(E) $+11°34'07''.92$	(W) $-17°53'20''.6$
ϕ	$+45°50'54''.92$	$+28°45'14''.4$
H (m)	1435 (1380 a.s.l.)	2427.6 (2370 a.s.l.)
Astronomic		
Λ	$+11°34'22''.14 \pm 0''.45$	$-17°53'37''.9 \pm 0''.45$
ϕ	$+45°50'36''.99 \pm 0''.41$	$+28°45'28''.3 \pm 0''.60$

Some practical considerations are worth mentioning. First, the astronomic horizon is not the visible one, even in open sea: the eye of the observer is usually located at a certain height above the surface of the water, so that the visible horizon is a minor circle depressed with respect to the astronomical one according to this height and to the horizontal refraction. See, for example, a discussion of the effect in Smart (1965). Second, the celestial North pole is, at the present epoch, very near the bright star α Ursae Minoris (or Polar Star), a second magnitude star, while the South pole, in the constellation of Octans and approximately $20°$ distant from the bright galaxy Large Magellanic Cloud, has no bright nearby star. Therefore, the celestial poles are not directly materialized. However, they can be identified with great accuracy by using a number of circumpolar stars that permit the accurate determination of the position of the meridian and of the astronomic latitude, as we will discuss in the next paragraph and in Chapter 3, provided the atmospheric refraction (see Chapter 11) is taken into consideration. Much more intriguing has, in the past, been the determination of the longitude of a given location. Many ways were devised (for instance the ephemerides of the Moon or of the moons of Jupiter, as proposed by Galileo in the 17th century), but finally the superb clocks produced during the 19th century allowed precise time to be kept, even on board the great ships. Since then, continuous improvements in clock technology and the broadcasting of radio signals by networks of ground stations and of satellites has largely eliminated the problem.

2.2 The Hour Angle and Declination System

The time variation of the coordinates due to the diurnal rotation of the Earth will now be considered. On the celestial sphere, the meridian cuts the celestial equator in a point M (see Figure 2.3). For any given star X, let us draw the great circle passing through X and the visible pole P.

This circle, which also passes through the other pole Q, is called the hour circle of X. Let X' be the intersection of the hour circle on the equator, and call hour angle of X, $HA(X)$, the arc MX', usually counted westward from M in $(^h\ ^m\ ^s)$ between 0^h and 24^h. Other choices are possible. For example, it is very intuitive to measure HA positive westward from 0^h to 12^h, and negative eastward from 0^h to -12^h. Upper culmination on the meridian is when $HA = 0^h$, lower culmination when $HA = 12^h$. The second coordinate of X is the arc $X'X$, counted in $(°\ '\ '')$ from $0°$ to $90°$, positive toward the North pole and negative toward the South pole. This second coordinate, indicated with $\delta(X)$: arc $X'X = \delta(X)$, is named declination of X. With the passage of time during the day, each star X will therefore describe its parallel of declination, namely the small circle $\delta(X) = $ const, by continuously increasing its HA. The Sun behaves in the same way, but in addition it has an Easterly motion of approximately $1°$/day along a great circle called the ecliptic (see later) with respect to the fixed stars. The fact that the Sun in its daily motion does not

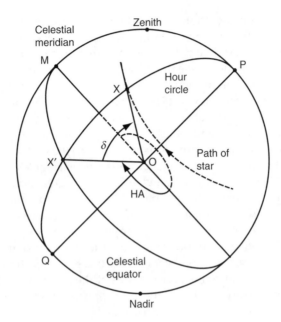

FIGURE 2.3
The hour angle and declination system.

move on a great circle has quite interesting consequences on the illumination of walls, but this argument will not be developed in this book. Similar but more complex considerations must be applied to the Moon, the planets, and other moving bodies such as comets and asteroids.

For a site at latitude ϕ in the Northern hemisphere, if $\delta(X) > 90° - \phi$, the star will never raise or set, being always above the horizon; these stars are called circumpolar stars. If instead $\delta(X) < -(90° - \phi)$ the star will never be visible above the local horizon. The same considerations apply to sites in the Southern hemisphere, with due account to the signs of ϕ and δ.

By means of circumpolar stars, we can determine ϕ even without knowing the declination of the star. Consider indeed a circumpolar star: the semi-sum of its altitudes h above the horizon in upper and lower culmination will immediately give ϕ; at the same time, the semi-difference will provide $\delta(X)$.

Be aware of the different aspect of the celestial sphere in the two hemispheres. Assume that you are in the Northern hemisphere, standing with your shoulders to the visible pole, namely with the South cardinal point in front, East to the left, and West to the right. All stars will describe their parallel of declination in the clockwise direction. If you are in the Southern hemisphere with the same attitude, the celestial North pole will be in front, and the stars will move in an anticlockwise direction, but of course always from East to West. In the same manner, the figures of the Moon and of the constellations will appear upside down.

2.3 The Equatorial System

The (HA, δ) system introduced in the previous paragraph had the virtue of fixing in time one of the two coordinates, namely the declination. To also make the second coordinate constant, a point on the equator is needed which remains as fixed as possible with respect to the stars. To define this point, let us consider the locus occupied by the Sun during its yearly motion, namely the great circle called ecliptic on which the Sun (indicated with ⊙) moves Eastward by approximately $1°/$day. The ecliptic is inclined to the equator by an angle ε (obliquity of the ecliptic) of approximately $23°27'$. The equator and ecliptic intersect each other in two opposite points called equinoxes: the vernal equinox is that point where the Sun transits at the beginning of the spring, around March 21st. The autumn equinox occurs six months later, around September 21st. On both points $\delta(⊙) = 0°$, but the derivative $\dot{\delta}(⊙)$ is positive in the first case and negative in the second. The vernal point is usually indicated with the astrological sign of Aries ♈, graphically approximated in the present text with the Greek letter γ (gamma), the autumn point with the astrological sign of Libra ♎ approximated with the Greek letter Ω (Omega).

The points on the ecliptic at $90°$ from the equinoxes are called solstices, respectively, summer solstice (around June 21st) and winter solstice (around December 22nd); the declination of the Sun in these points is $\delta_⊙ = \pm\varepsilon$. The great circles passing through the poles and the equinoxes or the solstices are called colures of the equinoxes or of the solstices, respectively. The poles of the ecliptic, indicated with E and E$'$, are points belonging to the colure of the solstices.

With reference to Figure 2.4, given a star X draw the great circle through P and X intersecting the equator in X$'$; as origin of the first angular coordinate choose the vernal equinox γ, and measure the angle γX$'$ in a direct sense: this angle is the right ascension of star X. The second angular coordinate of the equatorial system remains the declination $\delta(X)$ defined in the previous paragraph.

The right ascension of the star X is indicated with α, $\alpha(X) = $ arc γX$'$, and is usually is measured in (^{h} ^{m} ^{s}) from 0^h to 24^h. Notice the sense of α, opposite to that of HA. The right ascension can also be defined as the angle at the pole between the hour circle passing through X and the vernal colure.

For the Sun, at the equinoxes $\alpha_⊙ = 0^h$ and 12^h, respectively, at the solstices $\alpha_⊙ = 6^h$ and 18^h. The North pole of the ecliptic E has $\alpha(E) = 18^h$, $\delta(E) = 90° - \varepsilon$. Point E is in the constellation of Draco, near the gaseous nebula NGC 6543; the nearest bright star in the vicinity of E is ω Draconis, of the fourth visual magnitude, $3°$ away from it.

The right ascension will remain constant in time as much as γ remains fixed with respect to the stars. We will see in Chapter 5 that γ is subject to secular and periodic motions (general precession, nutation). However, on short time scales (say 1 year) the pair of coordinates (α, δ) will be almost constant.

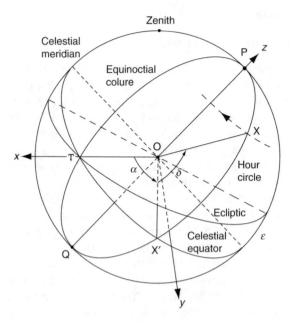

FIGURE 2.4
The equatorial system. The figure also shows a Cartesian (x, y, z) system having the celestial equator as xy plane, and the axis z directed to the celestial North pole P.

For longer periods, suitable and accurate correction formulae will be derived. The equatorial system is therefore the fundamental one for all precise descriptions of the celestial sphere, and it is employed by all major star catalogues.

Let us define a further angular quantity, namely the sidereal time (ST), as the hour angle of point γ:

$$ST = HA(\gamma)$$

Because of the rotation of the Earth, the sidereal time is a continuously varying angle between 0^h and 24^h. Taking into account the opposite sense of α and of HA, for any star X we have the fundamental relationship:

$$\alpha(X) = ST - HA(X) \tag{2.11}$$

which fixes the transformation between the hour angle and declination system and the equatorial system. The practical application of formula 2.11 requires due consideration to the adopted convention for HA, because by definition $0^h \le \alpha \le 24^h$. At any rate, when a star transits through the meridian in upper culmination, its right ascension coincides with the sidereal time. This is a very useful relation, which can be read in both ways: either we have a catalogue of fundamental stars giving accurate right ascensions, and so we measure ST by observing their upper transits, or we have a good knowledge of ST and we derive the right ascension of the transit stars. Let us examine this second possibility in more depth. ST is a quantity that varies with time (meant as the fundamental variable of all mechanical laws) in a fairly regular way; better yet, suppose that all deviations from uniformity are so small that we can disregard them. We can therefore build a clock whose reading coincides at each time with *ST*. For all practical applications we could then legitimately identify *ST* with a time, although the rigorous definition is the instantaneous angle on the celestial equator between the meridian and point γ. This approach was used by all astronomers of the past to build fundamental catalogues of stars. However, modern data show that the rotation of the Earth is not as uniform as it was supposed. Both the direction of the rotation axis in an inertial frame, and the angular velocity show well measurable fluctuations. This argument will be expounded in detail in later chapters; in the following paragraphs we will continue our discussion as if *ST* was for all purposes a uniform time.

Therefore, let us build a telescope with a mechanical mount having only one degree of freedom, that of elevation, while the optical axis is constrained as accurately as feasible in the meridian plane. The focal plane of the telescope is equipped with a high precision grid of wires, in order to accurately measure the instant of transit of the star. This telescope is known as a meridian circle or transit instrument, according to the several possible practical realizations. If we can identify the instant of transit of γ, and have our *ST* clock start from zero at that precise instant, by measuring the Zenith distance and the *ST* of passage of any star we derive its right ascension and declination at the date of observation. This procedure is easier said than done, all sort of systematic errors being possible, for example errors of the latitude of the observatory, of the position of the meridian, of the zero point and march of the clock, of the correction formulae for precession and nutation, etc. As an example, let us discuss one of these systematic errors, namely the dependence of the precision in right ascension from the declination. Indeed, a hypothetic polar star would never cross the meridian, while an equatorial star has the maximum linear speed in crossing the wires on the focal plane. Let s be the thickness of the wire; the time ΔT employed by the star to cross the wire will be:

$$\Delta T = \frac{Ks}{\cos \delta}$$

where K is an instrumental constant. Therefore, the error in ΔT is:

$$d\Delta T = Ks\left(\frac{\sin \delta}{\cos^2 \delta}\right)d\delta$$

namely with a systematic dependence from the declination.

The construction of a fundamental catalogue is indeed one of the most complex operations of all astronomy. We will see in Chapter 3 how *ST* can be directly linked to the position of the Sun on the ecliptic (in a somewhat pedantic sense, sidereal is a deceiving adjective: the Sun, not the stars, actually defines *ST*). Many stellar catalogues are simply of differential nature, giving positions relative to a set of fundamental stars. After the works carried out by many astronomers in the 17th, 18th, and 19th centuries (we recall the names of Flamsteed, Maskelyine, Bessel, and Auwers), in more recent time the so-called Fundamental Katalog FK was adopted, containing some 1500 bright stars. From 1964 to 1988 the standard catalogue was its fourth edition, FK4. The fifth revision, FK5, was published in 1988 (see Fricke et al., 1988). Positions and proper motions of 1535 bright stars were derived after a new determination of the origin of right ascension, with the adoption of the precession constants recommended by the International Astronomical Union (IAU) in 1976, and with the elimination of the elliptical aberration from the mean coordinates (as explained in Chapter 7). Furthermore, an extension of the FK5 was published, containing information on an additional 3117 secondary stars (most of them fainter than the primary stars, down to magnitude 9.5).

A new fundamental reference system, called the International Celestial Reference Frame (ICRF) and based on the position of a number of extra-galactic radio sources, has been adopted since 1997. The catalogue based on this frame, whose origin has been transferred to the barycenter of the Solar System, is called the International Celestial Reference System (ICRS). The catalogue of the European astrometric satellite Hipparcos has been tied to this system, and so have the ephemerides of the Solar System bodies published by the Jet Propulsion Laboratory. A fuller treatment of the ICRF will be given in Chapter 6.

For general astronomical applications, not limited to the construction of fundamental catalogs, the previous considerations have shown the advantages of building telescopes having an equatorial mount. Indeed, only the hour angle axis needs an accurate tracking rate, the declination axis being used for fast slewing and occasional corrections to the pointing. However, for all large telescopes, modern engineering has favored the Alt-Azimuth mount, because the structural flexures are much more controllable with vertical and horizontal axes. The disadvantage is the need to control, with great precision and continuously variable speed, not two but three axes (including the field rotation, as will be discussed in Chapter 3). Digital electronics has made it possible to solve this problem in an accurate and economic way. An example of a modern Alt-Az telescope is the already quoted 3.5 m TNG, shown in Figure 2.5.

2.4   The Ecliptic System

The system of ecliptic coordinates (see Figure 2.6) has the ecliptic as fundamental plane, inclined to the equator by the obliquity ε. The origin of the ecliptic longitudes is the same point γ, which belongs both to the ecliptic and to the equator. Ecliptic longitudes λ are given in ($^\circ\,{}'\,{}''$) between 0° and 360°, or in ($^h\,{}^m\,{}^s$) between 0^h and 24^h, increasing in the same sense of the right ascensions. Ecliptic latitudes β are given in ($^\circ\,{}'\,{}''$) between 0° and $\pm 90°$, as the declinations. All planets revolve around the Sun in the same direct sense

FIGURE 2.5
The 3.5 m Telescopio Nazionale Galileo (TNG). The vertical and horizontal axes are clearly seen. The field rotation is compensated by counter-rotation at the Nasmyth foci on the horizontal axis. (Photo by the author.)

(although they appear to have complex motions with respect to the fixed stars, for instance reversals of movement, and even stationary points), with orbital planes having small inclinations i to the ecliptic. The notable exceptions are Mercury ($i = 7°$) and Pluto-Charon ($i = 17°$). The obliquity of the ecliptic itself

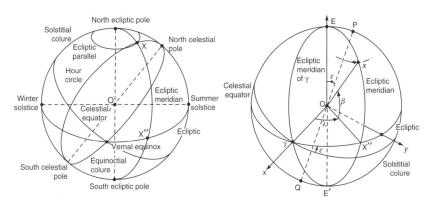

FIGURE 2.6
The ecliptic system. On the left, the ecliptic sphere, on the right the ecliptic coordinates longitude and latitude, and the Cartesian ecliptic axes.

is not absolutely fixed, the planetary perturbations cause a very slight decrease (at the present epoch) amounting to $0''.47$ per year. The ecliptic passes through an ensemble of stars that were organized by the ancient astronomers in constellations bearing the names of animals (these constellations define a band in the sky that was called Zodiac, from the Greek name of animals). The names given by the Latin poet Ausonius (IV century A.D.), are: Aries, Taurus, Gemini, Cancer, Leo, Virgo, Libra, Scorpio, Sagittarius, Capricornus, Acquarius, and Pisces. Those are also the names of the "astrological signs," that are fixed with respect to the calendar but not with respect to the stars, because of the precession of the equinox (see Chapter 4 through Chapter 6). Actually, there is a 13th constellation that perks through the Zodiac, namely Ophiucus, where the Sun passes in December, a fact which is mostly ignored by the astrologers.

For objects of the Solar System, an ecliptic Cartesian coordinate system (x, y, z) can be established, having its center at the Earth or at the Earth–Moon barycenter, or at the Sun's center or at the barycenter of the Solar System, according to the convenience. The x-axis is directed toward point γ, the y-axis at $90°$ in the ecliptic plane, and the z-axis directed toward the ecliptic North pole E.

2.5 The Galactic System

Another reference system, of a somewhat different nature than those of the previous frames where the Earth and the Sun played a fundamental role, must be considered. In the galactic system, the center coincides by definition with the observer, however, the fundamental plane is determined by the distribution in space of cosmic matter. Therefore, the practical construction of the system does not rely on measurements of directions, but on counts of stars or on the determination of hydrogen gas surface brightness in the various areas of the sky.

The old system of galactic coordinates was based on counts of stars, which confirmed the visual impression that we are at the center of a very flattened system with a plane of symmetry containing the Sun. The pioneering work of William Herschel at the end of the eighteenth century, and later that of his son John, must be remembered at this point. The Herschels set forth a coherent plan of stellar gauges over the entire celestial sphere, which was continued by many others astronomers after them, and finally led to the first galactic system (l^I, b^I). The longitudes l^I are counted in degrees between $0°$ and $360°$ in the direct sense. The latitudes b^I are counted in degrees between $0°$ and $±90°$. The North pole G of (l^I, b^I) is in the direction of the cluster of galaxies in Coma ($\alpha_G = 12^h.8$, $\delta_G = +27°.4$), almost in the plane of the ecliptic. This orthogonality between the ecliptic and the galactic plane has been the subject of many theoretical studies on the tidal effect of the Milky Way on the present structure of the Solar System.

The procedure of counting stars according to their position and apparent luminosity is by no means an easy task. Even applying all best statistical methods, the determination of the fundamental plane as the one passing through the areas of maximum density of stars is hampered by the presence of a strong interstellar absorption, whose amount increases precisely going into the plane (we shall examine the interstellar absorption again in Chapter 16). Immune to this absorption is instead the surface brightness of the interstellar gaseous hydrogen clouds, which are concentrated toward the galactic plane. Thanks to their low density and low temperature, they emit a strong spectral line in the radio-frequency domain, at $\lambda = 21$ cm (corresponding to 1420 MHz). It has been found that the galactic plane determined by the areas of maximum 21-cm brightness is inclined by approximately 3° to that of (l^I, b^I). Furthermore, a strong and point-like radio-source in Sagittarius, in the very direction of the Milky Way center, having equatorial coordinates $(\alpha_{GC} = 17^h42^m, \delta_{GC} = -28°55'$ at 1950.0), can be used as zero point for the longitudes. This radio-frequency galactic system, which became available around 1960, was initially named second galactic system (l^{II}, b^{II}). Its superiority over the old system was soon evident, so that by a resolution of IAU in 1976 the system based on star counts was discontinued, and the radio system was simply named (l, b). Should one need to use the old system, the Astronomical Observatory of Lund has published convenient transformation tables (see Notes). With more precision, referring to the equinox and equator at B1950.0 (see Chapter 4 for the definition of the Besselian year), the equatorial coordinates of the Galactic North pole G, and the position angle θ_{GC} of the galactic center from such a pole are:

$$\alpha_G = 12^h49^m, \quad \delta_G = +27°.4, \quad \theta_{GC} = 123° \quad (\text{B1950.0})$$

The position of the galactic system with reference to the equatorial one is shown in Figure 2.7.

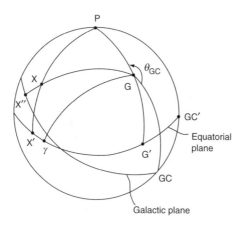

FIGURE 2.7
The system of galactic coordinates.

The galactic coordinates are never used for high precision positional work. Care must be exercised not to confuse galactic coordinates, whose center is the observer (but in practice the Sun), with galacto-centric (X, Y, Z) coordinates having the galactic center as the origin. The transformation between the two requires a model of the Milky Way, in particular the determination of the distance of the Sun from the center of the Galaxy, presently estimated at around 8 kiloparsecs.

To close this chapter, we mention that the distribution of the nearest galaxies, in particular those seen in Virgo, has been used to determine a super-galactic coordinate system (*L*, *B*); see, for example, deVaucouleurs and Peters, 1984.

Notes

1. There are other systems of coordinates used by astronomers and not covered in this book, for instance those on the surfaces of the Sun and of the planets, or those associated with the magnetic field of the Earth, or of Jupiter. See for instance in Hapgood, M., 1992, Space physics coordinate transformations: A user guide, *Planetary and Space Science*, **40**, pp. 711–717 (Hapgood, M., 1997, Space physics coordinate transformations: A user guide. *Planetary and Space Science*, **45**(8), p. 1047, Corrigendum). See also in: http://sspg1.bnsc.rl.ac.uk/Share/Coordinates/ct_home.htm and in: http://www.space-plasma.qmul.ac.uk/heliocoords/ by M. Franz.

2. Names of the constellations: The stars which are visible to the naked eye (numbering about 3000 in each hemisphere) have long since been arranged in constellations, namely in groups of stars defining broad areas over the celestial sphere, to be used as a first rough indication of direction. The names and borders of these groupings changed according to epoch and civilization, the constellations of the Chinese astronomers for instance, or of the Maya, being different from those of the Arabs. According to the present international convention (since 1922), there are 88 constellations in the sky with precisely defined borders, bearing the names of animals (Leo, Draco, Ursa Major, etc.), of mythological heroes (Cassiopea, Andromeda, etc.), of particular figures (Libra, Corona Borealis, etc.).

3. Names of the stars: The stars visible to the naked-eye are indicated by Greek letters followed by the abbreviated Latin name of the constellation, letter α indicating the brightest star of that constellation, β the second brightest, and so on until ζ, the last visible class (e.g., α UMa, β CMi, γ Dra). By telescopic observations, the lettering system was extended to cover the entire Greek alphabet. The visual brightness indicated by the Greek letter

is usually valid only inside a particular constellation, because keeping a uniform magnitude scale over the visible celestial hemisphere is a difficult task, impossible to achieve with the naked eye. The brightest naked-eye stars often have a proper name, deriving either from Greek or Arabian designations (e.g., the proper name of α Lyr is Vega, that of α Ori is Betelgeuse, that of β Ori is Rigel, that of β Per is Algol, and so on).
Two useful books in this regard are:
Allen, R.H., 1963, *Star Names, Their Lore and Meaning*, Dover Publication Inc., New York.
Bakich, M.E., 1995, *The Cambridge Guide to Constellations*, Cambridge University Press, Cambridge.

4. Counts of stars: see in Trumpler, R.J., Weaver, H., 1953, *Statistical Astronomy*, and in Binney, J., Merrifield, M., 1998, *Galactic Astronomy*.

5. Old and new galactic coordinates: the conversion between equatorial coordinates (at 1950.0), and the old and new galactic coordinates, was published in the Annals of the Observatory of Lund nr. 15, nr. 16 and nr. 17, 1962. A graphical representation of equatorial to new galactic coordinates transformation is given, e.g., by Kraus, J.D., 1966, *Radio Astronomy*.

3

Transformations of Coordinates

In this chapter, we consider several rules from transforming coordinates from one system to another. Two techniques will be used, that of matrix rotation and that of spherical trigonometry. In the majority of cases, the transformations will be rigid rotations around the origin. In other cases, a translation of origin must be added, for example, in passing from the geocentric to the heliocentric coordinates of a comet. Later on, we shall encounter phenomena that give rise to slight distortions of the celestial sphere, for example, the aberration and the gravitational deflection of light.

3.1 Transformations by Matrix Rotation

Consider two right-handed Cartesian orthogonal systems, say (x, y, z) and (X, Y, Z), having the same origin O. In order to transform the coordinates of any point P, (x, y, z), in one system to those in the other, (X, Y, Z), the following relationships can be used:

$$\begin{cases} X = x \cos xX + y \cos yX + z \cos zX \\ Y = x \cos xY + y \cos yY + z \cos zY \\ Z = x \cos xZ + y \cos yZ + z \cos zZ \end{cases} \tag{3.1}$$

or else, with matrix notation:

$$\begin{pmatrix} X \\ Y \\ Z \end{pmatrix} = \mathbf{R} \begin{pmatrix} x \\ y \\ z \end{pmatrix}, \quad \mathbf{R} = \begin{pmatrix} \cos xX & \cos yX & \cos zX \\ \cos xY & \cos yY & \cos zY \\ \cos xZ & \cos yZ & \cos zZ \end{pmatrix} \tag{3.2}$$

The distance r of P from O is clearly invariant under this rotation:

$$r^2 = x^2 + y^2 + z^2 = X^2 + Y^2 + Z^2$$

Let us consider the polar system (r, λ, β) defined by $O(x, y, z)$, and a rotated one (r, Λ, B) defined by $O(X, Y, Z)$

$$
\begin{cases}
x = r \cos \beta \cos \lambda \\
y = r \cos \beta \sin \lambda, \\
z = r \sin \beta
\end{cases}
\qquad
\begin{cases}
X = r \cos B \cos \Lambda \\
Y = r \cos B \sin \Lambda \\
Z = r \sin B
\end{cases}
\qquad (3.3)
$$

(as pointed out in Chapter 1, several authors prefer to use the complement of β as polar angle). By substituting in the previous matrix notation 3.4 we obtain:

$$
\begin{pmatrix}
\cos B \cos \Lambda \\
\cos B \sin \Lambda \\
\sin B
\end{pmatrix}
= \mathbf{R}
\begin{pmatrix}
\cos \beta \cos \lambda \\
\cos \beta \sin \lambda \\
\sin \beta
\end{pmatrix}
\qquad (3.4)
$$

where $\mathbf{R}$ must be specified for each case. Notice that all dependence on r has disappeared, so that those relations also apply on the sphere with unit radius. The inverse transformation is obtained by interchanging the role of (x, y, z) with (X, Y, Z), paying attention to maintain the positive sense on the angles. This implies that the matrix of the inverse rotation is the transpose of $\mathbf{R}$:

$$
\mathbf{R}^{-1} = {}^T\mathbf{R}, \quad ({}^T R_{ij} = R_{ji}), \quad \mathbf{R}^{-1}(\theta) = \mathbf{R}(-\theta), \quad (\mathbf{R}_i \mathbf{R}_j)^{-1} = \mathbf{R}_j^{-1} \mathbf{R}_i^{-1}
$$

Furthermore, a general rotation can always be represented as the result of three different successive rotations, $\mathbf{R}_1$ around the x-axis, $\mathbf{R}_2$ around the y-axis, $\mathbf{R}_3$ around the z-axis, $\mathbf{R} = \mathbf{R}_1 \mathbf{R}_2 \mathbf{R}_3$, with:

$$
\mathbf{R}_1(\phi_1) =
\begin{pmatrix}
1 & 0 & 0 \\
0 & \cos \phi_1 & \sin \phi_1 \\
0 & -\sin \phi_1 & \cos \phi_1
\end{pmatrix}, \quad
\mathbf{R}_2(\phi_2) =
\begin{pmatrix}
\cos \phi_2 & 0 & -\sin \phi_2 \\
0 & 1 & 0 \\
\sin \phi_2 & 0 & \cos \phi_2
\end{pmatrix},
$$

$$
\mathbf{R}_3(\phi_3) =
\begin{pmatrix}
\cos \phi_3 & \sin \phi_3 & 0 \\
-\sin \phi_2 & \cos \phi_3 & 0 \\
0 & 0 & 1
\end{pmatrix}
$$

As a first example, consider the transformation from equatorial coordinates (α, δ) to ecliptic coordinates (λ, β), by orienting the axes x and X from O toward the vernal point γ, the z-axis to the celestial North pole P, and the Z-axis to the ecliptic North pole E. The values of the angles are:

$$
xX = 0, \quad xY = \frac{\pi}{2}, \quad xZ = \frac{\pi}{2}, \quad yY = \varepsilon, \quad zZ = \varepsilon, \quad zY = \frac{3}{2}\pi + \varepsilon, \quad \text{etc.}
$$

where $\varepsilon \approx 23°27'$ is the obliquity of the ecliptic. The two systems are therefore connected by a rotation of ε around the x-axis, $\mathbf{R}_1(\varepsilon)$, or inversely

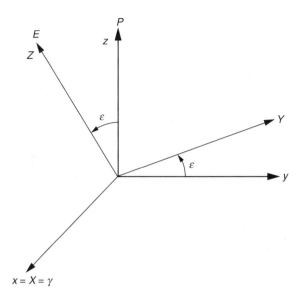

FIGURE 3.1
Transformation between equatorial and ecliptic coordinates.

of $-\varepsilon$ around the X-axis (see Figure 3.1).

$$\mathbf{R}_1(\varepsilon) = \begin{pmatrix} 1 & 0 & 0 \\ 0 & \cos\varepsilon & \sin\varepsilon \\ 0 & -\sin\varepsilon & \cos\varepsilon \end{pmatrix} \approx \begin{pmatrix} 1 & 0 & 0 \\ 0 & 0.9171 & 0.3987 \\ 0 & -0.3987 & 0.9171 \end{pmatrix}$$

The sought-for transformation is therefore:

$$\begin{cases} \cos\beta\cos\lambda = \cos\delta\cos\alpha \\ \cos\beta\sin\lambda = \cos\delta\sin\alpha\cos\varepsilon + \sin\delta\sin\varepsilon \\ \sin\beta = -\cos\delta\sin\alpha\sin\varepsilon + \sin\delta\cos\varepsilon \end{cases} \qquad (3.5)$$

The inverse is (pay attention to the signs):

$$\begin{cases} \cos\delta\cos\alpha = \cos\beta\cos\lambda \\ \cos\delta\sin\alpha = \cos\beta\sin\lambda\cos\varepsilon - \sin\beta\sin\varepsilon \\ \sin\delta = \cos\beta\sin\lambda\sin\varepsilon + \sin\beta\cos\varepsilon \end{cases} \qquad (3.6)$$

You will notice that three equations are needed to determine two angles and their signs (quadrants). We have already remarked that great care is needed in numerical calculations, especially in the proximity of the poles of the systems.

With the same technique we can transform Alt-Az (A, h) coordinates (with our adopted origin from the South; several authors prefer the Zenith distance z instead of the altitude h) in hour angle (HA) and declination (δ),

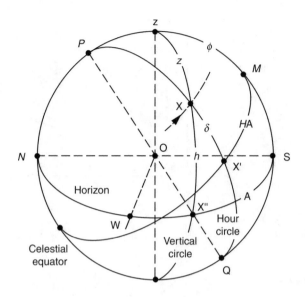

FIGURE 3.2
Transformation between Alt-Az and hour angle and declination of star X. Arc X″X is the elevation h, arc X′X the declination δ; arc MX′ is the hour angle HA, arc SX″ is the Azimuth A, arc NP = arc MZ is the latitude ϕ.

and then, by the knowledge of the sidereal time, to equatorial (α, δ). In this case, axes x and X will both point to W, axis y to S, axis z to Z, axis Y to M on the celestial equator, and axis Z to the celestial North pole P (see Figure 3.2). Clearly, the astronomical latitude ϕ of the site is needed.

In this case, the rotation matrix will be:

$$\mathbf{R} = \begin{pmatrix} 1 & 0 & 0 \\ 0 & \sin\phi & \cos\phi \\ 0 & -\cos\phi & \sin\phi \end{pmatrix} \tag{3.7}$$

However, by convention, the sense of the Cartesian angles is opposite to that of HA and A, both increasing in the retrograde sense, and therefore:

$$\begin{cases} \sin HA \cos\delta = \sin A \cos h \\ \cos HA \cos\delta = \cos A \cos h \sin\phi + \sin h \cos\phi \\ \sin\delta = -\cos A \cos h \cos\phi + \sin h \sin\phi \end{cases} \tag{3.8}$$

while the inverse transformation is:

$$\begin{cases} \cos h \sin A = \cos\delta \sin HA \\ \cos h \cos A = \cos\delta \cos HA \sin\phi - \sin\delta \cos\phi \\ \sin h = \cos\delta \cos HA \cos\phi + \sin\delta \sin\phi \end{cases} \tag{3.9}$$

Suppose now that we know the equatorial coordinates (α, δ) of a given star X, and the sidereal time (ST), so that HA is also immediately known. In order to point a telescope having an Alt-Az mount toward X, we need to calculate its (A, h) from Equation 3.9; the third equation will tell us if the star is visible above the horizon. The visibility limit $h = 0°$ is reached when:

$$\cos HA = -\tan \delta \tan \phi \qquad (3.10)$$

(hour angle of rising or setting). In the same way, the third of Equation 3.8 gives the Azimuth of rising or setting:

$$\cos A = -\sin \delta \sec \phi$$

Notice that only the equatorial stars $(\delta = 0°)$ raise and set exactly on the E and W points. For the circumpolar stars it is always true that $|\tan \delta \tan \phi| > 1$, so that they are visible above the horizon at all times. The two following relations also apply:

$$A = \arctan \frac{\sin HA \cos \delta}{\cos HA \cos \delta \sin \phi - \sin \delta \cos \phi} \qquad (3.11)$$

$$h = \arcsin(\cos \delta \cos HA \cos \phi + \sin \delta \sin \phi) \qquad (3.12)$$

For a circumpolar star having $|\delta| > |\phi|$, the Azimuth remains constrained between a maximum and a minimum value (occidental or oriental digression, or elongation), for which the following relations hold:

$$\cos HA = \tan \phi \cot \delta, \quad \sin h = \sin \phi / \sin \delta, \quad \sin A = -\cos \delta / \cos \phi$$

Notice that two solutions are possible for HA and A, one is for the occidental, the other for the oriental digression, as it can be easily discriminated.

It is useful also to derive the angular velocities $\dot{A}, \dot{h}$, by using the sidereal time as the time variable $t = ST$, and neglecting the effects of atmospheric refraction (plus other much smaller terms due to the motion of the equinox and of the pole):

$$\frac{dHA}{dt} = 1, \quad \dot{\delta} = 0, \quad \dot{h} \cos h = -\cos \delta \sin HA \cos \phi$$

From this and from the first of Equations 3.8 we obtain:

$$\dot{h} = -\cos \phi \sin A, \quad \dot{A} = \sin \phi + \cos A \tan h \cos \phi \qquad (3.13)$$

The velocity in altitude $\dot{h}$ is always restricted between ± 1 (namely $\pm 15°/$ sidereal hour), it is nil for a telescope at the geographic poles, and it is maximum for an equatorial telescope. More complex is the behavior of the Azimuthal velocity. At the horizon $\dot{A} = \sin \phi$, and therefore it is positive in the Northern hemisphere, and negative in the Southern, both at rise and set (and obviously stationary at the terrestrial equator). This result can be understood by remembering that we have defined the apparent sense of rotation of the celestial sphere by turning our shoulders to the visible pole.

In the particular case of a star transiting at the Zenith ($\delta = \phi$) the Azimuthal velocity becomes infinitely large when the star approaches the meridian; for this reason a telescope with an Alt-Az mount has a blind spot around the Zenith, a cone whose aperture can be made smaller than $1°$ with careful selection of the motors and associated controls.

For a circumpolar star, at the maximum digressions the velocity is all in altitude; this fact can be advantageously utilized for better determination of the meridian and of the latitude.

3.2 Transformations by Spherical Trigonometry

Spherical trigonometry is the second method of transformation. As an example, Figure 3.3 gives the elements necessary to carry out the transformation between equatorial and ecliptic coordinates. We would easily find Equation 3.8 and Equation 3.5 again, and their inverse.

More complex is the transformation from equatorial to galactic, so it is better to see it in detail. With reference to Figure 2.4, let P be the celestial North Pole, G the galactic center, and CG the galactic plane. Given a star X, from the spherical triangles we obtain:

$$\begin{cases} \cos b \sin(\theta_{GC} - l) = \cos \delta \sin(\alpha - \alpha_{GC}) \\ \cos b \cos(\theta_{GC} - l) = \cos \delta_G \sin \delta - \sin \delta_G \cos \delta \cos(\alpha - \alpha_G) \quad (3.14) \\ \sin b = \sin \delta_G \sin \delta + \cos \delta_G \cos \delta \cos(\alpha - \alpha_G) \end{cases}$$

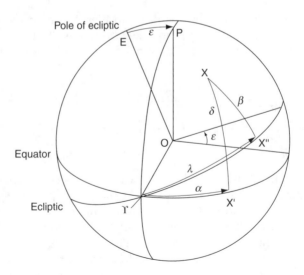

FIGURE 3.3
Transformation of equatorial to ecliptic coordinates.

and inversely:

$$\begin{cases} \cos \delta \sin(\alpha - \alpha_G) = \cos b \sin(\theta_{GC} - l) \\ \cos \delta \cos(\alpha - \alpha_G) = \sin b \cos \delta_G - \cos b \sin \delta_G \cos(\theta_{GC} - l) \quad (3.15) \\ \sin \delta = \sin b \sin \delta_G + \cos b \sin \delta_G \cos(\theta_{GC} - l) \end{cases}$$

Should one prefer the technique of matrix rotation, remember the equatorial coordinates of the three points:

$$\gamma(0,0), \quad G(192°.3, +27°.4), \quad GC(265°.6, -28°.9)$$

(at epoch B1950.0, as defined by IAU), and calculate the angular distances:

$$\cos \gamma G = \cos xZ = -0.86760, \quad \gamma G = xZ = 150°.2,$$
$$\cos \gamma GC = \cos xX = -0.06690, \quad \gamma GC = xX = 93°.9$$

and so on. Finally, the complete direct rotation matrix at epoch B1950.0 is:

$$\mathbf{R}_G = \begin{pmatrix} -0.06690 & +0.49273 & -0.86760 \\ -0.87276 & -0.45035 & -0.18838 \\ -0.48354 & +0.74459 & +0.46020 \end{pmatrix} \quad (3.16)$$

Notice that the equatorial coordinates of the star have to be precessed to B1950.0 before carrying out the transformation. Although not formally defined, and remembering the low precision usually needed for galactic coordinates, we can assume the following values for J2000:

$$G(192°.84, +27°.13) = (12^h51^m, +27°07'.7),$$
$$GC(266°.41, -28°.94) = (17^h45^m.6, -28°56'.2).$$

3.3 Some Examples and Applications

1. Let us apply the previous relationships to calculate the angular distance between two stars X_1 and X_2, a number that is independent of the particular system of coordinates. To be specific, consider equatorial coordinates, and draw the great circle passing through the two stars (see Figure 3.4).

   Irrespective of the position of γ, we have:

   $$\cos X_1 X_2 = \sin \delta_1 \sin \delta_2 + \cos \delta_1 \cos \delta_2 \cos \Delta\alpha, \quad \Delta\alpha = |\alpha_1 - \alpha_2| \quad (3.17)$$

   Let us now imagine viewing the celestial sphere from O. The great circle X_1P gives the direction of the celestial North pole through X_1; called position angle p of star X_2 with respect to star X_1 the angle $P\hat{X}_1X_2$ counted from the North toward East, which is also the angle

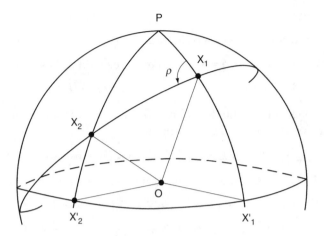

FIGURE 3.4
Computing the angular distance between two stars X_1 and X_2.

at the vertex X_1 of the spherical triangle X_1PX_2. Therefore:

$$\sin X_1X_2 \sin p = \cos \delta_2 \sin \Delta\alpha$$

$$\sin X_1X_2 \cos p = \cos \delta_1 \sin \delta_2 - \sin \delta_1 \cos \delta_2 \cos \Delta\alpha$$

In many applications, for example, binary stars, pairs of galaxies, etc., the angular distance between the two objects is so small that no error is committed by allowing:

$$\cos X_1X_2 = 1, \quad \sin X_1X_2 = X_1X_2 = s, \quad s \sin p = \cos \delta_1 \Delta\alpha,$$

$$s \cos p = \Delta\delta, \quad s = \sqrt{(\Delta\alpha \cos \delta_1)^2 + \Delta\delta^2}$$

2. Consider a telescope mechanically mounted in Alt-Az. Its field of view is in continuous rotation with variable angular speed, because the celestial sphere rotates around a direction that does not coincide with that of the mechanical axes. For a given star X, call parallactic angle q the angle $q = Z\hat{P}X$, which can be expressed as:

$$q = Z\hat{P}X = \frac{\sin A \cos \phi}{\cos \delta} \tag{3.18}$$

Its angular velocity (time derivative with $t = ST$) is:

$$\dot{q} = \frac{\cos \phi \cos A}{\cos \delta \cos q}\dot{A} = \frac{\cos \phi \cos A}{\cos h} \tag{3.19}$$

Field rotation is also encountered in equatorially mounted telescopes if part of the structure is fixed with respect to the ground, for example, in the so-called Coudè focus, where the light is brought by several mirrors to a large instrument on the floor of the observatory.

When the parallactic angle is 90° (namely when the hour and vertical circles through the star are perpendicular to each other), the star is said to be at digression, or elongation, as we have discussed before.

3. Apply the transformation between equatorial and ecliptic coordinates to the Sun (assuming that $\beta_\odot = 0°$, an approximation valid for the present purpose). From Equation 3.6 it is immediately seen that:

$$\sin \alpha_\odot = \tan \delta_\odot / \tan \varepsilon \qquad (3.20)$$

so that, with due consideration to the date (namely to the quadrant), the measurement of $\delta_\odot$ gives at any time the origin of the right ascension, namely the position of point γ. This consideration underlines the fundamental role played by the Sun in the definition of the sidereal time.

4. The astronomical night is defined as that period of time when the Sun is 18° or more below the local horizon (see Figure 3.5).

The period when the Sun has $0° < h_\odot < -6°$ is the civil twilight, when $-6° < h_\odot < -12°$ is the nautical twilight, and finally that $-6° < h_\odot < -18°$ is the astronomical twilight. The inset of the civil twilight can be ascertained by the appearance of first magnitude stars that of astronomical twilight by the visibility of 4th magnitude stars. Those durations obviously depend not only on the latitude of the site, but also on the season (namely on $\delta_\odot$). To determine the length of the above arcs, and ignoring the small corrections due to atmospheric refraction and diurnal Eastward motion of the Sun, the relation:

$$\sin h_\odot = \cos HA_\odot \cos \delta_\odot \cos \phi + \sin \delta_\odot \sin \phi$$

can be utilized. Table 3.1 shows examples of the duration of the astronomical night for sites at $\phi = 0°, 23°.5, 45°, 66°.6$ for the three

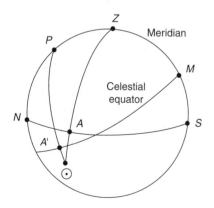

FIGURE 3.5
The astronomical night. The arc $A_\odot$ must be $\geq 18°$.

TABLE 3.1

Duration of the Astronomical Night

	$\delta_{\odot}$			$\delta_{\odot}$		
	$-\varepsilon$	$0°$	ε	$-\varepsilon$	$0°$	ε
$|\phi|$		$HA_{\odot}$ **(Hours)**		**Duration Astronomical** **Night (Hours)**		
$00°.0$	7.31	7.20	7.31	9.39	9.60	9.38
$23°.5$	6.69	7.31	8.25	10.62	9.38	7.50
$45°.0$	6.16	7.73	10.38	11.68	8.54	3.24
$66°.5$	5.41	9.39	—	13.18	5.24	—

cases $\delta_{\odot} = 0°$, $\pm\varepsilon$. It is customary to give the durations only to the minute. The total number of astronomical hours is larger for an observatory at lower latitudes; furthermore, it can be seen that, because of the different duration of the seasons (see also in Chapter 4 and Chapter 10), the Southern hemisphere has slightly more dark hours than a Northern one.

The legislation of several countries refer to the start and end of the civil night for obligations such as switching on and off the road lamps, the car lights, etc.

By the same formulae, one can calculate the duration of the arc where the Sun is always above the horizon.

Section A12 of the *Astronomical Almanac* provides the approximate universal time (UT) of rising or setting of a body having equatorial coordinates (α, δ) for every year, for any given place of latitude ϕ and longitude Λ:

$$UT = 0.997[\alpha - \Lambda \pm \cos^{-1}(-\tan\phi\tan\delta) - GMST \text{ at } 0^h \text{ UT}]$$

where GMST is the Greenwich mean sidereal time (see Chapter 4 for the definition of UT and GMST). The positive sign corresponds to setting, and the negative to raising. Those instants are calculated ignoring the horizontal atmospheric refraction, which amount to approximately $34'$ (see also Chapter 10). The rising or setting the center of the Sun is actually $51'$ below the astronomical horizon (the visible horizon depending, as already said, by the altitude of the observer on the sea level). The same considerations apply to the Moon, complicated, however, by the much smaller distance of this body (lunar parallax), so that the position of the observer on the geoid must be taken into account.

Exercises

1. *Duration of the day.* The angular dimension of the Sun is approximately $32'$. The duration of the day (meaning the period

of time when the Sun is above the horizon) is usually calculated with reference to the upper refracted limb of the solar disk, not to the center. Show that the correction to Equation 3.10 is equal to $-\sin 51' \sec \phi \sec \delta$. Calculate the duration of the day (arc of Sun above the horizon) for a site at $\phi \approx +46°$, respectively, around the summer and around the winter solstice (answer: 15^h44^m in summer, 8^h39^m in winter).

2. *Azimuth counted from the North.* Suppose that the Azimuth A is counted from N toward E; find the matrix rotation to derive (HA, δ) from (A, h); answer: $\mathbf{R} = \mathbf{R}_2(\phi - 90°)\mathbf{R}_3(180°)$.

   The inverse needed to obtain (A, h) from (HA, δ) is:

   $$\mathbf{R}^{-1} = \mathbf{R}_3(-180°)\mathbf{R}_2(90° - \phi)$$

   In practice, change sign to $\sin A$ and $\cos A$.

   In the same way, if the Zenith distance z is used instead of the altitude h, replace $\cos h$ with $\sin z$.

3. *Matrix conversion from (α, δ) to (HA, δ).* In this case the full machinery of matrix conversion is surely overabundant. However, when it is set up it will also work in simple cases. A star has equatorial coordinates $\alpha = 17^h20^m36^s.622$, $\delta = +40°00'00''.00$, find (HA, δ) when the sidereal time is $ST = 6^h10^m16^s.550$. The rotation angle bringing the hour angle in coincidence with the right ascension has module ST around the polar axis:

$$\mathbf{R}_3(ST) = \begin{pmatrix} \cos ST & \sin ST & 0 \\ -\sin ST & \cos ST & 0 \\ 0 & 0 & 1 \end{pmatrix} = \begin{pmatrix} -0.04482 & +0.99899 & 0 \\ -0.99899 & -0.04482 & 0 \\ 0 & 0 & 1 \end{pmatrix}$$

However, (HA, δ) is a left-handed system, (α, δ) is a right-handed one, so the correct transformation by Equation 3.2 is:

$$\begin{pmatrix} X \\ Y \\ Z \end{pmatrix}_{HA,\delta} = \begin{pmatrix} 1 & 0 & 0 \\ 0 & -1 & 0 \\ 0 & 0 & 0 \end{pmatrix} \mathbf{R}_3(ST) \begin{pmatrix} x \\ y \\ z \end{pmatrix}_{\alpha,\delta}$$

$$= \begin{pmatrix} 1 & 0 & 0 \\ 0 & -1 & 0 \\ 0 & 0 & 0 \end{pmatrix} \begin{pmatrix} \cos ST & \sin ST & 0 \\ -\sin ST & \cos ST & 0 \\ 0 & 0 & 1 \end{pmatrix} \begin{pmatrix} \cos \alpha \cos \delta \\ \sin \alpha \cos \delta \\ \sin \delta \end{pmatrix}$$

$$= \begin{pmatrix} -0.748103 \\ -0.164705 \\ +0.642817 \end{pmatrix}$$

Finally:

$$HA_1 = \arctan\frac{Y}{X} = \arctan 0.2201637 = 12°24'58''.91,$$

$$\delta = \arctan\frac{Z}{\sqrt{X^2 + Y^2}} = \arcsin Z = \arcsin 0.6428175 = 40°00'08''.05$$

However, HA_1 cannot be the correct answer, because X and Y are both negative, so we have to add $180°$ to HA_1, finally obtaining $HA = 192°24'58''.92 = 12^h49^m39^s.928$, a result which could have been immediately obtained by $(ST - \alpha)$.

The calculations made for this exercise were purposely carried out with only five significant digits, to show that the final numbers are slightly in error. Repeat with seven digits.

4. A given star X has $(A, h) = (317°.6, 32°.4)$ as observed from a site having $\phi = 45°51'$; determine its (HA, δ).

Disregarding the atmospheric refraction, from Equation 3.8 we immediately obtain $\delta(X) = -3°.0$, $HA(X) = 325°.3 = 21^h41^m$ $(= -34°.7 = -2^h19^m)$. Therefore, the star X will reach upper culmination after 2^h19^m.

5. From a site having $\phi = 28°.755$ observe a star X having $(HA, \delta) = (-9^h30^m, 85°18')$; determine its (A, z).

Notice that the star is circumpolar. Again, without considering the atmospheric refraction, from

$$\cos z = \sin \delta \sin \phi + \cos \delta \cos \phi \cos HA$$

get $z = 65°.01$ $(h = 24°.99)$; and then, from

$$\cos A = (\sin \delta - \cos z \sin \phi)/\sin z \cos \phi,$$

get $A = 184°.2$.

However, the star has a fairly high declination, so it would be advisable to make recourse to the expression of $\tan A$ from Equation 3.9 for better precision.

6. Assume a small error $d\alpha$, $d\delta$ in your knowledge of the equatorial coordinates of a given star. Determine the corresponding errors on the ecliptic coordinates if the error on ε can be considered negligible. Solve the inverse problem. Discuss the behavior of the errors in proximity of the ecliptic (respectively, celestial) poles. Hint: make recourse to the angle p between the declination and the latitude circles.

4

The Movements of the Earth and the Astronomical Times

In this chapter, some notions on the diurnal rotation and annual revolution of the Earth will be expounded, in order to allow a first understanding of the several definitions of time used in Astronomy. More explanations on the different time scales will be provided in Chapter 10, while the dynamics of the Earth's rotation and revolution will be discussed in Chapter 6 and Chapter 12, respectively. Further notions on the time scales based on the Moon are given in Chapter 14 and Chapter 15.

4.1 The Movements of the Earth

The Earth's diurnal rotation takes place around a polar axis whose direction, with respect to the distant stars (namely in an inertial frame of reference), will be considered in this chapter as invariable, and with absolutely constant angular velocity. In other words, the rotation is expressed by a constant vector Ω. Furthermore, the direction of Ω is assumed to coincide with the polar axis c of the ellipsoid which mathematically describes the Earth's figure. These assumptions are an approximation to the reality, but they are convenient for the present purposes.

The apparent direct movement of the Sun with respect to the fixed stars, amounting to approximately 1° per day, Eastward on the ecliptic, reflects the annual revolution of the Earth around the Sun. According to the first two Kepler's laws, expressed for the geocentric observer:

1. The apparent orbit of the Sun is an ellipse of eccentricity e and semimajor and semiminor axes a and b, having the Earth in one of the two foci. The equation of the ellipse is:

$$\frac{1}{r} = \frac{1}{p}[1 + e \cos v], \quad a = \frac{p}{1 - e^2}, \quad b = a\sqrt{1 - e^2} \quad (4.1)$$

The argument ν is named the true anomaly of the Sun: by definition, its initial direction $\nu = 0$ coincides with that of the semimajor axis a, at the instant when the Sun passes through the perigee Π. Given that the civil year begins when the longitude of the Sun is approximately 280°, while the longitude of the perigee at the present time is $\lambda_\Pi \approx 282°.97$ (with low precision, use the formula: $\lambda_\Pi = 282.940 + 0°.017\Delta t$, with Δt in years since 2000), this passage occurs a few days after the beginning of the year, for instance the 2nd of January in 2002.

2. The areal velocity (not the angular one!) of the Sun along the ecliptic is constant:

$$\dot{A} = \frac{dA}{dt} = \frac{1}{2}r^2\frac{d\nu}{dt} = \frac{C}{2} \tag{4.2}$$

where r is the distance Earth–Sun, and C (or $C/2$) is the so-called area-constant. These two movements provide the basis for two astronomical time scales, one connected with the diurnal rotation (the day), the other with the annual revolution (the year).

4.2   The Sidereal Time (*ST*)

In Chapter 2, the sidereal time (*ST*) as the hour angle of the vernal equinox γ: $ST = HA(\gamma)$ was defined. At each rotation of the Earth, *HA* increases by one sidereal day of 24 h. Notice that *HA* is an angle along the celestial equator, representing the position of the equinox with respect to the local meridian. However, the equinox γ itself is not directly visible as a point, its position being defined by the declination of the Sun $\delta_\odot$ through the relation 3.20:

$$\sin \alpha_\odot = \cot \varepsilon \tan \delta_\odot$$

In other words, *ST* is operationally referred to the Sun, not to the stars, so that the adjective "sidereal" is somewhat misleading. In practice, the operation of referring the equinox directly to the Sun is seldom carried out. In order to determine *ST*, it is much easier to utilize the upper meridian transit of a set of fundamental stars, whose right ascensions also define the origin of the equatorial system. A word of caution is in order here because each particular set of fundamental stars determines a slightly different equinox. Presently, the best realization of the fundamental catalogue is the already quoted ICRS (adopted by resolution of the IAU starting January 1st, 1998), whose right ascension origin can be taken to define the position of γ. At any rate, irrespective of its origin, the sidereal time is very uniform. Better yet, it is as uniform as the rotation of the Earth.

4.3 The Solar Time and the Equation of Time

A second rotational time scale can be defined by using the Sun, which for everyday life is certainly much more important than the equinox. Therefore, let us call solar day the interval of time between two successive upper culminations of the Sun on the meridian of a particular site, and solar time $T_\odot$ the hour angle of the Sun, augmented by 12 h so that the solar day starts at midnight, not at noon (this convention was adopted in 1925, however, for the three following years not all observatories conformed to the resolution, so that care must be taken when using the dates preceding 1928):

$$T_\odot = HA_\odot + 12^{\mathrm{h}}$$

This is the time indicated by a sundial (apart from the effects of the atmospheric refraction that can be ignored in this context) in that particular place. Notice that while the sidereal time derives only from the rotation of the Earth, the solar time reflects both the diurnal rotation and the yearly revolution: those two movements do not have any fundamental connection (apart a very slight influence through the constants of precession that here we may ignore). This independence is also at the root of the difficulties for building yearly calendars based on counting integer numbers of solar days, as will be discussed later.

Furthermore, we must pay attention that the Sun does not belong to the equator, but to the ecliptic, moving on it according to Kepler's first two laws; those two factors affect both the duration and the uniformity of the solar time. Regarding the duration, the Sun appears to move in a direct sense (Eastward) on the ecliptic by approximately $1°$ each day (more precisely, on average by $360°/365$ days $\approx 3^{\mathrm{m}}56^{\mathrm{s}}/$day) with respect to the fixed stars, and therefore also with respect to the equinox (at least in this approximation); these $3^{\mathrm{m}}56^{\mathrm{s}}$ represent the extra time the Sun takes to pass the following day in meridian with respect to the equinox. The solar day (indicated with j; more precisely, j is the mean solar day) is then, on average, $3^{\mathrm{m}}56^{\mathrm{s}}$ longer than the sidereal day. In the same manner, all units of solar time are correspondingly longer than the units of sidereal time having the same name.

Regarding the uniformity, the solar time $T_\odot$ is grossly nonuniform, as we show in detail. Given that $HA_\odot = ST - \alpha_\odot$, then:

$$T_\odot = HA_\odot + 12^{\mathrm{h}} = ST - \alpha_\odot + 12^{\mathrm{h}} \tag{4.3}$$

It is readily seen from Equation 4.3 that $T_\odot$ and $\alpha_\odot$ have the same degree of nonuniformity. Let $\lambda_\odot$, $\alpha_\odot$, $\delta_\odot$ be, respectively, the ecliptic longitude, right ascension, and declination of the Sun; from Equation 3.6 the following relations can easily be derived:

$$\sin \delta_\odot = \sin \lambda_\odot \sin \varepsilon, \quad \tan \alpha_\odot = \tan \lambda_\odot \cos \varepsilon \tag{4.4}$$

Taking the time derivative of the second equation and inserting the first one, we obtain:

$$\dot{\alpha}_\odot = \frac{\cos \varepsilon}{1 - \sin^2\lambda_\odot \sin^2\varepsilon} \dot{\lambda}_\odot = \frac{\cos \varepsilon}{\cos^2\delta_\odot} \dot{\lambda}_\odot \qquad (4.5)$$

Formula 4.5 comprises both the above mentioned effects, namely that of variable angular speed on the ecliptic and that of projection on the equator. To quantify the nonuniformity of $\dot{\alpha}_\odot$ we take into account both factors:

- According to Kepler's II law, the Sun has a daily motion greater at the perigee than at the apogee: $\dot{\lambda}_\odot \approx 61'.1 \approx 4^m4^s\,j^{-1}$ around the 4th of January, $\dot{\lambda}_\odot \approx 57'.2 \approx 3^m49^s\,j^{-1}$ around the 4th of July. This factor alone would produce a solar day 15^s longer at the beginning of January than at the beginning of July.

- The same motion of $\dot{\lambda}_\odot$ on the ecliptic corresponds to different arcs on the equator according to the declination, from a minimum value of $\dot{\lambda}_\odot\cos\varepsilon$ ($\approx 3^m37^s$) per day at the equinoxes, to a maximum value of $\dot{\lambda}_\odot/\cos\varepsilon$ ($\approx 4^m16^s$) at the solstices. Due to this effect of projection, a constant solar motion along the ecliptic would result in a day approximately 39^s longer at the equinoxes than at the solstices.

Therefore, the duration of the true solar day is continuously variable, for two different reasons which are out of phase and combine with each other with different signs; as a result, the longest solar day occurs around mid-December, and lasts approximately $24^h00^m30^s$, approximately 52^s longer than the shortest day which occurs a few days before the autumn equinox. Those seemingly small differences steadily accumulate with the passage of the days, reaching several minutes before changing sign, as we will discuss in the paragraph *Equation of Time E*.

In order to construct a uniform solar time, following Newcomb (1906), let us introduce two hypothetical Suns having constant angular velocity, namely a fictitious Sun $F_\odot$ moving on the ecliptic (called by some authors dynamic mean Sun), and a mean Sun $M_\odot$ moving on the equator. Both bodies move with the same constant daily motion $n = 3548''.3\,j^{-1}$. This value derives from the length of the tropical year (period of time between two consecutive passages of the Sun through point γ, see later a for a more precise definition).

By definition, the fictitious Sun $F_\odot$ coincides with the true Sun at perigee Π and apogee A. It trails behind the true Sun between perigee and apogee, and it precedes it between apogee and perigee. The longitude of $F_\odot$, $\lambda(F_\odot)$, is called the mean longitude of the Sun; it must not be confused with the longitude of the mean Sun. From the definition of $F_\odot$, it follows that the nonuniformity of its right ascension, namely the variation of $\dot{\alpha}(F_\odot)$ with the day along the year, is caused only by the projection effect due to its varying declination.

The difference between the ecliptic longitude of the true Sun and that of the fictitious Sun, namely the quantity $EC = \lambda_\odot - \lambda(F_\odot)$, is called the equation of the center (EC). It is a quantity which can be calculated from the equation of motion of the Sun in its orbit, as shown by the following considerations: the elements of the solar orbit are the eccentricity e, the instant t_0 of passage of the Sun through the perigee, the true anomaly v, and the mean anomaly $M = n(t - t_0)$, which is an auxiliary quantity uniformly increasing with time (it is convenient to express t in mean solar days j, so that $n = 236^s.555\,\mathrm{j}^{-1}$). Leaving the demonstration to Chapter 12 and Chapter 13, the following first-order relation can be derived:

$$\lambda_\odot = \lambda_\Pi + v \approx \lambda_\Pi + M + 2e \sin M \qquad (4.6)$$

where the last term expresses in an approximate manner the deviation from a circular orbit. By definition, the longitude of the fictitious Sun is a linear function of time:

$$\lambda(F_\odot) = \lambda_\Pi + n(t - t_0) = \lambda_\Pi + M \qquad (4.7)$$

Finally:

$$EC = \lambda_\odot - \lambda(F_\odot) = v - M \approx 2e \sin M = 0.03345R' \sin M = 115' \sin M \qquad (4.8)$$

At the first order, the equation of the center (EC) is therefore a periodic function of time, with a period of 12 months and an amplitude of approximately $115'$, namely 7^m40^s, roughly corresponding to the arc described by the Sun in 2 days. The phenomenon is so evident that Claudius Ptolemy could already ascertain it, although with an excessive value. Copernicus is credited with a determination very close to the true one.

Notice the different origins of the angles, the longitudes start from the vernal point γ, the anomalies from the perigee Π. From Equation 4.7, $M = \lambda(F_\odot) - \lambda_\Pi$, so that EC can also be written as:

$$EC = 2e \sin \lambda(F_\odot) \cos \lambda_\Pi - 2e \cos \lambda(F_\odot) \sin \lambda_\Pi + \cdots \qquad (4.9)$$

Let us take the time derivative of Equation 4.6, using the mean solar day j as unit of time:

$$\dot{\lambda}_\odot = n(1 + 2e \cos M + \cdots) = 3548''.3 + 118''.7 \cos M + \cdots \mathrm{j}^{-1} \qquad (4.10)$$

Notice that $\dot{\lambda}_\odot = n$ on two occasions, namely when the Sun passes through the semiminor axes of its orbit. Knowing the date when $M = 0°$ (e.g., January 2nd in 2002), we can easily calculate the variation of angular velocity along the ecliptic at each date.

To determine the relationship between $\alpha_\odot$ and $\lambda_\odot$ (a procedure called reduction to the equator), let us examine the second of the Equation 4.4, which according to Equation 1.10 can be written as:

$$\alpha_\odot = \lambda_\odot + \frac{\cos \varepsilon - 1}{\cos \varepsilon + 1} \sin 2\lambda_\odot + \frac{1}{2}\left(\frac{\cos \varepsilon - 1}{\cos \varepsilon + 1}\right)^2 \sin 4\lambda_\odot + \cdots$$

$$= \lambda_\odot - 148'.1 \sin 2\lambda_\odot + 3'.2 \sin 4\lambda_\odot + \cdots$$

Therefore, the right ascension of the true Sun is connected to its longitude by a series of multiples of $2\lambda_\odot$, with coefficients depending from ε. Inserting Equation 4.6 and Equation 4.9, we finally obtain:

$$\alpha_\odot \approx \lambda_\Pi + M + 2e \sin M - \tan^2 \frac{\varepsilon}{2} \sin 2(\lambda_\Pi + M) + \cdots \tag{4.11}$$

a series with periods of 12 months, 6 months, 4 months, etc. Taking the time derivative, we can find that $\dot\alpha(F_\odot) = n$ four times a year, when $\lambda(F_\odot) \approx 46°14'$, $133°46'$, $226°14'$, $313°46'$. The corresponding dates can be determined by knowing the date of the passage through the perigee.

When the fictitious Sun $F_\odot$ encounters the equator in γ coming from Π (later than the true Sun), let the mean Sun $M_\odot$ start from γ with the same uniform motion n. The two hypothetical Suns will coincide again at the autumn equinox; in this way, at any instant $\lambda(F_\odot) = \alpha(M_\odot)$. Finally, calculate the equation of time (E), namely the difference:

$$E = \alpha(M_\odot) - \alpha_\odot = -460^s.3 \sin n(t - t_0) + 592^s.2 \sin 2(\lambda_\Pi + M) + \cdots$$

$$= A \sin \lambda(F_\odot) + B \cos \lambda(F_\odot) + C \sin 2\lambda(F_\odot) + D \cos 2\lambda(F_\odot) \tag{4.12}$$

$$+ E \sin 3\lambda(F_\odot) + F \cos 3\lambda(F_\odot) + G \sin 4\lambda(F_\odot) + \cdots$$

The equation of time (E) is therefore a fairly complex, but well-known, function of time (see Figure 4.1).

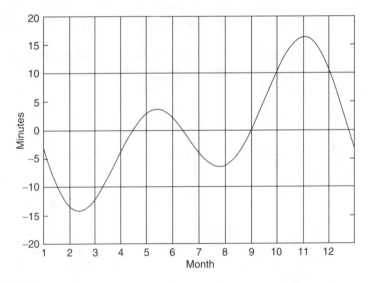

FIGURE 4.1
The equation of time E for J2000. The convention of sign is that adopted by the *Astronomical Almanac* (some texts use the opposite sign).

Its value is zero four times a year, namely at the middle of April, middle of June, beginning of September, and around the 25th of December; the maximum value of approximately $+16^m$ is reached in early November, the minimum value of approximately -14^m in middle February. Notice that the exact values at a particular date will vary by a few seconds from one year to the next, in a periodic behavior due to the presence of the leap year. For instance, in the year 2001 the values of the coefficients were: $A = -107^s.5$, $B = -428^s.5$, $C = +596^s.1$, $D = -2^s.1$, $E = +4^s.4$, $F = +19^s.3$, $G = -12^s.7$, $\lambda(F_\odot) = 279°.7 + 0.986d$, d being the number and fraction of days since January 1st, 2001. The precision of this expression is a few seconds.

For any particular site, the difference between the hour angles of the true and mean Sun will also equal the difference, changed in sign, between their two right ascensions:

$$HA_\odot - HA(M_\odot) = -\alpha_\odot + \alpha(M_\odot) = E$$

The hour angle of the mean Sun, augmented by 12^h in order to have the day starting at midnight, is called the local mean solar time $T(M_\odot)$:

$$T(M_\odot) = HA(M_\odot) + 12^h \tag{4.13}$$

The interval of time between two passages through the local meridian of the mean Sun is called mean solar day (the unit already indicated with j) which is divided in 24^h of 3600 sec of mean time (whose length is not the same of the sidereal second, as already said).

4.4 The Universal Time (UT)

In particular, the mean solar time at Greenwich is called universal time (UT):

$$UT = HA(M_\odot)(\text{Greenwich}) + 12^h \tag{4.14}$$

Sometimes, the letter Z is used to indicate Greenwich's mean time, for example 1433Z means 14h33m mean solar time at Greenwich.

For another site having longitude Λ, the local mean solar time at $UT = t$ is $T_\Lambda(M_\odot, UT = t) = t \pm \Lambda$, where the sign is $+$ if East of Greenwich, $-$ if West. It is absolutely equivalent to express the difference in longitude between two sites as difference in solar or sidereal time, because Λ is an angle.

4.5 The Tropical Year and the Rates of ST and UT

From their definitions, it follows that the sidereal time ST and the mean solar time $(T(M_\odot))$, including its particular case UT, have the same degree of uniformity of the Earth's rotation. However, the two times differ both in rate and in origin.

The ratio between the two rates can be easily determined. Let us call tropical year the interval of time between two consecutive passages of the mean Sun through the vernal equinox, in other words the interval of time needed for the longitude of the fictitious Sun to increase by 360° with respect to the equinox. At the level of precision of the present discussion, this length of time is the same as an increase of 360° of the right ascension of the mean Sun. However, we shall see in the following chapters that the equinox is not fixed among the stars, resulting in a slight difference between the two preceding definitions. The correct definition of the tropical year is the period of one complete revolution of the mean longitude of the Sun with respect to the dynamical equinox; this matter will be further discussed in connection with the Besselian year, and also in Chapter 10.

The value of the tropical year in mean solar days was determined by Newcomb (*op. cit.*) with utmost precision, thanks to its recording over several millennia; apart from a slight secular variation due to changes in the precession constants (see Chapter 10), Newcomb found:

$$1 \text{ tropical year} = 365^j.2421988\ldots = 365^j05^h48^m45^s.975\ldots$$

$$= 366.2421988\ldots \text{ sidereal days}$$

The difference of exactly one entire day using the solar or sidereal time units arises because after one tropical year exactly one more sidereal day has elapsed. Therefore:

$$\text{rate } ST = (1 + 1/365.2421988 = 1.002737909) \text{ rate } T(M_\odot)$$

$$\text{rate } T(M_\odot) = (1 - 1/366.2421988 = 1 - 0.002730434 = 0.997269566) \text{ rate } ST$$

$$24^h T(M_\odot) = 24^h3^m56^s.55537 \, ST, \quad 24^h ST = 23^h56^m04^s.09053 \, T(M_\odot)$$

$$1^s T(M_\odot) = 1^s.0027379 \, ST, \quad 1^s ST = 0^s.9972696 \, T(M_\odot)$$

Regarding the origin, we anticipate an approximate expression which will be completed in Chapter 10. At any given date T, the Greenwich sidereal time at $UT = 0^h$ (mean midnight) is:

$$ST_{\text{Greenwich}}(0^h UT) = 6^h41^m50^s.5481 + 8640184^s.812866T$$

where T (with its proper sign and all needed decimals) is in Julian centuries before or after the fundamental epoch January 1st, 2000 at 12^h UT (namely T is the number of days and fractions since that epoch, divided by 36,525).

For another site having longitude Λ, the local sidereal time at $UT = t$ is

$$ST_\Lambda(UT = t) = ST_{\text{Greenwich}}(0^h UT) \pm \Lambda + 1.0027379t$$

where the factor of the last term takes into account the different rates of the two time scales.

4.6 The Year and the Julian Calendar

The civil calendar adopted in many countries is based on the length of the tropical year, because the seasons follow the course of the Sun along the ecliptic. However, this length cannot be expressed by an integer number of days, not even by a rational fraction. Several remedies were adopted by the different cultures. In Rome, in approximately year 46 B.C., Julius Caesar agreed to the proposal of the astronomer Sosigenes of adding 1 day to the shortest month (February) each fourth year. In the Julian calendar, the fourth year is called bi-sextus, or leap year. In the first application of this rule, some 90 days had to be suppressed from the calendar. However, the situation was very confused at least for the following 30 years, until the times of Emperor Augustus. The extension of the Julian calendar into the past (namely before its adoption) is called the proleptic calendar. Because in Chronology year 0 does not exist, passing directly from 1 B.C. to 1 A.D., in performing calculations of intervals of time between two events happened one before and one after Christ, year 1 B.C. is year 0, year 2 B.C. is year -1, and so on. Year 0 is considered a leap year.

With the Julian reform, the duration of the year, averaged over a 4-year period, became exactly 365.25 mean solar days, and the Julian Century 36525 mean solar days. However, this round number is longer than the duration of the tropical year: after 1000 years the difference amounts to approximately 8 days. Some adjustments to the calendar were made during the Council of Nicea (325 A.D.). Finally, in 1582, Pope Gregorious XIII decreed to suppress 10 days, jumping from Thursday October 4 directly to Friday October 15. As a further element of the Gregorian reform, it was stated that only the secular years divisible by 400 would be leap years. Therefore, following the Gregorian reform, years 1600 and 2000 were leap years, but not 1700, 1800, and 1900: in a cycle of 400 years there are only 97 such leap years, and the average duration over each cycle is 365.2425 mean solar days. Because in 400 years there are 146,097 days, which is evenly divisible by 7, the Gregorian civil calendar exactly repeats at each cycle of 400 years. However good, the average value 365.2425 is still an approximation to the true value, so that the Gregorian calendar precedes the Sun by approximately one day every 2500 years. As a remedy, year 4000 could be considered a normal year, not a leap one, but no agreement has been reached.

4.7 The Besselian Year or Annus Fictus

Both the tropical and the Julian year are essentially durations, no precise origin being associated with their definitions. Following Bessel, the year starts when the longitude of the fictitious Sun, affected by aberration and referred to the mean equinox of date, is exactly $\lambda(F_\odot) = 280°$, and therefore

$\alpha(M_\odot) = 18^h40^m$. Such an instant, named epoch, is always within 1 day from midnight of the December 31st. In most applications, for instance in calculating the amount of precession, this slight difference between the start of the Besselian year and of the civil year is entirely negligible. The Besselian epoch is indicated by the notation B followed by the year, for example B1950.0; any other instant of time during that year is indicated by the fraction of year, for example B1950.45678.

The following year will start when again $\lambda(F_\odot) = 280°$, namely when $\alpha(M_\odot)$ has increased exactly by 360°. As we have indicated before, this interval of time is not quite the same as that of the tropical year, fixed by the mean longitude of the Sun with respect to the dynamical equinox. There is, therefore, a slight secular acceleration deriving from the different definition of $\lambda(F_\odot)$ and of $\alpha(M_\odot)$: the duration of the Besselian year is equal to that of the tropical year minus $0^s.148$ per tropical century since 1900; therefore B2000.0 started $7^s.4$ before the instant calculated from the duration of the tropical year.

In order to refer the Besselian year to the Julian calendar, it must be recalled that the fundamental epoch B1900.0 corresponds to 1900 January $0^d.813 = 1899$ December 31, 19^h31^m (notice the astronomical convention, year first, then month, day, hours, and the utilization of the 0 for the last day of the year). The calendar date of another epoch, say B1950.0, is obtained by considering 50 tropical years since then, namely 18262.110 days, or else 12.110 days more than 50 years of 365 days. Taking into account that 1900 was not a leap year, subtract 12 days, add 0.110 to 0.813 to finally obtain B1950.0 = 1950 January $0^d.923 = 1949$ December 31, 22^h09^m. This is the civil date of many stellar Catalogues such as the AGK3.

The convention of Bessel remained valid until 1984, when the IAU decreed to move the fundamental epoch to J2000.0 noon (not midnight!) = 2000 January $1^d.5$ UT (actually UT1, see Chapter 10), and to adopt Julian years of $365^j.25$ (or Julian centuries of 36525^j). Therefore, the beginning of year 1950.0 corresponds exactly to $18262^j.5$ days before the fundamental epoch, namely to 1950 January $1^d.0$, differing by 1^h51^m from B1950.0.

4.8 The Seasons

We have affirmed that the tropical year determines the succession of the seasons. Consider, for instance, the following values valid at epoch 1950.0:

$$\lambda = 282°04'30'' + M + 115' \sin M + \cdots, \quad M = 3548''.3(t - t_0),$$

$$\tag{4.15}$$

$$t_0 = 1950 \text{ January } 3.02$$

The seasons started when $\lambda_\odot = 0°$ (spring), $= 90°$ (summer), $= 180°$ (autumn), $= 270°$ (winter). Because no high precision is required, in order

TABLE 4.1

Dates of Start of the Seasons

Season	Start (1950)	Duration (days)	Start (2000)	Start (2096)
Spring	21.2 March	92.81	20.3 March	19.5 March
Summer	22.0 June	93.62	21.0 June	20.1 June
Autumn	23.1 September	89.82	22.7 September	21.9 September
Winter	22.4 December	89.00	21.5 December	20.9 December

to find the corresponding values of M and of t, we can ignore the term in $\sin M$, obtaining the values of Table 4.1.

The calculation has been repeated with the appropriate values of Equation 4.15 for the leap years 2000 and 2096 (when the start date will be the earliest of the 21st century).

The starting times retard 6^h each year, in a cycle of 4 years.

Table 4.1 shows that in the northern hemisphere the two warm seasons last 7 days longer than in the southern hemisphere; on the other hand, the Earth is closer to the Sun in the southern summer. Averaged over the globe, the sunlight falling on Earth at aphelion is approximately 7% less intense than at perihelion; however, the average temperature of the whole Earth at aphelion is approximately 2.3 °C higher than it is at perihelion: this happens because there is more land in the northern hemisphere and more seawater in the southern one. During the month of July, the northern hemisphere is tilted toward the Sun, and the Earth's overall temperature (averaged over both hemispheres) is slightly higher because the Sun shines mostly on continents, which have low heat capacity and warm up more easily. January is the coolest month because the Earth presents its water-dominated, high heat-capacity hemisphere to the Sun. Southern summer in January is therefore cooler than northern summer in July. In order to satisfy the thermal balance of the Earth as a whole, averaged over 1 year, efficient mechanisms of heat transport are required, such as regular winds and sea currents.

4.9 The Julian Date

In astronomy, it is customary to count the passage of time in mean solar days starting from an initial arbitrary date. Following the suggestion of Joseph Justus Scaliger (1583), the initial date is midday (not midnight!) of January 1st, 4713 B.C. (4713 B.C. = −4712). Such a system of dates is expressed in Julian Dates (JD). We follow the use of the *Astronomical Almanac*, which reserves the name of Julian Day to its integer part; be careful not to confuse the JD with a date in the Julian calendar, see also the Notes.

Thus, 1950 January 1st, 12^h UT, corresponds to JD = 2433283.0, and similarly:

$$B1950.0 = JD\ 2433282.423, \quad J2000.0 = JD\ 2451545.0.$$

(notice the coincide of the start of the astronomical year with the Julian Date). The inverse is:

$$\text{Julian epoch } J = J2000.0 + (JD - 2451545)/365.25$$

$$\text{Besselian epoch } B = B1900.0 + (JD - 2415020.31352)/365.2422$$

In order to avoid carrying too many decimals, and to start the day at midnight, a Modified Julian Date (MJD) has been introduced, having its zero date on 1858 Nov. 17.0:

$$MJD = JD - 240000.5$$

The JD (or MJD) scale furnishes a continuous reference of time; however, this scale is as uniform as the mean solar day itself. We will see in the following chapters (in particular in Chapters 6 and 10) that the duration of the day has a secular decrease, so that JD is not entirely satisfactory for dynamical purposes over intervals of centuries or millennia.

Notes

1. Gregorius XIII had the advice of several contemporary astronomers, in particular of Luigi Lilius, who also made a reform of the lunar calendar (important for the determination of Easter). The objectives and details of the new calendar were described in 1603 by Christoph Clavius in his letter *Romani Calendarii a Gregorio XIII P.M. restituti explicatio*. The Gregorian reform was not adopted immediately by all countries; for instance, it came to be used in the U.K. in 1752 (this is the reason for the confusion about the year of death of Galileo Galilei and of the birth of Isaac Newton being the same), and in Turkey only in 1927.

   For the historically oriented reader, the following two books can be recommended: Coyne, G.V., Hoskins, M.A., Pedersen, O., Eds., 1983, *Gregorian Reform of the Calendary*, Pontificia Academia Scientiarum; Heilbron, J.L., 1999, *The Sun in the Church*, Harvard University Press, Harvard.

   Notice that, according to many historians, J. J. Scaliger named Julian Day his system of counting days in honor of his father Julius, and not of Julius Caesar, so that the adjective Julian has nothing to do with the Julian calendar!

2. The determination of the date of Easter (the first Sunday after the full Moon happening the 21st of March or the first since that date, as established by the Nicea Council of 325 A.D.) was always a difficult computational problem, because for the Church the 21st of March is not the spring equinox in the astronomical significance. Therefore the date of Easter can vary between the 22nd of March and the 25th of April, although the most frequent interval is between March 25th and April 19th. Carl F. Gauss found a tabular expression which is almost always, but not strictly so, correct. See Meeus (1991), for Gauss's and other expressions.

3. The decision of reversing the sign of the terrestrial longitude Λ, taken by IAU in 1982, has been resisted by several authors, because it not only goes against the previous century-long practice, but also against the longitude system adopted for all other planets of the Solar System. Different space missions have adopted different sign conventions for their longitude systems, for instance the ongoing NASA missions to Mercury, Messenger, has adopted the opposite sign of Mariner 10. The same is true for previous and contemporary missions to Mars. Therefore, one has to be aware of this change before comparing the respective cartographies.

4. Another way of representing the equation of time is by means of the so-called analemma, an eight-shaped curve which is often found on sundials. For a vivid pictorial representation of the analemma see: http://www.uwm.edu/~kahl/Images/Weather/Other/analemma.html, and: http://vrum.chat.ru/Photo/Astro/analema.htm

Exercises

1. Calculate the mean anomaly of the Sun at the beginning of the year 2002.

2. An Almanac gives the ST(Greenwich) at 0 h UT of a given day, calculate the ST in a particular place of longitude Λ at any instant t of $T(M_\odot)$ of the same day.
   First solution: If you do not know UT, calculate the local ST at 0 h $T(M_\odot)$, which is equal to ST (Greenwich, 0^h UT) $+0.0027379\,\Lambda$ (for instance, for Asiago $0.0027379\,\Lambda = -7^s.58$). This term is often referred to as the local constant. Then calculate:

$$ST(t^h T(M_\odot)) = ST(\Lambda, 0^h T(M_\odot)) + 1.0027379t$$

Second solution: In your observatory there is already a clock giving t in UT (which is the common situation today), so that:

$$ST(t^h UT, \Lambda) = ST(\text{Greenwich}, 0^h UT) + 1.0027379(t^h UT) + \Lambda$$

(in the IAU convention, subtract West longitude, add East longitude). Inversely, if one has to calculate $T(M_\odot)$ starting from the local ST, the following relation will be used:

$$T(M_\odot) = 0.9972696(ST - ST(0^h T(M_\odot)))$$

3. Let a place be at longitude $75° = 5^h$ West, and the local sidereal time be $ST = 23^h 30^m$. The date is 1985, July 8. Find the corresponding UT.

   The first operation is to add the West longitude to find the Greenwich ST:

$$ST(\text{Greenwich}) = 4^h 30^m$$

   The Almanac gives, for that date, the value $ST(\text{Greenwich}, 0^h UT) = 19^h 03^m 34^s.376$, which must be subtracted from $4^h 40^m$, giving a time difference (in ST units) of $9^h 29^m 25^s.624$; multiply by 0.9972695663 and finally obtain UT $= 9^h 27^m 52^s.337$.

   The ST used in Exercises 1 and 2 is the mean sidereal time; to have the apparent one, the small correction named equation of the Equinox

$$EE = \Delta\psi \cos \varepsilon$$

   must be added. See also Chapter 5 and Chapter 10.

4. Calculate the Julian day corresponding to a calendar date.

   A formula used by many calculators, but valid only in the Gregorian calendar (after October 15, 1582), is the following:

$$JD(Y, M, D, H) = 367Y - INT\left[7 \times \frac{Y + INT((M+9)/12)}{4}\right]$$

$$+ INT\left(\frac{275M}{9}\right) + D + 1721013.5 + \frac{H}{24} + 0.5$$

$$\times SIG(190002.5 - 100Y - M) + 0.5$$

where INT is the function integer, and SIG is the function signum.

   After February 28, 1900 and until 2099, the last two terms can be omitted.

   For instance $JD(1996,1,1,0) = 2450083.500$, $JD(2004,1,1,4.75) = 2453009.250$.

5. Calculate the civil date corresponding to JD $= 2446108.5$, knowing that the JD at 0^h UT of January 1, 1900 was JD1900 $= 2415018.5$.
   Solution: The number of days elapsed since the initial epoch is ND $=$ JD $-$ JD1900 $= 31088$. The number of years NY elapsed since the initial date is: (NY) $=$ INT(JD $-$ JD1900)/365 $= 85$, where INT is the function integer, and the remainder is $0.1726 \times 365 = 63$ days. However, in 85 years there are 21 leap days, so that the number of days elapsed since the beginning of the year 1985 is actually $63 - 21 = 43$. Subtracting the 31 days of January, we finally obtain the wanted civil date, namely 1985, February 12, 0^h UT.

   To solve this problem in a general way, one formula is the following

   - Let $Z =$ INT(JD $+ 0.5$) and $F =$ (JD $+ 0.5$) $- Z$
   - If $Z < 2299161$ let $A = Z$
   - If $Z \geq 2299161$ determine: $A' =$ INT($Z - 1867216.25$)/36524.25, and let $A = Z + 1 + A' -$ INT($A'/4$)

   Then calculate:

   $$B = A + 1524, \quad C = \text{int}\left(\frac{B - 122.1}{365.25}\right), \quad D' = \text{INT}(365.25C),$$

   $$E = \text{INT}\left(\frac{B - D'}{30.6001}\right)$$

   The day of the month (with decimals) is then $D = B - D' -$ INT($30.6001E$) $+ F$.

   The number of the month is $M = E - 1$ if $E < 14$, $M = E - 13$ if $E = 14$ or 15. The year is $Y = C - 4716$ if $M > 2$, $C > -4715$ if $M = 1$ or $= 2$ (namely, January and February are considered the 13th and 14th months of the preceding year).

   For other formulae, and similar problems connected with the date and day of the week, see the *Almanac for Computers*, or the book by Meeus (1991).

5

The Movements of the Fundamental Planes

The equatorial and ecliptic coordinates are based on the plane of the celestial equator and the ecliptic, respectively, having a common origin in the vernal point γ: every movement of these planes with respect to the fixed stars will result in a variation of the equatorial and ecliptic coordinates with time. The study of these movements is the subject of the present chapter. The stars will provide an ideally fixed reference frame against which we will determine the complex movements of the equator, the ecliptic, and their intersections.

For fundamental dynamical reasons, in the inertial system the ecliptic plane is much more stable than the equator, whose movements are larger, and known with some residual imprecision even today. Those minute uncertainties, however small, are of great interest for the astronomer, because they somewhat hamper the precise knowledge of the system of motions and of the overall field of forces of the Milky Way. Hence, the efforts made, not only by geophysicists but also by astronomers, to know with better and better precision the movements of the terrestrial observer.

5.1 First Dynamical Considerations

The equatorial system (α, δ) is the only one used in high precision positional catalogues. However, it depends on the orientation in space of the Earth and its rotation (position of the equatorial plane with respect to the fixed stars, meridian, sidereal time), and on the revolution around the Sun (ecliptic, point γ). The motion of the Earth can be considered, at least in a first approximation, as a combination of two unrelated motions (although the constants of precession weakly tie one to the other, as we will see later), namely a translation of the center of mass and a rotation of the figure around an axis passing through the barycenter.

In a first, rough approximation, the barycenter of the Earth revolves around the Sun as a point-like particle subject to the gravitational pull of the Sun and of the other planets. More rigorously, it is the Earth–Moon (E–M) barycenter that follows Kepler's laws with respect to the barycenter of the Solar System. Therefore, it is preferable to identify the ecliptic with the orbital plane of this E–M barycenter, and free it from the periodic perturbations mostly due to Venus and Jupiter. The Sun is never more than $2''$ above or below this mean ecliptic, a fact that justifies the simple treatment given in the previous chapters.

Much more complex is the description of the orientation and rotation of the Earth, for several different reasons, among which we quote the following:

- The mass of the whole Earth is $M_\oplus = 5.976 \times 10^{27}$ g, the mass of the ocean is approximately $10^{-4} M_\oplus$, and that of the atmosphere approximately $10^{-7} M_\oplus$. Even considering the solid Earth alone, the distribution of its mass does not possess spherical symmetry (indeed, to the precision of the measurements, not even an azimuthal symmetry).

- The position of its rotation axis is influenced by the presence of the Moon and of the Sun (precession and nutation).

- Its rotation axis does not necessarily coincide with the minor axis of the geometrically best-fitting ellipsoid.

The variable orientation of the Earth's instantaneous rotation axis with respect to the fixed stars causes a variation of their equatorial coordinates. The variation of its position with respect to the axis of the geometric ellipsoid causes a small variation of the astronomical latitude and longitude of each observatory, and a wandering of the astronomical poles around the geodetic poles, which is confined to a circle approximately 20 m in diameter (free or Eulerian nutation). Furthermore, the solid Earth is not perfectly rigid, and the distribution of masses can change both at the surface (winds, tides, currents) and in the interior (earthquakes). Therefore, the rotation of the Earth cannot be uniform for arbitrarily long times; at the present epoch we witness a secular decrease of its angular velocity, with over-imposed periodic fluctuations and also abrupt changes. To be sure, all these complications are small, but well measurable with today's instruments.

The general dynamical problem presents great theoretical complexity; no less are the practical difficulties of disentangling the minute motions of the observers from the motions of the stars themselves, and to keep a uniform time system for centuries or millennia. However, the different effects have different amplitudes, and we are justified, at least for the present needs, to follow an almost historical description. At the end, we will sum one over the other the different effects, although the validity of this superposition will break down at very high precision (to set a limit, a precision of a few tenths of arcsec can be achieved; to reach the thousandth of arcsec, a more rigorous approach is needed).

In the following, we shall call epoch the initial time when all constants are assumed to be well known, and date the generic instant of observation. In several texts, the elementary time unit is the year set equal to 1, so that the word arc often means angular velocity. Here, for better clarity, we shall distinguish the two concepts, by explicit indication of the time and the differentials and derivatives. Notice that up to now we have consistently used the word motion to indicate an angular velocity, not a movement! Such convention will also be kept in the remainder of the present text.

Much of our treatment can be traced back to the books of Newcomb (1906) and Danjon (1980) quoted in the Bibliography. Since 1984, the IAU has enforced a profound revision of several fundamental constants and variables, including time. Therefore, we have provided both pre and post 1984 definitions and values in several places. Yet another revision has been applied by IAU starting in 2006, as will be briefly discussed in the Notes to this chapter and in Chapter 6.

5.2 The Precession of the Equinox

The dominant variation of the stellar coordinates was discovered in 129 B.C. by Hipparchus, comparing his own determination of the ecliptic coordinates of Spica (α Vir) with those derived 144 years earlier by Timocharis: while the ecliptic latitude had remained constant, the longitude had increased by approximately $2°$ (namely by $50''.4$ per year). The discrepancy of $2°$ was too large to be attributed to measurement errors. Soon after, the same variation of longitude was found on all stars. To explain it, Hipparchus imagined a rotation in direct (anticlockwise) sense of the whole sphere of the fixed stars around the ecliptic pole. The Sun would therefore encounter the vernal point γ each year at a somewhat earlier time. In the words of Hipparchus, the ingress of the Sun in γ would precede the previous one by the time taken to describe an arc of $50''.4$, hence the expression "precession of the equinox." Through the centuries, the constellation where the Sun is seen to ingress in γ changes; it was Aries at the time of Hipparchus, it is Pisces today, after another 2000 years it will be Aquarius.

Some 1600 years passed before Copernicus made the right interpretation of precession, it is the Earth's rotation axis that describes a retrograde cone of semiaperture ε around the ecliptic pole E, in a period of approximately 25,800 years ($= 360°/50''.4 \ y^{-1}$, often called a platonic year). The celestial pole P is therefore seen at each time on a point of the small circle distant ε from E, as shown in Figure 5.1. In other words, the parallel of ecliptic latitude $\beta = 90° - \varepsilon$ is the locus described by P in 25,800 years. The South celestial pole is seen to move on the corresponding circle having $\beta = -(90° - \varepsilon)$.

During this movement, the celestial poles will be seen in different constellations; today, the North pole approaches the bright star α UMi. The present distance of about $45'$ will decrease to a minimum value of $27'$ in 2102, and then it will progressively augment. As already said, there is no correspondingly bright star near the present position of the celestial South pole (the one used by the *Astronomical Almanac* as pole star for obtaining southern latitudes is σ Octanctis, of visual magnitude 5.5).

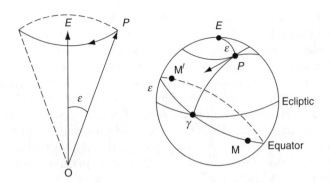

FIGURE 5.1
Two equivalent representations of the precession of Hipparchus. Left, the vector from the center of the Earth to the celestial North pole P describes a cone of fixed semiaperture ε around the ecliptic North pole E, in retrograde direction. Right, the instantaneous motion of the celestial pole is a vector tangent both to the small circle distant ε from the ecliptic pole and to the equinoxial colure. Arc $M\gamma$ = arc $\gamma M'$ = $90°$.

The precession correctly interpreted by Copernicus was explained on dynamical bases by Newton in his *Principia* (1687): the Earth cannot be a sphere, it must have a massive annulus around the equator. Therefore, the Moon, which is usually not on the equatorial plane, will exert a torque on the Earth's rotation axis. This torque causes a movement of the axis perpendicular to the instantaneous plane passing through the axis itself and the Moon. Quite often, an analogy is made with a spinning top; however, the sense of precession and rotation are opposite, the spinning top precesses and rotates in the same sense, the Earth precesses in the opposite sense of the diurnal rotation. The Sun exerts the same effect of the Moon; however, because the amplitude of the torque is proportional to the mass of the responsible body and to the inverse cube of its distance from the Earth's barycenter, the magnitude of the lunar effect is more than twice that of the Sun. Therefore, the term lunisolar precession is usually employed to describe the phenomenon discovered by Hipparchus.

However, Hipparchus's description and Newton's explanation must be only partially correct, for two reasons: first, the Moon's orbit is inclined by $5°9'$ to the ecliptic plane; second, the distance of the Moon, and to a far lesser extent that of the Sun, change during the lunar month and the tropical year, respectively. As a consequence, the true movement of the celestial poles cannot rigorously follow the small circles at a constant distance ε from the ecliptic poles: a series of cyclic terms of different amplitudes and periods, also affecting the instantaneous obliquity, must be present. These effects have a periodic nature however, they do not accumulate for centuries as the lunisolar precession does, so they went undetected until the advent of telescopic observations. The credit for their discovery goes to the English astronomer Bradley, in the eighteenth century, using a long series of measurements of the declination of the bright star γ Dra (not too distant from the ecliptic pole E). The declination of the star, once corrected for the lunisolar precession, appeared to increase by $18''$ from 1727 to 1736, and to decrease by the same amount from 1736 to 1745, as if the celestial pole has an oscillatory movement (Bradley called this nutation, the same term used for the oscillation of a ship's masts) of amplitude $\pm 9''$ and a period of 18.6 years around a mean position. The period is exactly the same of the retrogradation of the nodes of the lunar orbit on the ecliptic. Quite obviously the same effect had to be present on the right ascension of γ Dra, but Bradley could not measure it because of the insufficient stability of his clock. What Bradley measured is only the principal term of the nutation, many smaller terms with a variety of periods and amplitudes are present in the overall oscillation of the pole (see also Chapter 6).

Until now, we have tacitly assumed that the plane of the ecliptic is fixed with respect to the distant stars. This assumption is not entirely correct, as Cassini had indeed already suspected. The level of accuracy reached by eighteenth century observational astronomy, and the parallel theoretical progresses of Celestial Mechanics (Euler predicted the movement of the ecliptic under the influence of the planets, in particular of Venus and Jupiter) imposed to consider a moving ecliptic. Comparing the values of ε obtained by Copernicus, and then by Tycho Brahe at the end of the 16th century, with the determinations performed during the 19th century by many authors (e.g., Bessel, Struve, LeVerrier, Oppolzer, Newcomb), and finally with the value measured today ($\varepsilon \approx 23°27'$), a secular decrease of approximately $0''.5$ per year is well established. The comparison of the observed values with the planetary perturbation theory firstly expounded by Euler required a truly reliable set of planetary masses. Until the middle of the 20th century this was not the case, the very existence of Pluto was not known before 1938. Today, the masses of the planets have been very well determined, thanks to the accurate measurements of the accelerations of spacecraft that have navigated all the planets of the Solar System (except Pluto, which will be reached by the NASA mission New Horizons in 2015; see Notes). Modern theoretical developments have not only provided orbital elements correctly for millions of years, but have also discovered the decisive influence of the Moon on the stability of ε. The present-day diminution is actually part of a variation with an amplitude of approximately $\pm 1°$ and a period of 44,000 years (see Laskar et al., 1993). Planet Mars has no heavy moon. Therefore, its obliquity can change by much larger values, and so its climate can have extreme variations, in times of the order of 100,000 years.

It is easily understood that the influence of the planets on the obliquity of the ecliptic is a periodic effect, but the periods must be so long that their effects will accumulate for many centuries, just as the lunisolar precession does. We are therefore justified to call it planetary precession, and to add its magnitude to the lunisolar precession to obtain the general precession (because of its negative sign, the planetary precession makes the constant $50''.4$ become slightly smaller). A planetary nutation must also be present; however, the movements of the ecliptic with respect to the equator do not change the stellar declinations, but only the common origin of the right ascensions, so that the planetary nutation goes unnoticed in differential measurements.

5.3   The Movements of the Fundamental Planes

Let us consider in Figure 5.2 the two fundamental planes, celestial equator and ecliptic, at two dates t_1 and t_2 (with t_2 later than t_1 in order to fix the sense) intersecting in γ_1 and γ_2 with obliquities ε_1 and ε_2, respectively. Each element in this figure, e.g., the angle j, can be thought of as composed of two parts, a secular one and a short period one. During several decennia, or even centuries, the first part can be developed in a time series with only the first few terms of importance, so we can legitimately write:

$$j(t) = at + bt^2 + ct^3 + \cdots + n(t) - n(t_0) \qquad (5.1)$$

where the secular terms (precession) are zero at the initial epoch t_0, but not the short period ones (nutation). The instantaneous elements are the true elements, those freed by nutation are the mean elements (mean equator, mean equinox, mean obliquity, etc.).

As already pointed out, the definition of ecliptic requires some caution, in the literature there is some confusion of terminology. While the celestial equator is the same for any observer, the ecliptic is only approximately so, because of the finite distance of the Sun. From the knowledge of the shape of the Earth, the apparent ecliptic of a particular observer (the topocentric ecliptic) can be accurately referred to an ideal geocentric observer. However, this translation is not entirely satisfactory: we would be better to call ecliptic the plane passing for the E–M barycenter and freed from all periodic perturbations (due mostly to Venus and Jupiter, which amount to $0''.4$ and to $0'.2$, respectively). With respect to this ecliptic, the geocentric latitude of the Sun is not strictly zero, it can reach $\pm 0''.6$ according to the instantaneous position of the Moon, which changes with a period of approximately 29 days. Summing up the three different effects, the geocentric latitude of the Sun can amount to $\pm 1''.2$, an angle sufficiently small that we shall ignore it in almost all subsequent consideration. Furthermore, the instantaneous ecliptic is essentially never used, what is called ecliptic is rather the mean ecliptic in the above-mentioned sense of mean planes and mean coordinates. The mean equinox would then be the intersection of the mean equator with the mean

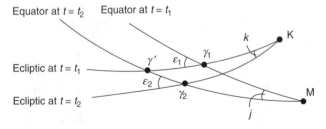

FIGURE 5.2
The movements of the fundamental planes.

ecliptic, and the true equinox the intersection of the true equator with the mean ecliptic. Finally, after the introduction of the ICRS, we could call mean equinox the zero point of that catalogue, thus apparently avoiding the problem. However, the ambiguity of the concept "ecliptic" remains, as pointed out by Fukushita (2003, see Notes).

5.4 First-Order Effects of the Precession on the Stellar Coordinates

As was remarked at the beginning, the moderate precision sought for in our treatment allows the consideration of each effect one after the other, and then to sum their amounts; therefore we shall consider first the lunisolar, then the planetary, and then the sum of the two, namely the general precession.

We shall first proceed in the pre 1984 way, when the elementary unit of time was the tropical year, and the role played by the vernal point appears more clearly. The procedure will allow the mean coordinates at any date to be derived, starting from the known mean coordinates at a given fundamental epoch. We have called lunisolar precession, without other adjectives, that due to a constant lunisolar torque on a rigid Earth, whose effect would be a strictly periodic rotation of the celestial pole P around the ecliptic one E, with constant velocity P_0 and obliquity ε, namely a progressive, uniform increase of all longitudes by:

$$P_0 \cos \varepsilon = \dot{\lambda} = \psi \approx +50''.37, \quad P_0 \approx +54''.91 \quad \text{(per tropical year)} \quad (5.2)$$

where ψ is the notation favored by many authors, and used in the following; it is evident that ψ is an arc per unit time (in this case, the tropical year). The value $P_0 = 54''.91$ derives for $1/3$ from the Sun and for $2/3$ from the Moon, it is a function of the orbital elements of the two forcing bodies, of the dynamical figure of the Earth (moments of inertia) and of the obliquity of the ecliptic. In principle, it could be determined by the theory, but not at the level of precision reached by the observations (see also Chapter 6). A variation of any of these elements causes a variation in P_0 and also in ψ, but for the moment we set aside this possibility. With reference to Figure 5.1 and Figure 5.2, in the assumption of a fixed ecliptic, after an elementary time $dt = 1$ tropical year the moving equator has performed an elementary rotation dj around the diameter MM'. During the same time, the celestial pole has moved from P to P+dP along a great circle which is perpendicular to the solstitial colure, and which is also the hour circle of the initial equinox γ_1, so that $dP = dj$.

The intersection between the ecliptic and the equator moves from γ_1 to γ_2, describing the elementary precession in longitude:

$$d\lambda = \psi \, dt = \frac{dj}{\sin \varepsilon} \approx +50''.37$$

The projection of this elementary arc on the moving equator amounts to the elementary lunisolar precession in right ascension:

$$d\mu = (\psi \, dt) \cos \varepsilon = \frac{dj}{\sin \varepsilon} \cos \varepsilon \approx +46''.21$$

The perpendicular component along the hour circle of γ_1 is the elementary lunisolar-precession in declination:

$$dj = (\psi \, dt) \sin \varepsilon \approx +20''.05$$

Returning to angular velocities, it is customary to put:

$$m = \psi \cos \varepsilon \ (m = +46''.21/y = +3^{s}.08/y),$$

$$n = \psi \sin \varepsilon \ (n = +20''.34/y = +1^{s}.34/y)$$

Let us take the differentials of the transformation between ecliptic and equatorial coordinates:

$$d\alpha = \frac{\partial \alpha}{\partial \lambda} d\lambda + \frac{\partial \alpha}{\partial \beta} d\beta + \frac{\partial \alpha}{\partial \varepsilon} d\varepsilon, \quad d\delta = \frac{\partial \delta}{\partial \lambda} d\lambda + \frac{\partial \delta}{\partial \beta} d\beta + \frac{\partial \delta}{\partial \varepsilon} d\varepsilon$$

and introduce the initial simplifying assumptions that $\dot{\beta} \equiv 0$, $\dot{\varepsilon} \equiv 0$. From Equation 3.7 we easily derive:

$$\dot{\alpha} = \dot{\lambda}(\cos \varepsilon + \sin \varepsilon \sin \alpha \tan \delta) = m + n \sin \alpha \tan \delta$$

$$\approx [3^{s}.08 + 1^{s}.34 \sin \alpha \tan \delta] \, y^{-1} \tag{5.3}$$

$$\dot{\delta} = \dot{\lambda} \sin \varepsilon \cos \alpha = n \cos \alpha \approx 20''.05 \cos \alpha \, y^{-1} \tag{5.4}$$

Equation 5.4 informs us that the declination increases with time in the whole hemisphere having the vernal point as its pole ($\cos \alpha \geq 0$), and decreases in the other one. However, point γ makes a full revolution on the ecliptic in approximately 25,800 years, so that the declination of a star will always be confined between a maximum and a minimum value. This systematic behavior gives the possibility of determining n with great precision.

More complex is the time derivative of the right ascension expressed by Equation 5.3: due to the greater importance of term m, the right ascension will increase on most part of the celestial sphere. However, there is a locus where:

$$\dot{\alpha} = m + n \sin \alpha \tan \delta = 0$$

This locus is a spherical triangle having as vertices the ecliptic and celestial poles, and those stars for which the angle is 90°. In each hemisphere, this locus is approximately a small circle of diameter ε passing through the two poles; inside those two small circles the right ascensions will decrease with time. In the Northern hemisphere this situation occurs for $12^{h} \leq \alpha \leq 24^{h}$; in the Southern one for $0^{h} \leq \alpha \leq 12^{h}$. Notice also that $\dot{\alpha}$ becomes very large in

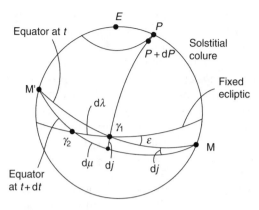

FIGURE 5.3

The elementary lunisolar precession. The equator performs an elementary rotation around the line MM'.

the proximity of the celestial poles, so that the calculations there need particular care.

It is instructive to derive the same results from a rotation of the Cartesian equatorial system having the x-axis toward γ and the z-axis toward the celestial North pole at the initial epoch t_0. At the time $t_0 + dt$, the position of the system is given by a retrograde rotation dj around Oy, followed by a retrograde rotation $d\mu$ around Oz. Therefore, the elementary rotation is:

$$\begin{cases} dx = -y\,d\mu - z\,dj \\ dy = x\,d\mu \\ dz = x\,dj \end{cases} \tag{5.5}$$

Let us apply it to the point representing the initial position of the star, namely to the direction:

$$\begin{cases} a = \cos \alpha \cos \delta \\ b = \sin \alpha \cos \delta \\ c = \sin \delta \end{cases} \tag{5.6}$$

After simple passages we obtain the result:

$$\begin{cases} d\alpha = d\mu + \sin \alpha \tan \delta\, dj \\ d\delta = \cos \alpha\, dj \end{cases},$$

$$\begin{cases} \dfrac{d\alpha}{dt} = \left(\dfrac{d\mu}{dt}\right)_{t_0} + \left(\dfrac{dj}{dt}\right)_{t_0} \sin \alpha \tan \delta\, dj \\ d\delta = \left(\dfrac{dj}{dt}\right)_{t_0} \cos \alpha \end{cases} \tag{5.7}$$

These expressions are similar to the previous ones, but they contain the explicit possibility that the "constants" m and n can vary with the epoch t_0.

Let us now consider the planetary precession, namely the effect of the gravitational perturbations of the planets on the ecliptic, and consequently on the precession rate of γ along the equator and on the value of the obliquity ε. The previous discussion has given us the mathematical possibility to fix the equatorial plane and its pole in the inertial frame. Therefore, we need only consider the small movement of the ecliptic pole around the so fixed celestial one. This movement can be represented by an elementary rotation of the ecliptic plane about a given line KK' (Figure 5.4), where K and K' are the poles of the arc EE'. In the notation of the *Astronomical Almanac*, the longitude of K, namely the arc γK, is indicated with Π_A; its value is approximately $174°.8$.

The elementary rotation of the ecliptic moves γ in a direct sense along the equator, by $g = 0''.13$ y^{-1}. The projection of this motion on the ecliptic amounts to $g \cos \varepsilon = 0''.11$ y^{-1}, which must by subtracted from the value of the lunisolar precession to obtain the value of the general precession $G = \psi - g \cos \varepsilon = +50''.26$ y^{-1}. As a consequence, the value of the precession in right ascension, m, slightly decreases to approximately $46''.07$ y$^{-1} = 3^s.07$ y^{-1}. The precession in declination, n, is not affected.

There is a second consequence of this rotation of the ecliptic around the line KK', namely the slight decrease of the obliquity itself. Again with reference to Figure 5.4, consider the elementary spherical triangle $\gamma K \gamma'$, where γ' is the position of γ on the fixed equator after 1 year, Π_A is the longitude of K and π_A is the angle in K (namely the inclination of the mobile ecliptic on the fixed one after 1 tropical year); from Equation 1.6 we obtain:

$$\sin \Pi_A \sin \pi_A = \sin g \sin(\varepsilon + \dot\varepsilon),$$

$$\cos \varepsilon \cos g = \sin g \cot \Pi_A + \sin \varepsilon \cot(\varepsilon + \dot\varepsilon)$$

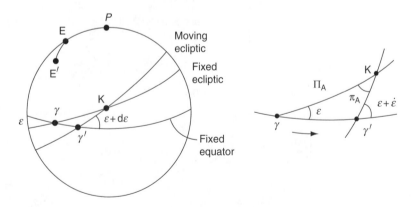

FIGURE 5.4
The planetary precession. On the right, the situation after 1 year. Notice that arc $\gamma K = \Pi_A$ is approximately $174°.8$, not as drawn.

where $g, \dot{\varepsilon}, \pi_A$ are three very small angles, so that:

$$g = \pi_A \sin \Pi_A \operatorname{cosec} \varepsilon, \quad \dot{\varepsilon} \approx \pi_A \cos \Pi_A \approx -0''.47 \, \mathrm{y}^{-1}$$

After Δt years, the inclination of the mobile ecliptic on the fixed one will be:

$$k = \pi_A \Delta t \approx \dot{\varepsilon} \Delta t / \cos \Pi_A$$

For instance, at the date J2001.5 the inclination had changed by $0''.7$ with respect to the mean ecliptic of epoch J2000.0.

Therefore, the obliquity of the ecliptic is not constant. As a consequence, none of the elements $\psi, m, n, G, g, \Pi_A, \pi_A$, would be really constant, even if P_0 did not vary (as indeed it does).

We come now to the post 1984 set of constants and procedures. Since 1984, the Julian year and the new system of constants adopted by IAU in 1976 (see also Kaplan, Notes) have been enforced in all Almanacs. Here are some of these values:

$$G = 50''.290966 + 0''.02222 \, T, \quad \varepsilon = 23°26'21''.448 - 0''.00468150 \, T$$

$$m = 46''.124362 + 0''.02793 \, T, \quad n = 20''.043109 - 0''.008533 \, T$$

$$\Pi_A = 174°.8764 + 0°.9137 \, T, \quad \dot{\varepsilon} = 0''.46815 - 0.0007 \, T,$$

$$g \cos \varepsilon = 0''.1055 - 0''.0189 \, T$$

where T is the number of Julian centuries of 36,525 days of 86,400 sec, starting from the fundamental epoch J2000. Notice that the ratio between the duration of the Julian year and that of the tropical year is only 1.00002136, and it does not entirely account for the strong revision of the value of G. Using centuries, the precessional constants become:

$$M = 1°.2812323 \, T + 0°.0003879 \, T^2 + 0°.0000101 \, T^3$$

$$N = 0°.5567530 \, T - 0°.0001185 \, T^2 - 0°.0000116 \, T^3$$

Furthermore, it has become customary to calculate the mean coordinates not for the beginning of the year, but for its midpoint.

Let us ignore these small variations for the moment, and set the task to derive the equatorial coordinates of a star, given by a catalog at the mean equinox J2000.5, at a date $t = t_0 + \Delta t = \mathrm{J}2000.5 + \Delta t$. In a first approximation, the following formulae will suffice:

$$\alpha_{2000.5+t} = \alpha_{2000.5} + (m_{2000.5} + n_{2000.5} \sin \alpha_{2000.5} \tan \delta_{2000.5}) \Delta t \qquad (5.8)$$

$$\delta_{2000.5+t} = \delta_{2000.5} + n_{2000.5} \sin \alpha_{2000.5} \, \Delta t \qquad (5.9)$$

with Δt in Julian years (of course, one has to pay attention to the different units used in right ascension and declination).

A small refinement will improve the precision if the time variation of m and n is known:

- Derive the values of the "constants" at the epoch $t_{1/2}$ intermediate between J2000.5 and t, and calculate the two other "constants" m', n':

$$t_{1/2} = \frac{(t_0 + t)}{2}, \quad m' = m_{1/2}(t - t_0), \quad n' = n_{1/2}(t - t_0)$$

- Using these values of m', n', derive from Equation 5.8 and Equation 5.9 the values of the coordinates $(\alpha_{1/2}, \delta_{1/2})$ at the intermediate epoch:

$$\alpha_{1/2} = \alpha_0 + \frac{1}{2}m' + \frac{1}{2}n' \sin\alpha_0 \tan\delta_0, \quad \delta_{1/2} = \delta_0 + \frac{1}{2}n' \cos\alpha_{1/2}$$

- Then, inserting $(\alpha_{1/2}, \delta_{1/2})$ again into Equation 5.8 and Equation 5.9, derive the final ones:

$$\alpha = \alpha_0 + m' + n' \sin\alpha_{1/2} \tan\delta_{1/2} \tag{5.10}$$

$$\delta = \delta_0 + n' \cos\alpha_{1/2} \tag{5.11}$$

These approximate methods, however, fail for a star close to the celestial North pole, or for long periods of time. In the first case we resort to the rigorous formulae expounded in the following (see also Exercises). In the second case, Equation 5.8 and Equation 5.9 can be regarded as the first terms of a development in series of $\alpha(t)$, $\delta(t)$. Adding the second derivatives as well we obtain:

$$
\begin{aligned}
\alpha_{t_0+t} &= \alpha_{t_0} + \dot{\alpha}_{t_0} t + \frac{1}{2!}\ddot{\alpha}_{t_0} t^2 + \cdots, \\
\delta_{t_0+t} &= \delta_{t_0} + \dot{\delta}_{t_0} t + \frac{1}{2!}\ddot{\delta}_{t_0} t^2 + \cdots
\end{aligned}
\tag{5.12}
$$

For full precision, the derivatives must contain also the variations of the "constants" (see Exercises).

We shall discuss again the problem of deriving the coordinates at a wanted date after having discussed the nutation.

5.5 The Nutation

We have called nutation the collection of the short period movements of the equator. The principal part, namely that discovered by Bradley, is due to the influence of the Moon, whose orbital plane is inclined by $i = 5°9'$ to that of the ecliptic (see Figure 5.5; here, we ignore some smaller movements of the Moon described in Chapter 14 and Chapter 15).

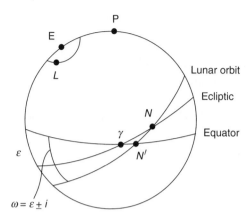

FIGURE 5.5
The lunar orbit and Bradley's nutation. L is the pole of the lunar orbit.

Let us call N the ascending node of this orbit on the ecliptic (ascending meaning that node where the lunar latitude passes from negative to positive values), and N' that on the equator. The two nodes are not fixed in inertial space: because of the solar perturbation on the lunar orbit, N precesses along the ecliptic in a retrograde sense, by approximately $191''$ each day (approximately three lunar diameters westward per lunation; the path of the Moon is a very complex one, a fact which must be carefully allowed for in studies such as the lunar occultations, described in Chapter 15), making an entire turn in 18.6 years. Correspondingly, the pole L of the lunar orbit describes a small circle of radius i around the ecliptic pole E, in the same period of time. As a consequence, the inclination ω of the lunar orbit on the celestial equator (and so the declination of the Moon) varies between approximately $\pm 18°.8$ and $\pm 28°.8$, according to the longitude of the node $\lambda(N)$, whose expression is:

$$\lambda(N) = 125°.04452 - 1934°.1363\ T + 0°0000.2971\ T^2 \tag{5.13}$$

if T is in Julian centuries from J2000.0. The variation of the inclination ω implies a variation of the perturbing torque exerted by the Moon on the geoid.

To see the effect on the declination of the Moon, consider the spherical triangle $N'\gamma N$; the angle in γ is ε, the angle in N' is $180° - \omega$, the arc γN is $\lambda(N)$, the angle in N is i. Therefore:

$$\cos \omega = \cos \varepsilon \cos i - \sin \varepsilon \sin i \cos \lambda(N), \quad \varepsilon - i \leq \omega \leq \varepsilon + i$$

When $\lambda(N) = 0°$, as in 1987.9 and again in 2006.4, the node coincides with γ, the inclination assumes the maximum possible value ($\omega = +28°36'$), and the declination reaches $\pm 28°36'$ during a lunation. When $\lambda(N) = 180°$, the node coincides with the autumn equinox, the inclination on the equator is the minimum possible ($\omega = +18°18'$), and the declination varies between $\pm 18°18'$.

Now, consider the ascending node N' on the equator and its right ascension $\alpha(N')$, namely the arc $\gamma N'$. From the same spherical triangle $N'\gamma N$ (see again Figure 5.5) one can see that:

$$\sin \gamma N' = \sin \alpha(N') = \sin \gamma N \frac{\sin i}{\sin \omega} = \sin \lambda(N) \frac{\sin i}{\sin \omega},$$

$$|\alpha(N')| \leq \arcsin\left(\frac{\sin i}{\sin \omega}\right) \approx 13°.0$$

Therefore, N' oscillates along the equator around γ in 18.6 years, with an amplitude of approximately $\pm 13°$.

Summarizing these considerations, the nutation discovered by Bradley can be described in the following way: the instantaneous movement of the celestial North pole is no longer along the great circle Pγ, but along PN$'$, so that the nutation periodically changes not only the origin of the longitudes, but also the obliquity, with amplitudes:

$$\Delta \lambda = -17''.2 \sin \lambda(N), \quad \Delta \varepsilon = 9''.2 \cos \lambda(N)$$

This movement can be visualized as the instantaneous true pole P describing a retrograde cone around the mean pole P_m, which in turn describes a cone of aperture ε around E. Imagine looking at this movement from the outside of the celestial sphere, as in Figure 5.6. On the plane tangent to the celestial sphere through P_m, the locus occupied by P is an ellipse, of semimajor axis $\Delta \eta = 9''.2$ and semi-minor axis $\Delta \xi = 17''.2 \sin \varepsilon = 6''.9$, described with a period of 18.6 years in the retrograde sense. By dynamical arguments, d'Alembert showed that the ratio of the two axes must be equal to $(\cos 2\varepsilon / \cos \varepsilon)$.

The complete phenomenon of nutation contains many other terms of smaller but nonnegligible amplitudes and different periods; the second most important term in longitude has an amplitude of $1''.32$ and a period of 183 days, that in obliquity has an amplitude $0''.57$ and the same period (these values slightly change with the epoch).

Conventionally, the nutation terms have been divided in two groups, those having long periods (>90 days) and those having short periods (<35 days), which have often been treated separately in the derivation of the true coordinates. This argument is further expanded in the Notes and in Chapter 6. In the following, we will indicate the complete nutation in longitude with $\Delta \psi$, that in obliquity with $\Delta \varepsilon$.

The nutation can also be regarded as a retrograde rotation R_3 of the equatorial mean Cartesian frame (x_M, y_M, z_M) around the z-axis by $\Delta \psi \cos \varepsilon$, followed by a direct rotation R_2 around the y-axis by $\Delta \psi \sin \varepsilon$; the nutation in obliquity is then a retrograde rotation R_1 around the x-axis of amplitude $\Delta \varepsilon$. Using the same procedure as seen in Equation 5.6 and Equation 5.7, we can

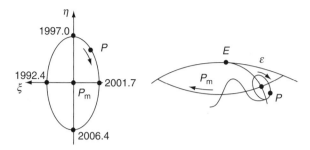

FIGURE 5.6
Nutation of the instantaneous pole P around the mean pole P_m. The η-axis points toward the ecliptic pole E.

easily show that the nutation in longitude has the same structure of the lunisolar precession, and therefore will cause a variation of the equatorial coordinates given by:

$$\Delta\alpha = \Delta\psi\,(\cos\varepsilon + \sin\varepsilon\sin\alpha\tan\delta), \quad \Delta\delta = \Delta\psi\,(\sin\varepsilon\cos\alpha) \quad (5.14)$$

The term in obliquity causes a variation of ε, but it does not affect the position of γ. By taking the derivatives of Equation 3.6, after a few simple calculations we obtain:

$$\Delta\alpha = -\Delta\varepsilon\cos\alpha\tan\delta, \quad \Delta\delta = \Delta\varepsilon\sin\alpha$$

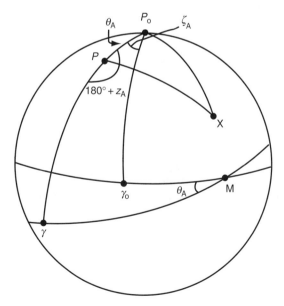

FIGURE 5.7
Newcomb's precessional angles. Rigorously, arcs P_0P and $P\gamma$ are not part of the great circle $P_0\gamma$.

Therefore, the total effect of the nutation is:

$$\begin{cases} \Delta\alpha = \Delta\psi\,(\cos\varepsilon + \sin\varepsilon\sin\alpha\tan\delta) - \Delta\varepsilon\cos\alpha\tan\delta \\ \Delta\delta = \Delta\psi\sin\varepsilon\cos\alpha + \Delta\varepsilon\sin\alpha \end{cases} \tag{5.15}$$

whose numerical values depend on the particular date. For instance, for the year 2000 and to a precision of $1''$, the *Astronomical Almanac* gives the following numerical expressions:

$$\begin{cases} \Delta\psi = -0°.0048\sin(125°.1 - 0.053d) - 0°.0004\sin(198°.0 + 1.971d) \\ \Delta\varepsilon = +0°.0026\cos(125°.1 - 0.053d) + 0°.0002\cos(198°.0 + 1.971d) \end{cases} \tag{5.16}$$

being $d = \text{JD} - 2451543.5$ (2451544.0 is the JD at Greenwich noon on January 0, 2000).

5.6 Approximate Formulae for General Precession and Nutation

Let us complete the problem of allowing for precession and nutation, again at the first order only. We already know how to precess the mean coordinates (α_0, δ_0) from the mean equinox at the epoch t_0 of the catalog to the mean coordinates (α, δ) at the beginning of year t of the date of observation, using Equation 5.8 and Equation 5.9, or a more refined procedure. Call Δt the remaining fraction of the year. Now, we have to add the general precession for this fraction, and the nutation terms given in Equation 5.15, obtaining:

$$\Delta\alpha = (m + n\sin\alpha\tan\delta)\Delta t + \Delta\psi\,(\cos\varepsilon + \sin\varepsilon\sin\alpha\tan\delta) - \Delta\varepsilon\cos\alpha\tan\delta$$

$$= m\left(\Delta t + \frac{\Delta\psi}{m}\cos\varepsilon\right) + n\left(\Delta t + \frac{\Delta\psi}{n}\sin\varepsilon\right)\sin\alpha\tan\delta - \Delta\varepsilon\cos\alpha\tan\delta \tag{5.17}$$

After some manipulation, we obtain an expression for the right ascension:

$$\Delta\alpha = A'(m + n\sin\alpha\tan\delta) + B'\cos\alpha\tan\delta + E' = A'a' + B'b' + E' \tag{5.18}$$

A', B', E' are named Bessel's daily numbers; their values are rapidly changing functions of the date (but not of the star's coordinates), and their values can be found tabulated in all major almanacs. The quantities a' and b', which depend only on the star coordinates, are called Besselian star's constants. In a similar way for the declination:

$$\Delta\delta = A'\,n\cos\alpha - B'\sin\alpha = A'a'' + B'b'' \tag{5.19}$$

Indeed, the Besselian star's constants (a', b', a'', b'') are also a slowly varying function of the epoch, because of the slow variation of m and n. Notice that the precise definition and units of Bessel's daily numbers and star constants varies in the different almanacs (see for instance Section B22 of the *Astronomical Almanac*). Furthermore, the expression 5.18 and the expression 5.19 become more complicated if higher precision is required, or if the star is close to the celestial pole. Still another way of computing the combined effect of nutation and precession is by means of the so-called independent day numbers f, g, G:

$$\alpha = \alpha_0 + f + g \sin(G + \alpha_0) \tan \delta_0$$

$$\delta = \delta_0 + g \cos(G + \alpha_0)$$

We refer to the *Astronomical Almanac* for their values, which are of course a function of the date. At this stage, we have obtained coordinates that are called the "true" coordinates of the star, but they are not yet the observed ones, because other phenomena contribute to the observed values, namely the aberration, the gravitational deflection of light, the parallax, the proper motions, the atmospheric refraction, which we will discuss in later chapters (see also Notes).

5.7 Newcomb's Rotation Formulae for Precession

After having discussed several approximate formulae, we describe the rigorous procedure introduced by Newcomb to derive the mean coordinates at an arbitrary date t from an initial epoch t_0. In the following, the notation will be that used by the *Astronomical Almanac*. Let us consider star X (see Figure 5.7) and let P_0 and P be two successive positions occupied by the celestial North pole at times t_0 and t. In the spherical triangle $P_0 PX$ consider the arc θ_A and the angles ζ_A, z_A. The angle ζ_A will be very small for small $(t - t_0)$, and so will be the angle z_A, because the arc $(P_0 P + P\gamma)$ differs very little from a great circle; furthermore, $\zeta_A \approx z_A$. The arc θ_A is not exactly the path described by the true pole, which is actually a somewhat irregular curve, but this fact is irrelevant here. At the second order in T, the elements (ζ_A, z_A, θ_A) are given by:

$$\begin{cases} \zeta_A = 0°.6402633T + 0°.0000839T^2 \\ z_A = \zeta_A + 0°.0002197T^2 \\ \theta_A = 0°.5567376T - 0°.0001183T^2 \end{cases}$$

if T is given in tropical centuries since B1950.0, or by:

$$\begin{cases} \zeta_A = 0°.6406161T + 0°.0000839T^2 \\ z_A = \zeta_A + 0°.000202T^2 \\ \theta_A = 0°.5567530T - 0°.0001185T^2 \end{cases} \tag{5.20}$$

if T is given in Julian centuries since J2000.0. As we remarked previously, the two sets do not give exactly the same results, due to the strong revisions of the constants.

From Equation 1.5 and Equation 1.6 we can derive the following rigorous formulae:

$$\begin{cases} \cos \delta \sin(\alpha - z_A) = \cos \delta_0 \sin(\alpha_0 + \zeta_A) \\ \cos \delta \cos(\alpha - z_A) = \cos \theta_A \cos \delta_0 \cos(\alpha_0 + \zeta_A) - \sin \theta_A \sin \delta_0 \quad (5.21) \\ \sin \delta = \sin \theta_A \cos \delta_0 \cos(\alpha_0 + \varsigma_A) + \cos \theta_A \sin \delta_0 \end{cases}$$

and their inverse expressions:

$$\begin{cases} \sin(\alpha_0 + \zeta_A) \cos \delta_0 = \sin(\alpha - z_A) \cos \delta \\ \cos(\alpha_0 + \zeta_A) \cos \delta_0 = \cos \theta_A \cos \delta \cos(\alpha_0 - z_A) + \sin \theta_A \sin \delta \quad (5.22) \\ \sin \delta_0 = -\sin \theta_A \cos \delta \cos(\alpha_0 - z_A) + \cos \theta_A \sin \delta \end{cases}$$

The same transformation (5.21) and the transformation (5.22) can be expressed as a rotation matrix $\mathbf{P}$ applied to an initial Cartesian system (x_0, y_0, z_0) to derive (x, y, z) and vice versa, namely $\mathbf{r} = \mathbf{P}\mathbf{r}_0$ or $\mathbf{r}_0 = \mathbf{P}^{-1}\mathbf{r}$. The elements of the matrices can be found by the above equations; for instance:

$$P_{11} = -\sin \zeta_A \sin z_A + \cos \zeta_A \cos z_A \cos \theta_A$$

For brevity, we omit the complete expressions here, which can be found in the *Astronomical Almanac*. Numerically, the diagonal elements are very close to one, all others are very close to zero, so that calculations must be carried out with at least seven significant decimal digits. We have already remarked that in the current practice the elements are not given at the beginning, but at the middle of the year; so, for instance, the values of the elements at J2000.5 are: $\zeta_A = z_A = +11''.53$, $\theta_A = +10''.02$.

To allow for nutation, the rotation $\mathbf{P}$ will be followed by rotation $\mathbf{R}_N$:

$$\mathbf{R}_N = \begin{pmatrix} 1 & -\Delta\psi \cos \varepsilon & -\Delta\psi \sin \varepsilon \\ \Delta\psi \cos \varepsilon & 1 & -\Delta\varepsilon \\ \Delta\psi \sin \varepsilon & \Delta\varepsilon & 1 \end{pmatrix} \quad (5.23)$$

whose numerical values have been given in Equation 5.16.

Although the method is accurate from the numerical point of view, it cannot be forgotten that the instantaneous values of the constants are not known with infinite precision, and, actually, some correction to their values is always possible with an *a posteriori* analysis of the observations.

5.8 Precession and Position Angles

Precession and nutation are rigid rotations of the celestial sphere, and as such they do not alter the angular distance between the stars. Therefore, the observed shape of a constellation, or of a nebula, will not be altered by the passage of the centuries (they will change because of the proper motions, but this is a different effect described in Chapter 8). However, the position angle p between two objects defined in Chapter 3, will change because it is measured from the variable direction of the North celestial pole. After some manipulation, it can be seen that:

$$\Delta p / \Delta t = n \sin \alpha \sec \delta = 0°.0056 \sin \alpha \sec \delta \text{ per year} \qquad (5.24)$$

Therefore, the position angle must be given in conjunction with an epoch. Its variation is very large in the proximity of the celestial poles. In a similar way, the differential coordinates of the two nearby objects ($\alpha_1 \approx \alpha_2 \approx \alpha$, $\delta_1 \approx \delta_2 \approx \delta$) will change by:

$$\frac{d\Delta\alpha}{dt} \approx n \sec \delta [\Delta\alpha \cos \alpha \sin d\delta + \Delta\delta \sin \alpha \sec \delta], \qquad \frac{d\Delta\delta}{dt} \approx -n \Delta\alpha \sin \alpha$$

Notes

1. The best method Hipparchus and the ancient astronomers had available in order to measure the positions of the stars without precise clocks, was to refer their positions to the Sun, using as intermediary a bright planet such as Venus, which can be seen in full daylight. An ingenious variant of the method is to take advantage of a lunar eclipse, when the Sun is at 180° from the Moon, to measure the position of a star near the ecliptic as Spica is, with respect to the center of the Earth's shadow on the Moon.

2. This chapter has discussed the procedures needed to derive the coordinates at a wanted date of observations from the mean coordinates given by a catalog at a certain epoch. Allowing for precession and nutation furnishes the so-called true coordinates at that date. However, these are not yet the apparent coordinates at the date, other effects must be added that will be discussed in the following chapters, such as diurnal and annual aberration, gravitational deflection of light, diurnal and annual parallax, proper motions, atmospheric refractions. We shall proceed in this book by superimposing these different effects, so that the order in which the different corrections will be applied is immaterial. This procedure is sufficient for most purposes, but it is not rigorous; see for instance the discussion by Soma, M., Aoki, S., 1990, Transformation from FK4 system to FK5 system, *Astronomy and Astrophysics*, **240**, p. 150. Standardized computer programs are available that follow generally agreed procedures, see for instance Wallace, P., 1994, The SLALIB library, *Astronomical Data Analysis Software and Systems III*, ASP Conference Series, D.R. Crabtree, R.J. Hanisch, J. Barnes, Eds., Vol. 61. A low precision, but very instructive,

graphical representation of the precession over the celestial sphere is given for instance in the book *Radio Astronomy* by J.D. Kraus (1966).

3. We have not included in the present discussion a small effect due to general relativity, and known as geodesic precession P_g. The inertial reference of frame in the neighborhood of the geocentric observer has a slight rotation with respect to the heliocentric frame, of the order of $+0''.0192$ per year, which is a small correction absorbed in ψ. Rigorously then: $\psi = P_0 \cos \varepsilon - P_g$. See also Chapter 9.

4. Among the many papers useful to follow the evolution of the determination of the precessional constant, see: Fricke, W., 1971, A rediscussion of Newcomb's determination of precession, *Astronomy and Astrophysics*, **13**, p. 298. The paper by Lieske, J.H., Lederle, T., Fricke, W., Morando, B., 1977, Expression for the precession quantities based upon the IAU (1976) system of astronomical constants, *Astronomy and Astrophysics*, **58**, p. 1, contains cross references to the notations used by several authors. The *1976 IAU Resolution on Astronomical Constants* is given, e.g., by Kaplan, G.H., 1981, *US Naval Observatory Circular No. 163*.

5. A most important paper is: by Fukushima, T., 2003, A new precession formula, *The Astronomy Journal*, **126**, pp. 494–534. This paper is based on the dynamical treatment given by J.G. Williams in 1994 (see Chapter 6); in particular, a careful discussion is given of the different meanings of ecliptic. Another discussion of the motions of the ecliptic is given by Harada, W., Fukushima, T., 2004, A new determination of planetary precession, *The Astronomical Journal*, **127**, pp. 531–538. This paper also contains an interesting nonlinear method of harmonic analysis.

6. Methods for Solar System objects. Allowance for precession and nutation in the case of objects of the Solar System presents several peculiarities: their Cartesian coordinates, velocities and accelerations are usually known, (often with such high precision to require a fully relativistic treatment); topocentric, geocentric, heliocentric, barycentric systems might be required; their positions are rapidly changing with respect to the background of the "fixed" stars. Therefore, we must postpone a fuller treatment to later chapters, after having discussed their orbits, the planetary aberration and the different time systems. However, some considerations can be made here: from the dynamical point of view, the really significant plane is that of the ecliptic, whose stability is much greater than that of the equator. Therefore, roughly speaking, of all elements characterizing the position in space of a planet, only those connected with the position of the equinox on the ecliptic will be affected by precession. We shall have to worry, for instance, about the longitude of the ascending node, very much less about the inclination. See the paper by Simon, J.L., Bretagnon, P., Chapront, J.,

Chapront-Touze, M., Francou, G., Laskar, J., 1994, Numerical expressions for precession formulae and mean elements for the Moon and the planets, *Astronomy and Astrophysics*, **282**, p. 663.

7. It will be useful to underline that precession and nutation do not alter the geographic coordinates, nor the position of the cardinal points on the horizon, a fact all too often overlooked by science fiction writers.

8. Information about NASA's mission to planet Pluto and Kuiper Belt Objects can be found in the web sites:
http://Pluto.jhuapl.edu/
http://www.nasa.gov/mission_pages/newhorizons/main/index.html

Liftoff occurred Jan. 19, 2006. Flyby studies of Pluto and Charon will be carried out in 2015. Successively, from 2016 to 2020, the spacecraft will visit the Kuiper Belt.

Exercises

1. Apply Equation 1.5 and Equation 5.7 to a body of the solar system, in order to transform its Cartesian coordinates (x, y, z) at the epoch t_0 to another date $t_0 + \Delta t$. Provided Δt is not too large, so that the precession constants do not change much, and ignoring the real movement of the body between the two dates, one obtains:

$$\begin{cases} x(t_0 + \Delta t) = x - (my + nz)\Delta t \\ y(t_0 + \Delta t) = y + mx\Delta t \\ z(t_0 + \Delta t) = z + nx\Delta t \end{cases}$$

where the constants must be expressed in radians per year ($m = 0.0002234$, $n = 0.000097$).

2. Calculate the second derivatives in Equation 5.12. Call τ the initial epoch, and write:

$$\begin{cases} \alpha = \alpha_\tau + a_1\Delta t + a_2\Delta t^2 + \cdots \\ \delta = \delta_\tau + b_1\Delta t + b_2\Delta t^2 + \cdots \end{cases}$$

After simple passages, we have:

$$a_1 = \left(\frac{d\alpha}{dt}\right)_\tau = m_\tau + n_\tau \sin \alpha_\tau \tan \delta_\tau$$

$$a_2 = \frac{1}{2}\left(\frac{d^2\alpha}{dt^2}\right)_\tau = \frac{1}{2}\left[\left(\frac{dm}{dt}\right)_\tau + \left(\frac{dn}{dt}\right)_\tau \sin\alpha_\tau \tan\delta_\tau\right]$$

$$+ \frac{1}{4}n_\tau^2\left[1 + 2\tan^2\delta_\tau\right]\sin 2\alpha_\tau + \frac{1}{2}m_\tau n_\tau \cos\alpha_\tau \tan\delta_\tau$$

$$b_1 = \left(\frac{d\delta}{dt}\right)_\tau = n_\tau \cos\alpha_\tau,$$

$$b_2 = \frac{1}{2}\left(\frac{d^2\delta}{dt^2}\right)_\tau = \frac{1}{2}\left[\left(\frac{dn}{dt}\right)_\tau \cos\alpha_\tau - n_\tau \sin\alpha_\tau\left(m_\tau \sin\alpha_\tau \tan\delta_\tau\right)\right]$$

3. Derive the mean coordinates of Polaris for the year 2100, knowing its J2000.0 position ($2^h31^m46^s.3$, $+89°15'50''.6$). Ignore the proper motion of the star.

This exercise has been adapted from Newcomb and Danjon (*op. cit.*). The time interval is very long, and the star is very close to the celestial North Pole, therefore we must use the formula 5.20 and the formula (5.21), with $T = 1$. The result is: J2100.0 ($5^h53^m14^s.9$, $+89°32'26''.3$).

The result given by Danjon, who, however, started from B1900.0 and with the pre-1894 constants, is B2100.0 ($5^h53^m36^s.7$, $+89°32'22''.7$).

6

Dynamics of Earth's Rotation

This chapter is dedicated to a very simplified dynamical analysis of the Earth's lunisolar precession and nutation and free (Eulerian) nutation. The discussion of the Earth's yearly revolution around the Sun is postponed to Chapter 12. Since Bradley's measurements, many mathematicians have accepted the challenge of explaining these observations, in particular d'Alembert, Clairaut, and Euler. Let us start by examining the intuitive explanation of the lunisolar precession put forward by Newton in his *Principia*. We shall give first an approximate description, sufficient to clarify the basis of the overall phenomenon, and then a more quantitative analysis.

6.1 Newton's Lunisolar Precession

We have already seen that the figure of the Earth is well approximated by an ellipsoid of rotation, of equatorial axis $a_\oplus$ and polar axis c; for the moment, the rotational axis is supposed to coincide with c. The reason for this shape is the centrifugal force due to the diurnal rotation. This figure can be thought of as an inner sphere (not necessarily homogeneous, but having density with radial symmetry), plus an equatorial bulge. Consider the gravitational attraction exerted by the point-like Sun S on each element of mass of such supposedly *rigid* body (see Figure 6.1); the attraction is proportional to the mass of the Sun and to the inverse square of the distance between the particular element and the Sun. The overall effect on the barycenter of the Earth is responsible for the annual revolution, and will not be considered further in this chapter.

Consider two mass points inside the spherical inner body, such as A and B in Figure 6.1. The force in A will be slightly different from the force in B, because of different distance and direction to S, but this slight imbalance will be perfectly compensated by the two symmetric points A′, B′. However, this compensation is not possible for points of the equatorial bulge, for example V, W. The net result will be a torque parallel to CS, perpendicular to the plane of the figure and directed inside. As is well known, a gyroscopic axis under

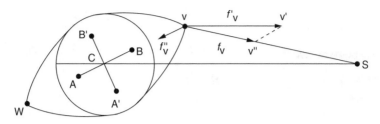

FIGURE 6.1
Solar precession, winter solstice. The nonspherical shape of the Earth has been greatly exaggerated.

the influence of a force will tend to orient itself in the plane orthogonal to that containing axis and force. Therefore, the combination of the angular momentum due to the diurnal rotation and of the torque will cause a gyroscopic movement of the Earth axis in the plane containing c and the perpendicular to CS. The solar torque is zero at the equinoxes; it would also be zero if the Sun reached declinations of $\pm 90°$, and it would be maximum if it reached declinations of $\pm 45°$, but these conditions never happen.

The same reasoning can be applied to the Moon, taking into account the different mass, distance, and position with respect to the equator.

Now, we shall show that the gyroscopic motion forced by the external body S has amplitude proportional to the mass of S and to the *inverse cube of its distance* CS, not to the inverse square. The explanation is intuitively simple, the imbalance can be considered due to the *gradient* of the gravitational force across the figure of the Earth. For this reason, the effect of the Moon is 2.2 times greater than that of the Sun. To see this conclusion in a more quantitative form, refer again to Figure 6.1. Calling r_S the distance between the barycenter of the Earth and of the forcing body S, we get:

$$r_V^2 = a_\oplus^2 + r_S^2 - 2a_\oplus r_S \cos \delta_S, \quad r_W^2 = a_\oplus^2 + r_S^2 + 2a_\oplus r_S \cos \delta_S$$

In the limit $a_\oplus / r_S \ll 1$ (certainly true for the Sun; for the Moon: $a_\oplus / r_\text{☾} \approx 1/60$):

$$r_V \approx r_S\left(1 - \frac{a_\oplus}{r_S} \cos \delta_S\right), \quad r_W \approx r_S\left(1 + \frac{a_\oplus}{r_S} \cos \delta_S\right)$$

Now call f'_V the component of the gravitational force f_V parallel to CS (and similarly f'_W):

$$f'_V/f_V = r_S/r_V, \quad f'_W/f_W = r_S/r_W, \quad \Delta f' = f'_V - f'_W = -K\frac{M_S}{r_S^3} a_\oplus \cos \delta_S$$

Therefore, the key factors for the amplitude of the torque are the mass of S and the inverse cube of its distance, the radius of the Earth, the mass of the equatorial bulge which enters in the still unspecified constant K.

Regarding the mass contained in the equatorial bulge, a rough estimate can be obtained by calculating the volume of the ellipsoid and subtracting that of the inner sphere: calling $\rho_\oplus$ the mean density, we would have:

$$m_{bulge} \approx \frac{8}{3}\pi\rho_\oplus c^3 \frac{a_\oplus - c}{c} \approx 2f M_\oplus \approx \frac{2}{300}M_\oplus \approx f_{bulge} M_\oplus$$

where f is the flattening factor defined in Chapter 2. However, the value $f_{bulge} \approx 2/300$ is certainly overestimated, because the density of the upper mantle of the Earth is smaller than the average (3.3 instead of 5.5), and because at any instant the effective mass is less than that of the whole bulge.

Regarding the order of magnitude of the precessional velocity, from the basic equations of precessional motions we recall that the torque is proportional to the product the moment of inertia C of the forced body (Earth) around its polar axis, times its rotational (diurnal) velocity Ω, times the precessional velocity ω_{prec}, so that:

$$\omega_{prec} \approx \frac{\Delta F \cdot a_\oplus}{C \cdot \Omega} \approx G \frac{m_{bulge} \cdot a_\oplus \cdot a_\oplus}{M_\oplus a_\oplus^2 \cdot \Omega} \frac{M_S}{r_S^3} \approx G \frac{f_{bulge}}{\Omega} \frac{M_S}{r_S^3}$$

(notice the cancellation of δ_S, due to the presence of sin δ_S on both terms). In general, a precessional parameter α can be defined as a function of the angular velocity Ω and dynamical ellipticity σ of the forced body, and of the mass and distance of the forcing body S:

$$\alpha = G \frac{\sigma}{\Omega} \frac{M_S}{r_S^3}$$

In Table 6.1 we provide an evaluation of the gravitational accelerations and precessional effects F_P exerted by several bodies of the Solar System at their minimum distance from the Earth.

Examine first the fourth and fifth columns, where the distance enters with the second power. The gravitational acceleration exerted by the Sun is 6.1×10^{-4} the gravity at the surface of the Earth (indicated with $g_\oplus$), and the gravitational attraction by the Moon is 200 times smaller than that of the Sun. The attractions exerted by Venus and Jupiter are two orders of magnitude

TABLE 6.1

Gravitational and Precessional Effects of Bodies of the Solar System on the Earth

Body	Mass (in Earth Masses $M_\oplus$)	Minimum Distance (in Earth Radii $a_\oplus$)	$g/g_\oplus$	$(g/g_\oplus)_\odot$	$F_{pr}/F_\odot$
Sun	3.33×10^5	2.354×10^4	6.1×10^{-4}	1	1
Moon	1.23×10^{-2}	60.2	3.4×10^{-6}	5.6×10^{-3}	2.19
Venus	8.15×10^{-1}	6.49×10^3	1.9×10^{-8}	3.2×10^{-5}	1.2×10^{-4}
Mars	1.08×10^{-1}	1.23×10^4	7.2×10^{-10}	1.2×10^{-6}	2.3×10^{-6}
Jupiter	3.18×10^2	9.85×10^4	3.3×10^{-8}	5.4×10^{-5}	1.3×10^{-5}
Saturn	9.52×10^1	2.00×10^5	2.4×10^{-9}	3.9×10^{-6}	4.6×10^{-7}

below that of the Moon. The last column, where the distance enters with its cube, shows that the precessional effect of the Moon is 2.19 times higher than that of the Sun, whereas those of Venus and Jupiter are, respectively, four and five orders of magnitude below the solar one. The values in the last column enter also into the discussion of the sea and solid Earth tides.

6.2   The Lunisolar Torque

In this section we examine the influence of the astronomical coordinates of the Sun and of the Moon on precession and nutation. Let us consider first the Sun (see Figure 6.2).

At the summer solstice, the moment of the force is oriented toward the equinox γ, and therefore the Earth's axis will tend to fall toward γ; the same will happen six months later, at the winter solstice. On the other hand, at the equinoxes the moment will be zero; thus we can consider the solar moment $K_\odot$ as composed of two vectors of constant and equal modules, one always directed toward γ, and one rotating in the equatorial plane with a period of 6 (not 12) months.

The first component forces the North Pole P to constantly fall toward γ. The instantaneous velocity of P will therefore be a uniform retrograde vector tangent both to the hour circle of γ and to the small circle of radius ε centered on the ecliptic pole E. The modulus of the solar vector is approximately 3.19 times smaller than the lunisolar precession, namely $15''.8 \sin \varepsilon \approx 6''.3$ per year, which would also be the elementary rotation of the equator around MM' in Figure 5.1. The vernal point itself is forced to a retrograde motion on the ecliptic of $15''.8$ per year. This would be the very slow *mean solar* precession.

The second component, of variable direction on the equatorial plane, is responsible for a movement of the *true* North Pole P along a circle around the *mean* pole, of very small amplitude. This solar nutation produces a periodic oscillation of γ about the mean position with a semiamplitude of $1''.3$, and a periodic variation of the obliquity with total amplitude of $0''.55$ (the obliquity being maximal at the equinoxes and minimal at the solstices).

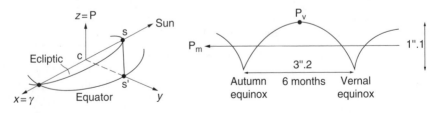

FIGURE 6.2
The solar precession.

The celestial pole would therefore appear to describe a cycloid around the mean pole, with arcs $3''.2$ long and $1''.1$ high, every six months.

In analytical terms, the previous discussion showed that the solar torque has amplitude $K_\odot$ given by:

$$K_\odot = \Delta f' a_\oplus \sin \delta_\odot = k' \sin \delta_\odot \cos \delta_\odot = k \sin 2\delta_\odot$$

At every date, the vector $\mathbf{K}_\odot$ is orthogonal both to the polar axis and to the line joining the Earth and the Sun, therefore pointing to direction $(\alpha_\odot - 90°)$ in the equatorial plane; in the geocentric equatorial Cartesian reference system (x, y, z) its components will be:

$$\begin{cases} K_x = k \sin 2\delta_\odot \cos(\alpha_\odot - 90°) = 2k \sin \delta_\odot \cos \delta_\odot \sin \alpha_\odot \\ K_y = k \sin 2\delta_\odot \sin(\alpha_\odot - 90°) = -2k \sin \delta_\odot \cos \delta_\odot \cos \alpha_\odot \\ K_z = 0 \end{cases} \quad (6.1)$$

or else, indicating as usual with $\lambda_\odot$ the longitude of the Sun:

$$\begin{cases} K_x = k \sin \varepsilon \cos \varepsilon(1 - \cos 2\lambda_\odot) \\ K_y = k \sin \varepsilon(-\sin 2\lambda_\odot) = k \cos \varepsilon \tan \varepsilon(-\sin 2\lambda_\odot) \\ K_z = 0 \end{cases} \quad (6.2)$$

Given that $\mathbf{K}_\odot$ cannot alter the module of the diurnal rotation velocity vector $\mathbf{\Omega}$, but only its direction, its components will be directly proportional to the Cartesian components of the velocity of the pole, $K_x \propto \dot{x}$, etc. Now, assume that $\lambda_\odot$ is a uniform function of time (this assumption is not strictly correct, as discussed in Chapter 4), namely that $\dot{\lambda}_\odot = 2\pi$ if the time unit is the tropical year. Then, by integrating Equation 6.2 starting from the initial point $\lambda_\odot = 0$, we get:

$$\begin{cases} x \operatorname{cosec} \varepsilon = P_\odot \cos \varepsilon \left(t - \dfrac{1}{2\dot{\lambda}_\odot} \sin 2\lambda_\odot \right) = p_\odot \left(t - \dfrac{1}{2\dot{\lambda}_\odot} \sin 2\lambda_\odot \right) \\ y = \dfrac{1}{2\dot{\lambda}_\odot} P_\odot \cos \varepsilon \tan \varepsilon \cos 2\lambda_\odot = \dfrac{p_\odot}{2\dot{\lambda}_\odot} \tan \varepsilon \cos 2\lambda_\odot \\ z = 1 \end{cases} \quad (6.3)$$

where $P_\odot$ is a "constant" whose value ($17''.2$ per tropical year) depends on the angular momentum of the Earth and on the average solar torque. The projection of $P_\odot$ on the ecliptic, $p_\odot = P_\odot \cos \varepsilon$, is the solar contribution to the total precession (its value is still less constant than that of $P_\odot$ because of the variable obliquity). Disregarding these minute variations, the term $x \cdot \operatorname{cosec} \varepsilon$ represents the motion in longitude $\Delta\lambda$ of the North Pole, and the term in y the variation of the obliquity $\Delta\varepsilon$; the z-component stays constant. We have confirmed, by this simplified analytical treatment, the existence of a progressive term in $\Delta\lambda$, increasing as $p_\odot t$, and of two periodic terms in $\Delta\lambda$ and in $\Delta\varepsilon$, both with a period of six months. We can easily see that there will be other periodic terms in the solar precession. A small annual term arises because the Earth–Sun distance has a slight variation between the aphelion and the perihelion. Furthermore, two periodic motions of period T_1 and T_2 couple together producing:

$$\frac{1}{T_+} = \frac{1}{T_1} + \frac{1}{T_2}, \quad \frac{1}{T_-} = \frac{1}{T_1} - \frac{1}{T_2} \tag{6.4}$$

so that two other periods of 4 and 12 months (and others of smaller amplitude) will be present in the complete solar nutation expression.

The above treatment cannot be extended so easily to the lunar case, because of the much greater intricacies of its orbit, of the larger precessional torque, and of the larger variation in declination caused by the inclination of $5°9'$ of the orbital plane on the ecliptic. However, one conclusion can be reached almost immediately: owing to the inverse proportionality of the amplitude to λ, which varies 13 times more rapidly than that of the Sun (13 lunar months in one year), the period of the lunar semimonth (13.7 days) has amplitude smaller than that of the six-month period in the solar one. Instead, the very long time (18.6 years, 6789 days) associated with the retrogradation of the nodes will be extremely important, say approximately $18.6/2.19 \approx 8.5$ times larger than the semiannual solar term. Any variation of the other orbital elements will give rise to corresponding terms in the lunar nutation, and to their coupling frequencies, so that the total number of measurable terms rapidly grows.

The 1980 IAU Theory of Nutation has been computed by determining the nutation in longitude and obliquity of a rigid Earth (Kinoshita, 1997), and introducing some modifications to allow for the nonrigidity. Table 6.2 highlights the nine terms with amplitude larger than $0''.03$ (the values are at J2000.0, all have minute variations).

Although the 1980 theory contains 106 terms in longitude and in obliquity, it still fails to represent the observations fully. Starting in 2006, the more accurate IAU 2000A precession and nutation theory will be implemented.

TABLE 6.2

High-Amplitude Terms in the 1980 IAU Theory of Nutation

Term Number	Period (days)	Longitude ($''$)	Obliquity ($''$)
1	6798.4	− 17.1996	9.2025
2	3399.2	0.2062	− 0.0895
9	182.6	− 1.3187	0.5736
10	365.3	0.1426	0.0054
11	121.7	− 0.0517	0.0224
31	13.7	− 0.2274	0.0977
32	27.6	0.0712	− 0.0007
33	13.6	− 0.0386	0.0200
34	9.1	− 0.0301	0.0129

6.3 The Precessional Potential

Let us consider a rigid Earth as an ellipsoidal body of center C. Let $C(X,Y,Z)$ be the geocentric equatorial Cartesian system connected with the body (and therefore rotating in the inertial frame), a, b, c the three principal semiaxes, and A, B, C the three corresponding moments of inertia:

$$A = \int (Y^2 + Z^2)\, dM, \quad B = \int (X^2 + Z^2)\, dM, \quad C = \int (X^2 + Y^2)\, dM \quad (6.5)$$

where dM is the mass element. The integration must be performed over the volume of the body. For symmetry reasons it is:

$$\int dM = M_{\oplus}, \quad \int X\, dM = \int Y\, dM = \int Z\, dM = \int XY\, dM = \cdots = \int X^2 Y\, dM = \cdots$$

$$= \int X^3\, dM = \cdots = 0$$

For our simplified model of homogeneous and rotationally symmetric ellipsoid we would have:

$$A = \frac{1}{5} M_{\oplus}(b^2 + c^2), \quad B = \frac{1}{5} M_{\oplus}(a^2 + c^2), \quad C = \frac{1}{5} M_{\oplus}(a^2 + b^2)$$

$$a = b = a_{\oplus}, \quad A = B, \quad \frac{C}{A} = \frac{2a_{\oplus}^2}{a_{\oplus}^2 + c^2} \approx 1.003, \quad \frac{C - A}{A} \approx \frac{C - A}{C} \approx 0.003$$

With reference to Figure 6.3, let Q be a massive point external to the Earth, at a distance r from C much greater than $a_{\oplus}$. Let ν be the angle PCQ.

For a generic point P inside the Earth, of elementary mass dM, distant R from C and r_{PQ} from Q, we have:

$$r_{PQ}^2 = R^2 + r^2 - 2rR \cos \nu = r^2 \left(1 + \frac{R^2 - 2rR \cos \nu}{r^2} \right)$$

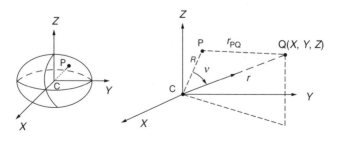

FIGURE 6.3
The precessional potential. Q is a generic body outside the Earth (in the present case, the Sun or the Moon), P a point-like mass inside it.

where $0 \leq R \leq a_\oplus$. With the simplification $a_\oplus << r_{PQ}, r$, we apply the series expansion seen in Equation 1.12:

$$\frac{1}{r_{PQ}} = \frac{1}{r}\left[1 + \frac{R}{r}\cos v + \frac{R^2}{2r^2}(3\cos^2 v - 1) + \frac{R^3}{2r^3}(5\cos^3 v - 3\cos v) + \cdots\right]$$

$$= \frac{1}{r}\left[1 + \frac{R}{r}P_1(\cos v) + \left(\frac{R}{r}\right)^2 P_2(\cos v) + \cdots\right] \tag{6.6}$$

where P_n are Legendre's polynomials of order n, defined by the general expression:

$$P_n(x) = \frac{1}{2^n n!}\frac{d^n}{dx^n}[(x^2 - 1)^n] \tag{6.7}$$

$$P_0(x) = 1, \quad P_1(x) = x, \quad P_2(x) = \frac{1}{2}(3x^2 - 1),$$

$$P_3(x) = \frac{1}{2}(5x^3 - 3x), \quad P_4(x) = \frac{1}{8}(35x^4 - 30x^2 + 3), \ldots$$

The elementary potential energy U in P due to Q is:

$$dU = GM_Q\frac{dM}{r_{PQ}}$$

which can be integrated over the whole body of the Earth to give:

$$U = GM_Q\int\frac{dM}{r_{PQ}} = G\frac{M_Q}{r}\left(U_0 + \frac{U_1}{r} + \frac{U_2}{2r^2} + \cdots\right) \tag{6.8}$$

In the same manner, the potential of the Earth on a distant point Q can be expressed as:

$$U = \frac{GM_\oplus}{r}\left[1 - \sum_{n=2}^{\infty}\left(\frac{a_\oplus}{r}\right)^n J_n P_n(\sin\phi')\right] \tag{6.9}$$

where ϕ' is the geocentric latitude of Q.

The constants J_n can be derived with high precision from the observations of the orbits of artificial satellites ($J_2 = 1.08 \times 10^{-3}$, $J_3 = -2.51 \times 10^{-6}$, $J_4 = -1.60 \times 10^{-6}$, etc.). This expression in terms of spherical harmonics, first derived by Laplace, will be used in Chapter 14 to examine how the nonspherical Earth influences the motion of a nearby artificial satellite.

The first term U_0 in Equation 6.8 is the mass of the Earth; the second term U_1 vanishes because of the definition of the barycenter, and all the following uneven terms U_3, U_5, etc. must be zero for symmetry reasons; therefore, in the present approximations we need only to consider U_2:

$$U_2 = \int(3\cos^2 v - 1)R^2 dM = A + B + C - 3I \approx 2A + C - 3I \tag{6.10}$$

where I is the moment of inertia of the Earth around the direction to Q. Therefore, to the order $(1/r)^3$:

$$U = GM_Q\left(\frac{M_\oplus}{r} + \frac{2A + C - 3I}{2r^3}\right)$$

Let (X_Q, Y_Q, Z_Q) be the coordinates of Q in the Earth-fixed reference system, and (l, m, n) its direction cosines:

$$l = \frac{X_Q}{r}, \quad m = \frac{Y_Q}{r}, \quad n = \frac{Z_Q}{r}, \quad l^2 + m^2 + n^2 = 1$$

Then:

$$I = Al^2 + Bm^2 + Cn^2 = A + (B - A)m^2 + (C - A)n^2 \approx A + (C - A)n^2$$

$$U \approx G\frac{M_Q}{r}\left(M_\oplus + \frac{(C - A)}{2r^2} - 3\frac{(C - A)}{2r^2}\frac{Z_Q^2}{r^2}\right)$$

Only the third term will be effective in precessional and nutational phenomena, being the only one that explicitly depends on the position of the forcing body Q with respect to the equator. Finally, Q being very distant from C, with a slight imprecision we can also put: $Z_Q/r = \cos\delta_Q$, so that:

$$U_{\text{prec}} \approx -3GM_Q\frac{(C - A)}{2r^3}\sin^2\delta_Q$$

Adding together Sun and Moon:

$$U_{\text{prec}} \approx -3G\frac{(C - A)}{2}\left[\frac{M_\leftmoon}{r_\leftmoon^3}\sin^2\delta_\leftmoon + \frac{M_\odot}{r_\odot^3}\sin^2\delta_\odot\right] \tag{6.11}$$

Let us take into account Kepler's third law (see Chapter 12):

$$P_Q^2(M_Q + M_\oplus) = \frac{4\pi^2}{n_Q^2}(M_Q + M_\oplus) = \frac{4\pi^2}{G}a_Q^3$$

where: P_Q is the period of revolution of Q (either Sun or Moon) around the Earth; n_Q its mean motion; and a_Q the semimajor axis of its orbit. Approximating the real orbits of Sun and Moon with circular ones, the cube of the instantaneous distances in Equation 6.11 can be substituted by the square of the mean motions, so that:

$$U_{\text{prec}} \approx -3\frac{(C - A)}{2}\left[n_\leftmoon^2\frac{M_\leftmoon}{(M_\leftmoon + M_\oplus)}\sin^2\delta_\leftmoon + n_\odot^2\frac{M_\odot}{(M_\odot + M_\oplus)}\sin^2\delta_\odot\right]$$

$$\approx -3\frac{(C - A)}{2}\left[n_\leftmoon^2\frac{M_\leftmoon}{(M_\leftmoon + M_\oplus)}\sin^2\delta_\leftmoon + n_\odot^2\sin^2\delta_\odot\right] \tag{6.12}$$

where the mass of the Sun has disappeared because it greatly exceeds that of the Earth.

Although many approximations were made in deriving Equation 6.12, some fundamental points have nevertheless been brought to light: namely, the dependence on the declination and mean motion of the forcing bodies, on the moments of inertia of the Earth, whose mechanical ellipticity can be derived without any model of the interior, and on the mass of the Moon, but not of the Sun. Before the advent of the Space Age (more precisely, before the spacecraft Ranger 5 had flown very near the Moon, in October 1962), the value of the lunar mass ($\approx M_\oplus/80$) was estimated from the constants of precession and nutation (see also Notes).

6.4 The Earth's Free Rotation

Let us compare the precessional energy given by Equation 6.12 with that of the diurnal rotation T:

$$T = \frac{1}{2}C\,\Omega^2, \quad \Omega = 2\pi/(\text{sidereal day}) \approx 7.292 \times 10^{-5} \ \text{sec}^{-1}$$

U_{prec} can be maximized by taking the highest value of the declination of Sun and Moon, hence:

$$\left|\frac{U_{\text{prec}}}{T}\right| \le 3\frac{(C-A)}{C}\left(\frac{n_\odot}{\Omega}\right)^2\left[1 + 0.25\frac{M_{\mathrm{D}}}{M_\oplus}\left(\frac{n_{\mathrm{D}}}{n_\odot}\right)^2\right] \tag{6.13}$$

where $n_{\mathrm{D}}/n_\odot \approx 1/13, n_\odot/\Omega \approx 1/366.25$, and therefore $U_{\text{prec}}/T \le 1 \times 10^{-7}$, a very modest fraction indeed. Therefore, in this paragraph we will examine the rotation of a free Earth, as if the Moon and the Sun were not forcing the precession and nutation.

Let us perform again the dynamical analysis of the forced lunisolar precession, with a more formal procedure. Consider two geocentric reference systems, the inertial ecliptic one (X_0, Y_0, Z_0), and the equatorial one (X, Y, Z) fixed in the body, and thus rotating with respect to the inertial system. For simplicity, we orient the axes (X, Y, Z) toward the principal axes of inertia. Let $\Omega(\Omega_1, \Omega_2, \Omega_3)$ be the angular velocity, resolved along the three orthogonal components, and I the inertia tensor, reduced to the diagonal form (A, B, C) by the choice of axes directions. The rotational energy of the body, and its total angular momentum are:

$$T = \frac{1}{2}(A\Omega_1^2 + B\Omega_2^2 + C\Omega_3^2),$$

$$\mathbf{M}(M_1, M_2, M_3) = \mathbf{M}(A\Omega_1, B\Omega_2, C\Omega_3) = \int \mathbf{R} \times (\mathbf{\Omega} \times \mathbf{R})dm$$

where the integral is extended over the mass of the body. Notice that the directions of $\mathbf{M}$ and $\mathbf{\Omega}$ do not necessarily coincide, unless the body has spherical symmetry, because the direction of $\mathbf{\Omega}$ need not coincide with that of the principal axis of inertia.

However, by assuming the spheroidal shape, with $A = B < C$, it can be shown that at all times the directions of $\mathbf{M}$, $\mathbf{\Omega}$, and axis Z will be in the same plane. In the absence of external forces, the direction of $\mathbf{M}$ is fixed in the inertial space, so that the inertial observer sees a precession of the rotation axis around $\mathbf{M}$, with angular velocity $\omega_{prec} = M/A$.

In the real case of external forces due to Moon and Sun, the vector equations of motions in the inertial system centered in the barycenter will be:

$$\frac{d\mathbf{P}}{dt} = \mathbf{F}, \quad \frac{d\mathbf{M}}{dt} = \mathbf{K}$$

where: $\mathbf{P}$ is the total linear momentum; $\mathbf{F}$ are the total external forces (namely, those of lunar and solar origin); and $\mathbf{K}$ their total moment. In the rotating body-fixed system (X,Y,Z), we can connect the three components of $\mathbf{\Omega}$ to the three components of $\mathbf{K}$ through the following Euler's equations:

$$\begin{cases} A\dfrac{d\Omega_1}{dt} + (C - B)\Omega_2\Omega_3 = K_1 \\[2mm] B\dfrac{d\Omega_2}{dt} + (A - C)\Omega_1\Omega_3 = K_2 \\[2mm] C\dfrac{d\Omega_3}{dt} + (B - A)\Omega_1\Omega_2 = K_3 \end{cases} \qquad (6.14)$$

In the absence of external forces, $K_{1,2,3} = 0$ and with azimuthal symmetry $A = B$, the system reduces to:

$$\begin{cases} A\dfrac{d\Omega_1}{dt} + (C - A)\Omega_2\Omega_3 = 0 \\[2mm] A\dfrac{d\Omega_2}{dt} - (C - A)\Omega_1\Omega_3 = 0 \\[2mm] C\dfrac{d\Omega_3}{dt} = 0 \end{cases}$$

After an elementary time dt, the rotating system (X,Y,Z) has a different orientation with respect to the inertial system (X_0,Y_0,Z_0) (see Figure 6.4), which can be resolved in an elementary rotation with angular velocities $(\Omega_1, \Omega_2, \Omega_3)$.

We have defined the plane XY as the geocentric equator, and the plane X_0Y_0 as the fixed ecliptic. The two planes intersect along the line of nodes, the ascending one being indicated with N. The instantaneous position of the rotating frame in the inertial system will thus be specified by three Eulerian angles (θ, ψ, φ). In the astronomical designation, N is the initial vernal equinox γ; θ and its time derivative express the obliquity of the ecliptic and its variation, and therefore the nutation; the time derivative of ψ is connected

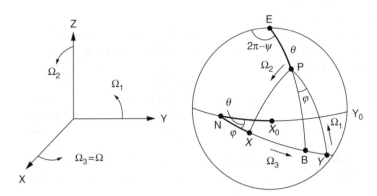

FIGURE 6.4
The Eulerian rotation angles. E, ecliptic pole; P, celestial North Pole; N, ascending node of the ecliptic on the equator; θ = angle EP = angle X_0NX = nutation angle; $2\pi - \psi$ = angle NX_0 = angle Y_0EP = precession angle; φ = angle NX = angle YPB = angle of diurnal rotation.

with the precession in longitude. According to the previous discussions, the derivatives are composed by secular and periodic terms, with coefficients function of the mechanical ellipticity of the Earth, of the orbital elements of the Sun and of the Moon, and of their masses (see also Notes). The angle φ expresses the diurnal rotation of the Earth.

The time derivatives of the Eulerian angles can be written in terms of the angular velocities $(\Omega_1, \Omega_2, \Omega_3)$. We omit the demonstration of these relationships, because of the many elaborated passages required to obtain them, and only recall the result:

$$\begin{cases} \dot{\theta} = \Omega_2 \sin\varphi - \Omega_1 \cos\varphi \\ \dot{\psi} \sin\theta = \Omega_2 \cos\varphi + \Omega_1 \sin\varphi \\ \dot{\varphi} = \Omega_3 + \cot\theta(\Omega_1 \sin\varphi + \Omega_2 \cos\varphi) \end{cases},$$

$$\begin{cases} \Omega_1 = \dot{\psi}\sin\theta\sin\varphi - \dot{\theta}\cos\varphi \\ \Omega_2 = \dot{\psi}\sin\theta\cos\varphi + \dot{\theta}\sin\varphi \\ \Omega_3 = \dot{\varphi} - \dot{\psi}\cos\theta \end{cases}$$

(6.15)

For the Earth, both Ω_1 and Ω_2 are both much smaller than Ω_3, and $\Omega_3 \approx \Omega$ (conditions not necessarily true in the general case). In the inertial system, the rotational energy T is thus given by:

$$2T = A(\dot{\theta}^2 + \dot{\psi}^2 \sin^2\theta) + C(\dot{\varphi} - \dot{\psi}\cos\theta)^2 \approx C\Omega^2$$

Using the Lagrangian formalism, namely:

$$\frac{\mathrm{d}}{\mathrm{d}t}\left(\frac{\partial T}{\partial\dot{\theta}}\right) - \frac{\partial T}{\partial\theta} = \frac{\partial U}{\partial\theta}$$

for θ and for the other variables, the following relations can be obtained:

$$
\begin{cases}
A\dot\Omega_1 + (C - A)\Omega_2\Omega_3 = \dfrac{\sin\varphi}{\sin\theta}\left(\cos\theta\,\dfrac{\partial U}{\partial\varphi} + \dfrac{\partial U}{\partial\psi}\right) - \cos\varphi\,\dfrac{\partial U}{\partial\theta} \\[2ex]
A\dot\Omega_2 + (A - C)\Omega_1\Omega_3 = \dfrac{\cos\varphi}{\sin\theta}\left(\cos\theta\,\dfrac{\partial U}{\partial\varphi} + \dfrac{\partial U}{\partial\psi}\right) + \sin\varphi\,\dfrac{\partial U}{\partial\theta} \\[2ex]
C\dot\Omega_3 = \dfrac{\partial U}{\partial\varphi}
\end{cases}
$$

where U is the total potential due to the external bodies. Notice that the assumption of symmetry permits $\partial U/\partial\varphi = 0$ in the last equation; Ω_3 is constant even in the case of forced precession and nutation (to be sure, if all other simplifying assumptions are fulfilled).

In the hypothetical case of the free Earth, U and its derivatives can be ignored, so that:

$$
\begin{cases}
A\dot\Omega_1 + (C - A)\Omega_2\Omega_3 = 0 \\
B\dot\Omega_2 + (A - C)\Omega_1\Omega_3 = 0 \\
C\dot\Omega_3 = 0
\end{cases}
\tag{6.16}
$$

The component Ω_3 of the angular velocity remains obviously constant, as before, while (remember, $B = A$):

$$
\frac{d\Omega_1}{dt} = -\Omega_3\frac{C - A}{A}\Omega_2 = -\omega\Omega_2, \qquad \frac{d\Omega_2}{dt} = \Omega_3\frac{C - A}{A}\Omega_2 = \omega\Omega_1
$$

The constant

$$
\omega = \Omega_3\frac{C - A}{A} \approx 0.003
$$

corresponds to a period of about 303 sidereal days. Finally, with an appropriate choice of the initial time, one can arrive at the result:

$$
\Omega_1 = \chi\,\Omega_3\cos\frac{C - A}{A}\Omega_3 t, \qquad \Omega_2 = \chi\,\Omega_3\sin\frac{C - A}{A}\Omega_3 t
\tag{6.17}
$$

where χ is a constant which has to be determined by the observations.

To understand the previous results, consider in Figure 6.5 the three vectors $\boldsymbol{\Omega}$, $\mathbf{M}$, $\mathbf{P_e}$, the latter being the unit vector in the direction of the pole of the spheroid, namely along the Z-axis (this representation often goes under the name of the French mathematician Poinsot). Let us normalize $\boldsymbol{\Omega}$ by Ω_3, and $\mathbf{M}$ by $C\Omega_3$:

$$
\boldsymbol{\Omega}/\Omega_3 = \left(\frac{\Omega_1}{\Omega_3}, \frac{\Omega_2}{\Omega_3}, 1\right),
$$

$$
\mathbf{M}/C\Omega_3 = \left(\frac{A}{C}\frac{\Omega_1}{\Omega_3}, \frac{A}{C}\frac{\Omega_2}{\Omega_3}, 1\right) \approx \left(0.997\frac{\Omega_1}{\Omega_3}, 0.997\frac{\Omega_1}{\Omega_3}, 1\right)
$$

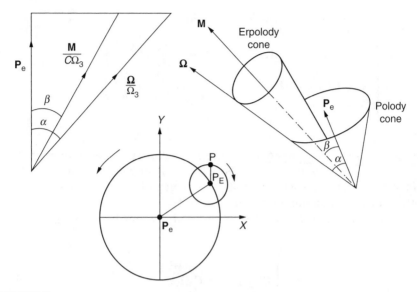

FIGURE 6.5
The rigid body rotation in the Poinsot representation. $\mathbf{P_e}$, unit vector to the pole of the geoid (axis Z); P_E, Eulerian position of the rotation pole. The observations prove that angle $\mathbf{P_e}P_E \approx 0''.3$, so that $\alpha - \beta \approx 0''.001$ ($\mathbf{M}$, $\mathbf{\Omega}$, and P_E essentially coincide). The direction $\mathbf{P_e}X$ defines the terrestrial meridian of Greenwich. The true instantaneous rotation pole is in P: the difference $P_E P$ is due to the forced lunisolar precession, varying approximately between 0 and $0''.02$ (Oppolzer's terms). The astronomical meridian of Greenwich passes through P.

The two vectors $\mathbf{\Omega}/\Omega_3$ and $\mathbf{M}/C\Omega_3$ are very similar in direction and amplitude. Furthermore, they must also be almost equal to $\mathbf{P_e}$, so that only fairly refined measurements will be able to distinguish between them. The theory cannot fix the value of the angle β in Figure 6.5, it must be determined by the observations, but then it must be: $[(\alpha - \beta)/\beta] = [(C - A)/A] \approx 1/303$.

Recalling that the three vectors must be at all times coplanar, we conclude that the inertial observer notices a diurnal rotation of the plane containing $\mathbf{\Omega}$ and $\mathbf{P_e}$ around $\mathbf{M}$; more precisely, the frequency will be:

$$\omega_{fp} = \frac{M}{I_1} = \left(1 + \frac{C - A}{A}\right)\Omega_3 \approx 1.003\Omega_3$$

The body-fixed observer, instead, in the time $1/\omega$ will notice a rotation of $\mathbf{\Omega}$ around $\mathbf{P_e}$ (cone of polody) and around $\mathbf{M}$ (cone of erpolody). The observations show (see next paragraph) that the angle between $\mathbf{\Omega}$ and $\mathbf{P_e}$ is approximately $0''.3$, so that the angle between $\mathbf{\Omega}$ and $\mathbf{M}$ must be approximately 303 times smaller, namely of the order of $0''.001$. Projected on the terrestrial surface at the geographic poles, these angles correspond to

9 m and 3 cm, respectively. Therefore, for the terrestrial observer the dominant motion is the polody.

We have already seen that the angular frequency of the polody should be, according to the above results:

$$\omega \approx \frac{C - A}{A} \, \Omega_3 \approx 0.003 \, \Omega_3$$

or else, a period of approximately 303 sidereal days (10 months).

This free-body nutation is superimposed to the forced lunisolar precession and nutation, whose presence forces the motion of **M** with respect to the fixed stars, and slightly alters the previous considerations by a minute variation of the apertures of the cones (the difference P_eP in Figure 6.5, Oppolzer's terms), at the level of few hundredths of arcsec.

In conclusion, if what is exposed is correct, from the Earth's surface we should observe the diurnal rotation axis to perform, every 10 months, a cone around the geodetic North Pole, with an amplitude σ that we cannot determine a priori (being a function of unknown initial conditions), but that hopefully is measurable. Figure 6.6 shows, in a graphical form, the difference between forced and free precession, also called *Chandler's wobble*.

The immediate observational consequence of this wobble of the rotation axis would be a periodic alteration of the astronomical latitude of each observatory, so that the phenomenon is also called *variation of the astronomical latitudes*. Obviously, the longitudes would be affected in the same way; however, with the intrinsic difficulties associated with all determinations of the zero point that we have discussed in several places.

Therefore, the astronomical latitudes ϕ should vary according to:

$$\phi = \phi_0 + \phi_1 \cos(\omega_{pol} t + q)$$

Many astronomers have attempted to detect the phenomenon, until finally, in 1888, Kustner measured variations at the level of $0''.3$. Soon after, Chandler was able to demonstrate the existence of two different periods, one of approximately 366 days (one sidereal year), and one of approximately 435 days. The instantaneous pole would be seen by an observer

FIGURE 6.6

The wobble of the rotation axis P around the geodetic pole P_e. L is a site with zenith Z. Left, the forced lunisolar precession in the approximation of the rotational pole coinciding with the pole of the figure produces a slow variation of the equatorial coordinates of the star S. Right, the slight difference between the two poles produces small variations of the latitude of L.

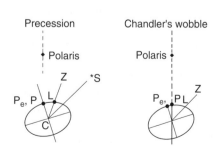

looking at it from above to perform two counterclockwise revolutions around the principal moment of inertia, with periods of 1 and 1.2 years, respectively. The first term is clearly of meteorological origin, the seasonal variation of the weight of the snow in the polar caps, the atmospheric loading of the regular winds, etc., but the duration of the second came as a real surprise, being much longer than the expected free Eulerian period. Therefore, one major assumption must be in error in the previous reasoning: Newcomb identified the responsible factor in the nonrigidity of the Earth. The Earth behaves more like an elastic than a rigid body, and indeed, its elasticity is surprisingly good, being comparable to that of a steel sphere. Precise computations to quantify the effects of elasticity are very difficult to make, it is however understandable that if the spheroid is elastic, only a fraction of its mass will be effective at any given time, lowering C and augmenting the frequency ω. Another way of looking at this phenomenon is to think that part of the equatorial elastic bulge will be readjusted by the centrifugal force around the rotation axis. In the same manner, the constants of the forced precession and nutation will be affected by the elasticity.

As a result of the discovery by Kustner and Chandler, an International Service for the Latitudes (ILS) was instituted at the close of the nineteenth century. Several stations, all having latitude $39°08'$, were chosen in Italy, Japan, U.S.A., and Russia, and equipped with instruments such as meridian circles, Zenith tubes, and Danjon's astrolabe. The ILS brought to light a complex wandering of the poles, which is usually represented in a Cartesian system centered in a mean pole and having the X-axis pointing towards Greenwich and the Y-axis towards west. The superposition of the two revolutions approximately averages to zero every 6 years; for this reason averages were made over that interval (initially from 1900 to 1905). (See Dick et al., 2000 for a detailed account of the first years of the ILS.)

The results of the ILS were in a sense disappointing, first of all because the instrumental and human resources were never really adequate to perform the difficult task. Furthermore, a great problem was the very definition of the reference and time systems: until much more precise clocks than astronomical phenomena became available, it was never possible to disentangle the minute perturbations affecting the meridian and the vertical of a particular site. An entirely different approach was necessary, which became possible in 1970 with the introduction of atomic clocks, with the very long baseline radio-interferometers (VLBI), and with laser ranging of dedicated satellites (e.g., the LAGEOS) and of the Moon.

Figure 6.7 shows a recent result obtained with the VLBI, together with a reanalysis of the historical drift of the mean pole.

In addition to the variation of latitudes and longitudes, such modern techniques have also confirmed, beyond doubt, the existence and import-ance of the relative movements of the continents, an idea put forward by

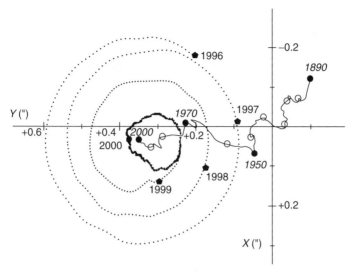

FIGURE 6.7

VLBI determination of the wandering of the poles from 1996 to 2000, together with a reanalysis of the historical data on the mean pole movement from 1890 to 2000.

Wegener in 1911. The distance between two observatories can change at the level of a few centimeters a year, a movement which is reflected in the relative displacements of the celestial sources of a few millarcseconds per year (see Figure 6.8).

At this level of precision, the concept of rotation of the Earth, as measured by stations on moving plaques, becomes extremely intriguing, and so do all

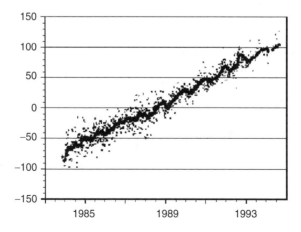

FIGURE 6.8

The increase of the relative distance between two VLBI stations (Westford in U.S.A. and Wetzell in Germany) from 1983 to 1995. The ordinate scale is in millimeters.

TABLE 6.3

The Main Movements of the Terrestrial Axis

In the inertial reference system

Lunisolar precession, discovered by Hipparchus, second century B.C., amplitude
$\sim 23°.5$, period $\sim 26{,}000$ years

Lunar nutation in obliquity, discovered by Bradley in 1737, amplitude $\sim 9''.2$, period
18.6 years. To this major term, several smaller components of lunisolar, geophysical,
and meteorological origin are overimposed

Secular decrease of the obliquity due to planetary perturbations, theoretically predicted
by Euler in 1748, amplitude $\sim 47''$/century

With respect to the body of the Earth

Pseudo-Eulerian wobble, predicted by Euler in 1767, measured by Kustner and then by
Chandler in 1891, period ~ 14 months, amplitude $\sim 0''.3$

Seasonal wobble, evidenced by Chandler, period 1 sidereal year, amplitude $\sim 0''.2$

Forced wobble, theoretically predicted by Oppolzer in 1856, amplitude $\sim 0''.02$, several
periods

Secular displacement of the pole

---

astronomical timescales. Further data on the length of the day, and of its
variations in the last 2500 years, will be provided in Chapter 10.

It is useful to summarize in Table 6.3 the main movements of the terrestrial
axis we have discussed so far.

6.5 Recent Developments

The International Earth Rotation Service (IERS) has selected a catalogue
of 608 extragalactic radio-sources (212 of which are considered defining
sources) that constitute the International Celestial Reference Frame (ICRF)
to be used for VLBI observations. Since the same ICRF sources are used
to study the movements of the pole, this reference frame does not
depend on the accuracy of the models of precession and nutation. In
1991, the IAU decided that the future standard celestial reference frame
would be centered in the barycenter of the Solar System, and realized by
an ensemble of very distant point-like extragalactic radio-sources, in such
a way to be as consistent as possible with the FK5, and with a full
account of the relativistic metrics. The extragalactic distances would
insure for centuries to come an essentially rotation-free system (at the
level of ± 20 microarcsec). This fundamental catalog, named ICRS
(International Celestial Reference System), has officially replaced the
FK5 since 1997; it is based on the equatorial coordinates at J2000.0 of the
608 radio-sources of the ICRF. The origin of the Right Ascension is
implicitly determined, establishing that the quasi-stellar object (Quasar)
3C 273 has $\alpha(\text{J}2000.0) = 12^h29^m06^s.6997$. This quasar is at cosmological

distance, and is also the only one bright enough in visible wavelengths (visual magnitude ≈ 13) to have been accessible to the satellite Hipparcos. Therefore, the positional system of Hipparcos could be referred to the ICRS, and so were the ephemeris of the Solar System bodies calculated by the Jet Propulsion Laboratory (designated JPL DE 403). Owing to the fact that Hipparcos contains all 1540 FK5 stars, the position of the FK5 pole and equinox are known in the ICRF to better than a few milliarcsec: the systematic difference of the equinox FK5-ICRS is 87 ± 10 milliarcsec. (See also, in the Bibliography, Walter and Sovers' *Astrometry of Fundamental Catalogues*.)

Given the continuously growing precision of time and Earth rotation angles (ERAs), in 2000 the IAU decided to adopt several resolutions (including also a revision of nomenclature), which will be enforced from 2006. In addition to the already quoted new precession and nutation theory, a more precise connection between the UT and Earth rotation angle (ERA) will be adopted. (See *The Astronomical Almanac* for 2006.)

Notes

- Smart, M. (*Celestial Mechanics*, Bibliography) gave a detailed discussion of Euler's Equation 6.15 covering the entire problems of general precession and nutation. It was shown there that:

$$-U_{\text{prec}} = \frac{3}{2}(C - A)n_\odot^2 \left\{ f\left(\frac{M_{\leftmoon}}{M_\odot}, r_{\leftmoon}, i, e, e_{\leftmoon} \right) \sin^2\theta \right.$$
$$\left. + \left[(g_1 \cos\psi - g_2 \sin\psi) \sin\theta\cos\theta + H\sin^2\theta \right] t + V \right\}$$

with f, g_1, g_2, H, V functions of the orbital elements of the Sun and of the Moon, and t the time since the initial epoch. From this, the expressions for $\psi(t)$ and $\theta(t)$ can be derived:

$$\psi(t) = at + bt^2 + \psi_{\text{per}} = 50''37 - 0''.0001t^2 - 17''.23 \sin\Lambda_{\leftmoon} + \cdots$$

$$\theta(t) = \theta_0 + ct^2 + \theta_{\text{per}} = 23°.26 + \frac{1}{2}5''.6 \times 10^{-6}t^2 + 9''.21 \cos\Lambda_{\leftmoon} + \cdots$$

where $\Lambda_{\leftmoon}$ is the longitude of the ascending node of the lunar orbits. These results justify the approximate discussion and validate the results obtained with Equation 6.12.

See also: Kinoshita, H., 1977, Theory of the rotation of the rigid earth, *Celestial Mechanics*, **15**, p. 277.

- Another excellent examination of this intricate problem is the paper by Williams, J.G., 1994, Contributions to the earth's obliquity rate, precession and nutation, *The Astronomical Journal*, **108**, pp. 711–724.

- The stabilization of the Earth's obliquity by the Moon is discussed in Laskar, J., Joutel, F., Robutel, P., 1993, *Nature*, **361**, p. 615.
- For a pre-Hipparcos discussion of the planetary, stellar, and extragalactic frames see:
  Folkner, W.M., Charlot, P., Finger, M.H., Williams, J.C., Sovers, O.J., Newhall, X.X., Standish, E.M., 1994, Determination of the extragalactic–planetary frame tie from joint analysis of radio interferometers and lunar ranging measurements, *Astronomy and Astrophysics*, **287**, p. 279.
- A recent publication about the polar motion is:
  Dick, S., McCarthy, D., Luzum, M., 2000, Polar motion: historical and scientific problems, IAU Coll. 178, ASP Conference Series Vol. 208.
- For a description of the VLBI see:
  http://giub.geod.uni-bonn.de/vlbi/, which contains several useful links.
- For a description of the Lageos program see:
  http://www.asi.it/html/eng/asicgs/cgs.html,
  http://www.earth.nasa.gov/history/lageos/lageos1.html
- For laser lunar ranging see:
  http://physics.ucsd.edu/~tmurphy/apollo/apollo.html
- The ensemble of celestial and terrestrial assumptions made by the IERS are exposed in the *IERS Technical Note* 21, July 1996 (Central Bureau IERS, Observatoire de Paris, France).
  Useful Web sites:
  http://www.iers.org/
  http://hpiers.obspm.fr/
  http://syrte.obspm.fr/iaucom19/ (IAU Commission 19 Rotation of the Earth)
  http://syrte.obspm.fr/iaudiv1/ (Division I 2000–2003 Fundamental Astronomy)
  http://www.usno.navy.mil/
  These sites contain the latest IERS and IAU resolutions, graphs, and tables of useful and recent data, and also products such as accurate subroutines for several astrometric applications.
- The original determination of the optical and radio position of 3C 273B in the FK4 frame is given by: Hazard, C., Sutton, J., Argue, A.N., Kenworthy, C.M., Morrison, L.V., Murray, C.A., 1971, 3C 273B — Coincidence of radio and optical positions, *Nature*, **233**, p. 89.
- The influence of tides on the rotation of the Earth is described in many books (see Bibliography) and articles, for instance:
  Brosche, P.U., Seiler, U., Suendermann, J., Wuensch, J., 1989, Periodic changes in the Earth's rotation due to oceanic tides, *Astronomy and Astrophysics*, **220**, pp. 318–320.

Ray, R.D., Steinberg, D.J., Chao, B.F., Cartwright, D.E., 1994, Diurnal and semidiurnal variations in the Earth's rotation rate induced by oceanic tides, *Science*, **264**, pp. 830–832.

Exercises

Derive the gravity at the surface of a rotating ellipsoidal body of mass m and radius R (Clairaut's formula).

Let U be the potential of the body gravitational field, and ω its angular velocity around the polar axis. On any equipotential equilibrium surface, the quantity U':

$$U' = U + \frac{1}{2}\omega^2 r^2 \cos^2\phi$$

where $\mathbf{r}$ is the radius vector (with $r \approx R$) and ϕ its angle with the equator, will be constant on that surface. From Equation 6.9 and following equations the relation:

$$U' = G\frac{m}{r} - G\frac{C-A}{2r^3}(3\sin^2\phi - 1) + \frac{1}{2}\omega^2 r^2 \cos^2\phi$$

is easily derived. Let $r = R(1 - \eta)$, η being a small quantity. Neglecting higher-order terms, it can be proven that:

$$\eta = \left[3\frac{C-A}{2mR^2} + \frac{\omega^2 R^3}{2Gm}\right]\sin^2\phi$$

Let σ indicate the ratio between centrifugal force to gravitational force at the equator:

$$\sigma = \omega^2 R^3 / Gm$$

and recall the equations for the oblate spheroid and of its flattening f given in Chapter 2. Expanding the expression of r^2 by the binomial theorem and retaining at the end only the terms in f, we finally get for the equilibrium surface:

$$r = R\left[1 - \left(3\frac{C-A}{2mR^2} + \frac{1}{2}\sigma\right)\sin^2\phi\right]$$

By measuring f and σ we can thus derive $(C - A)$. We can now derive U' to get the surface gravity:

$$g = -\frac{\partial U'}{\partial r} = G\frac{m}{r^2} - 3G\frac{C-A}{2r^4}\left(3\sin^2\phi - 1\right) - \omega^2 r(1 - \sin^2\phi)$$

and finally:

$$g = g_0\left[1 + \left(\frac{5}{2}\sigma - f\right)\sin^2\phi\right]\tag{6.18}$$

where g_0 is the gravity at the equator:

$$g_0 = G\frac{m}{R}\left[1 + f - \frac{3}{2}\sigma\right]$$

Equation 6.18 is Clairaut's formula. At the equator of the Earth, $g_0 = 978.049$ cm/sec^2 (measured) and 981.43 cm/sec^2 (corrected for rotation). At the second order, the more precise formula (Airy, Callandreau) can be applied:

$$g = 978.049\left(1 - 0.005288\,\sin^2\phi - 0.000006\,\sin^2 2\phi\right)$$

7

Aberration of Light

As seen in the previous chapters, precession and nutation are phenomena due to the variable orientation of the observer's system with respect to the fixed stars. The aberration, instead, is an effect due to the finite velocity of light, and to the varying motion of the observer with respect to the celestial source.

The finiteness of the velocity of light (indicated as usual by c) was suspected by many philosophers and physicists. Galileo Galilei suggested a method to measure it, but he probably never carried out the experiment. At the end of the 17th century, Oleg Roemer, then working in Paris, obtained the first reliable indication of the value of c by purely astronomical means. In 1727, Bradley discovered that the declination of γ Dra (a star not too distant from the ecliptic pole) had a periodic variation during the yearly revolution of the terrestrial observer due to the same cause, the finite velocity of light. The value of c derived from these astronomical observations was confirmed more than a century later, around 1850, by Fizeau and Foucault with the well-known terrestrial experiments employing toothed wheels and rotating mirrors.

7.1 The Solar Aberration

In Paris, Roemer had taken up the task of comparing the observed times of the eclipses of the Medicean moons of Jupiter with the tables compiled by Cassini. Those times could have had great practical importance for the determination of the longitude of ships (another idea of Galileo). The observations proved that Cassini's ephemerides were quite accurate at the quadratures of Jupiter with the Sun, retarding, however, or anticipating, from the observations by approximately 8 min near opposition or conjunction (see Figure 7.1). The required correction therefore had the form:

$$\Delta t = \pm \tau_a \cos(\lambda_\odot - \lambda_J)$$

where $\lambda_\odot$, λ_J are, respectively, the ecliptic longitudes of the Sun and of Jupiter, and $\tau_a \approx 8$ min.

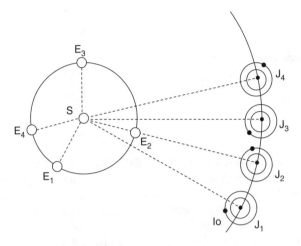

FIGURE 7.1
Successive positions of Jupiter and Io as seen from the Earth. Position 4 (conjunction) is not
directly observable.

This time interval was interpreted by Roemer as the time needed by the
light to cross the astronomical unit (AU), for the epoch a bold idea indeed.
Cassini had determined in 1672 the fairly precise value of 9″.5 for the solar
parallax. A value for the velocity of light could be derived, ranging from
230,000 to 360,000 km/sec according to the value of the light time (Roemer
had indeed obtained the excessive value of 11 min).

A discussion mostly limited to first order in the quantity V/c follows:
although the derived formulae will not be accurate to better than 0″.01, the
advantage of the approximate procedure is to separate all the various factors
entering in the aberration, and to allow the possibility of summing their
effects one over the other. Should greater accuracy be needed, e.g., in
reducing the data of the astrometric satellite Hipparcos, more precise
formulae have to be employed.

The Earth describes around the Sun an orbit that for the moment is taken
as circular, with radius $a_\odot = 1$ AU, and with uniform velocity vector $\mathbf{V}$,
whose direction is perpendicular to the radius vector, and whose modulus is
given by:

$$V = 2\pi \frac{a_\odot}{P} = na_\odot$$

P being the sidereal year, and n ($\approx 3548″.2/$day) the so-called *mean motion*.

The light will cross the AU in time τ_a (a quantity referred to as aberration
time, or equation of light, or light time for the unit distance). For the
geocentric observer, in those 8 min the Sun will have moved from its
apparent position (when the light left it) to the geometrically correct but
unobservable position corresponding to the arrival of the light. The angular

distance between the two positions is:

$$K = 2\pi \frac{a_\odot}{Pc} = n\tau_a = \frac{V}{c} \qquad \text{(radians)}$$

a quantity named *constant of solar aberration*. In round numbers:

$$V \approx 30 \text{ km/sec}, \quad K = V/c \approx 0.0001 \approx 20''.6$$

Under those simplifying hypotheses, angle K is constant during the year, even in sign. Notice that the solar aberration does not change the definition of the equinoxes as intersections among equator and ecliptic; it delays, however, the ingress in the equinoxes of the aberrated Sun with respect to the geometrical Sun, a difference important to remember in Newcomb's definition of the longitude of the mean *geometric* Sun (see Chapter 10). Furthermore, the solar aberration as previously described is not observable directly on the Sun, but it can be noticed from a number of external effects, such as velocity curves of binary stars, light curves of variable stars, etc. For instance, the radial velocities of the stars exhibit a yearly component due to the projection of the Earth velocity in the direction of each star; with this method, in 1928 Harold Spencer Jones obtained the value $K = 20''.475$.

The above considerations can be refined by taking into account that the orbit of the Earth is slightly elliptical ($e \approx 0.0167$), with the Sun not in the center but in one of the two foci. The modulus V of the velocity will therefore change with the date, being maximum at perihelion and minimum at aphelion. Therefore, it is useful to consider the vector $\mathbf{V}$ as the sum of two components: $\mathbf{V}_t$ perpendicular to the major axis of the ellipse (a direction known as *line of apses*), and $\mathbf{V}_r$ perpendicular to the radius vector. With simple geometric considerations it is seen that the moduli of both components are constant, and that constant is also the *direction* of $\mathbf{V}_t$:

$$V_r = \frac{na_\odot}{\sqrt{1-e^2}}, \quad V_t = \frac{na_\odot e}{\sqrt{1-e^2}} \approx \frac{1}{60} V_r$$

where $a_\odot$ is now the semimajor axis of the Earth's orbit, as shown in Figure 7.2.

Therefore, the velocity that properly enters in the definition of the solar aberration is $\mathbf{V}_r$, so that the constant of aberration is:

$$K = \frac{V_r}{c} = \frac{na_\odot}{c\sqrt{1-e^2}} \approx 20''.495$$

(the quantity K is not really constant over long time intervals, because of the secular change of e). The component $\mathbf{V}_t$ is responsible for the so-called *elliptical aberration* $K_e \approx 0''.343 \approx 0^s.023$, which changes from day to day and in principle is observable through the *equation of time* (see Chapter 10).

In total, the difference in ecliptic longitude between the aberrated (observable) and the geometric Sun is:

$$\lambda_\odot - \lambda = -K[1 - e\cos(\lambda_\odot - \lambda_\Pi)]$$

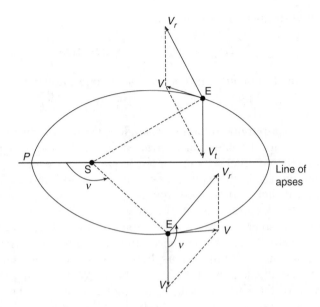

FIGURE 7.2
A representation of the elliptical motion of the Earth. P is the perihelion (at 180° from the perigee Π).

λ_{Π} being the longitude of the perigee (at 180° from the longitude of the perihelion; at the present epoch $\lambda_{\Pi} \approx 18^h 58^m$, see also Chapter 4). A corresponding equation must be applied to the difference in Right Ascensions. The *Astronomical Almanac* publishes in Section C the *geometric* ecliptic coordinates referred to as the mean equinox of date. The apparent longitude is then given by the low-precision formula:

$$\lambda_{app} = \lambda_{geom} + \Delta\psi - 20''.496/R$$

where $\Delta\psi$ is the nutation term for the date, and R the distance in AU. To this level of precision, the latitude is not affected.

In these considerations we have not yet discussed the very small effects due to the latitude of the Sun with respect to the dynamical ecliptic and to the velocity of the Earth with respect to the barycenter of the Earth–Moon system, approximately 12.5 m/sec. These components will be discussed later.

The value of K has been revised from time to time, according to the different determinations of a and c; for instance, in the first part of the 20th century the following values were adopted:

$$a_{\odot} = 149.504.000 \text{ km}, \quad c = 299.774 \text{ km/sec}, \quad e = 0.0167301 \text{ (1950)},$$

$$K = 20''.487 = 1^s.366, \quad Ke = 0''.343 = 0^s.023$$

The International Astronomical Union adopted, in 1976, a new set of constants that have been enforced since 1984:

$$a_\odot = 149.597.870 \text{ km}, \quad c = 299.792.458 \text{ km/sec} = 173.14 \text{ AU/day},$$

$$e = 0.01670790 \ (2000), \quad K = 20''.49552 = 1^s.366, \quad Ke = 0''.342 = 0^s.023$$

Using these values, the light time for unit distance becomes $\tau_a = 499^s.004782 = 8^m 19^s.004782$, corresponding to a solar parallax $\pi_\odot = 8''.794148$ (see also Chapter 8).

7.2 The Annual Aberration

As previously highlighted, the annual aberration affecting the stars was discovered by Bradley by observing with his meridian circle the second magnitude star γ Dra, in an attempt to measure its parallax. During the year, the declination of the star was seen to oscillate by $\pm 20''.5$ around a mean position, reaching the maximum deviation at the solar opposition or conjunction. The motion was too large to be attributable to a distance effect, and the dates were 3 months out of phase with those expected from the annual parallax; furthermore, Bradley was struck by the close numerical coincidence with the solar aberration constant, and thus suspected that the cause was the same, namely the finite velocity of light. Obviously the apparent Right Ascension of the star had to be affected in the same way, but that effect could not be measured by Bradley due to the lack of precision of his clocks.

Figure 7.3 shows a series of positions of γ Dra from 1920 to 1941, during a complete revolution of the nodes of the lunar orbit. One can see a small processional effect (small because of the proximity of the star to the ecliptic pole), the nutation discovered by Bradley himself (the sinusoid with period 18.6 years), and finally the annual variation due to aberration. Bradley's discovery conclusively proved the correctness of Roemer's hypothesis, gave a direct way to determine K, and provided a much more precise value of c.

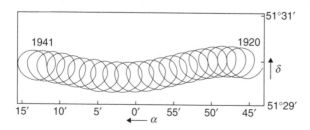

FIGURE 7.3

A series of positions of γ Dra from 1920 to 1941. (Adapted from Danjon, A., 1980, *Astronomie Generale*, Librairie Blanchard, Paris.)

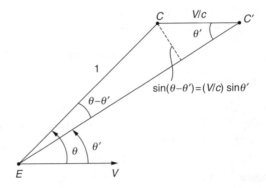

FIGURE 7.4
The intuitive explanation of aberration, in the prerelativistic approximation where $EC = 1$, $EC' > 1$. The Earth's velocity vector V points to the apex in the plane of the ecliptic, $90°$ from the Sun.

An intuitive way of understanding the yearly aberration, based on the Galilean transformation of velocities, and in the simplified hypothesis of circular orbit, is the following (Figure 7.4): let C be the center of the objective of the telescope, E the intersection of the optical axis with the focal plane, so that the line EC is the direction of sight. The Earth's velocity vector $\mathbf{V}$ points toward an instantaneous direction named *apex of motion* which is on the ecliptic at $90°$ from the Sun. Let θ be the angle between the line of sight and the apex, in the plane defined by the two directions. During the time Δt employed by the light to travel the distance CE, the Earth moves by $V\Delta t = (V/c)CE$. Therefore, the telescope must be pointed in direction θ', not θ, inclining it toward the direction of the apex. From the figure, it is easily seen that:

$$\sin(\theta - \theta') = \sin \Delta\theta = \frac{V}{c}\sin \theta' \qquad (7.1)$$

$V/c \sin \theta'$ being the component of the Earth velocity perpendicular to the apparent direction θ' of the star.

Since $V \ll c$, the sine of the difference approximately equals the difference:

$$\sin \Delta\theta \approx \Delta\theta = \theta - \theta' \approx \frac{V}{c}\sin \theta' \qquad \theta \approx \theta' + \frac{V}{c}\sin \theta' \qquad (7.2)$$

Using this approximate formula, the geometrical direction θ of the star, at the time when the light reached the observer (not at the time it left the source), can be derived from the apparent direction θ'. The difference (geometric minus apparent) is zero in the direction of the apex and of the antapex (at $180°$ from the apex).

With vector notation, let **n** be the unit vector in direction θ, **n**$'$ the unit vector in direction θ'; the Galilean addition of velocities gives:

$$\mathbf{n}' = \frac{\mathbf{n} + \dfrac{\mathbf{V}}{c}}{\left| \mathbf{n} + \dfrac{\mathbf{V}}{c} \right|} \tag{7.3}$$

Taking the scalar part of **n** $\wedge$ **n**$'$ then:

$$\sin \Delta\theta = \frac{\dfrac{V}{c}\sin\theta}{\sqrt{1 + 2\dfrac{V}{c} + \left(\dfrac{V}{c}\right)^2}} = \frac{V}{c}\sin\theta - \frac{1}{2}\left(\frac{V}{c}\right)^2\sin 2\theta + \cdots \tag{7.4}$$

which is the prerelativistic second-order approximation of the aberrational amount. Notice the dependence of the second term from $\sin 2\theta$, and that the apparent direction has been expressed as a function of the geometric one. The maximum value of the second-order term is $0''.001$.

The aberration effect we have described is independent of the wavelength, the focal length of the telescope, the distance of the star, and also its velocity with respect to the terrestrial observer. The aberration will be absolutely identical for both components of a binary system of stars or of galaxies (apart from that due to the slight difference in relative positions). One might also wonder what the correct value of c to be used in observations with ground telescopes is, either that in air or that in vacuum (the two velocities differ in the visible range by some 67 km/sec in normal conditions of temperature and pressure, a difference well measurable). The correct answer is the velocity in vacuum, because the atmosphere partakes of the same translational motion of the Earth barycenter, and no further aberration is introduced by its presence (the atmospheric refraction is one of the main factors limiting the precision of positional measurements, including the determination of the aberration, but this is an entirely different effect). The observational proof was obtained by the Astronomer Royal, George Biddell Airy in 1872, by filling his telescope with water: the aberration amount did not change.

In the simplifying hypothesis of the terrestrial circular orbit, the velocity vector **V** rotates during the year by $360°$ in the plane of the ecliptic with constant modulus, always pointing to $90°$ from the Sun. Consequently, the yearly aberration of a star having ecliptic latitude β will appear as an apparent elliptical motion with a semimajor axis parallel to the ecliptic and equal to K, and a semiminor axis perpendicular to the ecliptic and equal to $K \sin\beta$; this ellipse degenerates in a circle at the ecliptic poles, in a segment on the ecliptic itself. Notice that the star will never be seen in its geometrical position, except for an ecliptical star, twice a year.

The dimensions of this ellipse are the same for all celestial bodies having the same ecliptic latitude (whether they are planets, stars, galaxies, quasars, etc.), and do not reflect the ellipticity of the Earth's orbit, although finally the cause is the yearly motion. In other words, the aberration cannot be noticed by differential measurements of stars in a small area of sky. Over large angles, however, it will cause a (small) distortion of the celestial sphere, distinctively different from precession and nutation, which are rigid rotations of the sphere. Any other periodic motion of the terrestrial observer, for instance the diurnal rotation, or the motion around the Earth–Moon barycenter, will cause a corresponding periodic phenomenon of aberration, suitably scaled for its velocity value. Allowing for all these aberrational effects will result in referencing the position of each star to an ideal heliocentric observer (better yet, to an observer fixed in the barycenter of the Solar System). However, even the latter will be in secular motion with respect to the fixed star system, so that the true coordinates and the true aspect of the celestial sphere are to a certain extent unknown.

Before discussing the effect of the annual aberration on the equatorial coordinates (intuitively, the ellipse in ecliptic coordinates will remain the same in the equatorial ones, it will simply rotate according to the direction of the star), it is better to see the modifications introduced by special relativity (Einstein, 1905). For simplicity the discussion will be carried out in two dimensions, but it will emphasize all the important elements.

7.3 Lorentz Transformations

Let $O(x, y)$ and $O'(x', y')$ be two inertial reference systems, the first fixed with respect to the distant stars, the second in uniform motion with respect to the first with velocity $\mathbf{V}$ along x. The time is t in $O(x,y)$ and t' in $O'(x',y')$. Without loss of generality, we can assume that the two systems coincide at $t = t' = 0$. The Galilean transformations for positions and times are:

$$x' = x - Vt, \quad y' = y, \quad t' = t$$

Special relativity tells us that the correct transformations (named after Lorentz), are instead:

$$x' = \frac{x - Vt}{\sqrt{1 - \left(\dfrac{V}{c}\right)^2}}, \quad y' = y, \quad t' = \frac{t - \dfrac{Vx}{c^2}}{\sqrt{1 - \left(\dfrac{V}{c}\right)^2}}$$

Resorting again to the vector notation, the Lorentz transformations give, for the unit vectors $\mathbf{n}$ in direction θ and $\mathbf{n}'$ in direction θ', the expression:

$$\mathbf{n}' = \frac{\sqrt{1 - \left(\dfrac{V}{c}\right)^2}\,\mathbf{n} + \dfrac{\mathbf{V}}{c} + \left(\mathbf{n}\dfrac{\mathbf{V}}{c}\right)\left(\dfrac{\mathbf{V}}{c}\right)\Big/\left(1 + \sqrt{1 - \left(\dfrac{V}{c}\right)^2}\right)}{1 + \mathbf{n}\dfrac{\mathbf{V}}{c}} \tag{7.5}$$

Taking the modulus of the vector product $\mathbf{n} \wedge \mathbf{n}'$, and expanding in powers of (V/c), we obtain:

$$\sin \Delta \theta_{\text{Rel}} = \frac{\left(\dfrac{V}{c}\right)\sin \theta + \dfrac{1}{2\sqrt{1 - \left(\dfrac{V}{c}\right)^2}}\left(\dfrac{V}{c}\right)^2 \sin 2\theta}{1 + \dfrac{V}{c}\cos \theta}$$

$$= \frac{V}{c}\sin \theta - \frac{1}{4}\left(\frac{V}{c}\right)^2 \sin 2\theta + \cdots \tag{7.6}$$

At this level of approximation, the relativistic aberration differs from the Galilean expression 7.4 by $1/4(V/c)^2 \sin 2\theta$.

Therefore, the elementary formula 7.2 is approximated to terms of the order of $(V/c)^2$ for two distinct reasons:

1. Neglect of higher order terms in the trigonometric expansions
2. Incorrect transformation rules

Numerically, the Galilean and relativistic expressions give the same results for the annual aberration to within $0''.0005$. However, there is a profound conceptual difference: even in the vacuum above the terrestrial atmosphere, in the Galilean relativity aberration appears only in the direction of the group velocity, not in that of the phase velocity. According to special relativity, aberration is present in both (see the books by Møller (1962) and by Bohm (1965)).

7.4 Effects of Annual Aberration on the Stellar Coordinates

In order to determine the main components of the annual variation of the apparent ecliptic coordinates due to aberration, let us proceed for a moment with the circular approximation of Earth's orbit. On the celestial sphere, the plane defined by the observer, by the terrestrial apex T' and by the geometric position X of the star (of ecliptic coordinates λ, β), translates in a great circle passing for X and T' (see Figure 7.5).

The apparent position of the star on this great circle is X', displaced from X toward T' by the amount $XX' = K \sin \theta$; T' in its turn is on the ecliptic at $90°$ behind the Sun, whose true (geometric) longitude at the date is $\lambda_\odot$. Therefore, the displacement is always towards the Sun. The smallness of K allows use of plane trigonometry in the small triangle $XX'U$; after simple

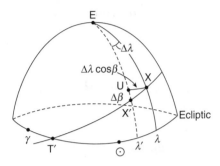

FIGURE 7.5
The displacement of the apparent position due to annual aberration in ecliptic coordinates, in the approximation of the Earth's circular orbit. E is the pole of the ecliptic. T' is the apex of the Earth's motion, 90° from the Sun.

passages, one gets:

$$\Delta x = (\lambda' - \lambda)\cos \beta = \Delta\lambda \cos \beta = -K \cos(\lambda_\odot - \lambda)$$
$$\Delta y = (\beta' - \beta) = \Delta\beta = -K \sin(\lambda_\odot - \lambda)\sin \beta$$

(7.7)

which are the parametric equations of an ellipse with semimajor axis K and semiminor axis $K \sin \beta$.

During the course of the year, the star traces an elliptical locus having the true star in its center. This is the annual aberration ellipse. The star describes the ellipse in retrograde sense if $\beta > 0$, in direct sense if $\beta < 0$, and passes through the end points of the semimajor axis when in conjunction with (westerly end point) or opposition (easterly end point) to the Sun, of the semiminor axis when at the quadratures (at the westerly quadrature S is nearer to the ecliptic). For a star at the ecliptic pole the ellipse degenerates in a circle, on the ecliptic on a segment: this is the only case when we can see the star in its true position, twice a year.

Let us now add the effect of the slight ellipticity e of the terrestrial orbit, namely the small and constant velocity component of amplitude Ke perpendicular to the semimajor axis. First of all, as already discussed for the solar aberration, the value of constant K in the annual aberration must be understood as:

$$K = \frac{V_r}{c} = \frac{na}{c\sqrt{1 - e^2}} \approx 20''.495.$$

Second, the effect of the perpendicular component is as follows: the geometric position of the star is not exactly at the center of the ellipse of aberration, but displaced with respect to it by $0''.343$, in a direction whose longitude is $\lambda = \lambda_\Pi - 90°$, namely at 90° from the geocentric longitude of the perigee of the Sun. This (almost) constant displacement, called *elliptic aberration*, when projected in longitude and latitude, amounts, respectively, to $\Delta\lambda_e = -eK \cos \beta \cos(\lambda_\Pi - \lambda)$, and $\Delta\beta_e = -eK \sin(\lambda_\Pi - \lambda) \sin \beta$; these two so-called *E-terms* must be added with their signs to the already determined $\Delta\lambda$, $\Delta\beta$. Some authors prefer using the longitude of the perihelion ($\lambda_\oplus = 180° - \lambda_\Pi$), thus reversing the signs of the two terms.

The value of this elliptic constant has, however, a slight secular variation, for two different reasons: because the eccentricity decreases with the centuries and because the Earth's orbit slowly processes in its plane (for this reason λ_{Π} increases by $11''.63/$year, performing a full sidereal revolution in approximately 110,000 years). Certainly, those are very small changes, the resulting variation of the position of the geometric star with respect to the center of the ellipse is of the order of $0''.02$ in 1000 years. Traditionally, the terms of the elliptic aberrations were not explicitly calculated in the reduction of the coordinates, they were masked in the published mean positions at each epoch. This habit was discontinued with the FK5 and in subsequent catalogs, in order not to introduce any systematic error, however, small in the coordinates. According to the IAU resolutions taken in 1976, stellar aberration is to be computed from the total velocity of the Earth referred to the barycenter of the Solar System, and mean catalog places are not to contain elliptic terms of aberration.

We now calculate the corresponding variations in equatorial coordinates. The aberration ellipse will maintain its shape, but its orientation will change in the new reference system. Ignoring for the moment the elliptic terms, from the general transformation rules given in Chapter 3 we derive the apparent geocentric coordinates α', δ' from the geometric geocentric coordinates α, δ through the expressions (see Figure 7.6):

$$
\begin{cases}
\Delta\alpha = \alpha' - \alpha = -K\dfrac{\sin\alpha\sin\lambda_\odot + \cos\varepsilon\cos\alpha\cos\lambda_\odot}{\cos\delta} = \dfrac{1}{c}\dfrac{-\dot{X}\sin\alpha + \dot{Y}\cos\alpha}{\cos\delta} \\[3mm]
\Delta\delta = \delta' - \delta = -K(\sin\varepsilon\cos\delta\cos\lambda_\odot + \cos\alpha\sin\delta\sin\lambda_\odot \\
\qquad\qquad\qquad - \cos\varepsilon\sin\alpha\sin\delta\cos\lambda_\odot) \\[3mm]
\qquad\quad = \dfrac{1}{c}(-\dot{X}\cos\alpha\sin\delta - \dot{Y}\sin\alpha\sin\delta + \dot{Z}\cos\delta)
\end{cases}
\tag{7.8}
$$

where $(\dot{X},\dot{Y},\dot{Z})$ are the components of the Earth's velocity vector, whose approximate values in AU/day are:

$$\dot{X} = +0.0172\sin\lambda_\odot, \quad \dot{Y} = -0.0158\cos\lambda_\odot, \quad \dot{Z} = -0.0068\cos\lambda_\odot$$

Precise values are found in the *Astronomical Almanac*.

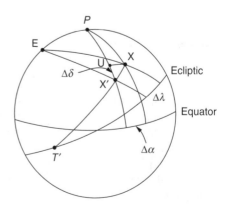

FIGURE 7.6
Aberration in the transformation from ecliptic to equatorial coordinates.

In order to remove the elliptic aberration, we must add the (almost) constant terms:

$$\Delta\alpha = -Ke[\sin \lambda_{\Pi} \sin \alpha + \cos \lambda_{\Pi} \cos \varepsilon \cos \alpha]/\cos \delta,$$

$$\Delta\delta = -Ke[\sin \lambda_{\Pi} \cos \alpha + \cos \lambda_{\Pi} \cos \varepsilon \sin \alpha]\sin \delta + \cos \lambda_{\Pi} \sin \varepsilon \cos \delta$$

for the FK5 and all catalogs based on it. The corrections therefore take the form:

$$\alpha' - \alpha = Cc + Dd, \quad \delta' - \delta = Cc' + Dd'$$

in which C, D depend on the sun's longitude and therefore on the date, while c, c', d, d' depend on the coordinates of the star and on the obliquity of the ecliptic ε. Or else:

$$\Delta\alpha = \frac{1}{15 \cos \delta}(C \cos \alpha + D \sin \alpha) = \frac{1}{15 \cos \delta}h \sin(H + \alpha)$$

$$\Delta\delta = C(\tan \varepsilon \cos \delta - \text{sen } \alpha \text{ sen } \delta) + D \cos \alpha \text{ sen } \delta = h \cos(H + \alpha) + i \cos \delta$$

where:

$$h \cos H = -K \text{ sen } \lambda_{\odot}, \quad h \text{ sen } H = -K \cos \varepsilon \cos \lambda_{\odot}, \quad i = -K \text{ sen } \varepsilon \cos \lambda_{\odot}$$

C and D are Bessel's day numbers, and H, h, i are the independent day numbers. Their values are found tabulated for each day in the *Astronomical Almanac*. Notice the formal similarity with the expression of the nutation, although the physical bases are so different.

We have already highlighted that the aberration introduces a slight distortion of the celestial sphere; the distance s between two stars, and their position angle p, will be altered by an amount that can be calculated in advance, and removed before the astrometric reduction of a plate is carried out. To give an order of magnitude, over an arc of 1° the maximum effect of the annual aberration is $0^s.02/\cos \delta$ in α, $0''.3$ in δ. Bessel's formulae provide a convenient way to carry out this differential calculation:

$$\Delta s = s[h \cos(H + \alpha) \cos \delta - i \sin \delta], \quad \Delta p = h \sin(H + \alpha) \tan \delta$$

7.5 The Diurnal Aberration

The diurnal rotation velocity will be responsible for a similar effect, of much smaller amplitude, and is dependent on the geocentric latitude ϕ' of the observer. Indeed, the diurnal velocity is approximately 0.46 km/sec at the equator (the angular velocity is $\omega \approx 7.292 \times 10^{-5}/\text{sec}$); its apex is on the equatorial plane, at 90° from the meridian and towards east, therefore with equatorial coordinates ($\alpha_{\oplus} = \text{LST} + 6^h$, $\delta_{\oplus} = 0°$), where LST is the local sidereal time. The diurnal aberration ellipse is thus parallel to the equatorial system, and its smallness permits the difference (apparent − geometric) to be treated with the first-order formulae; for an observer in geocentric latitude ϕ' at distance ρ km from the Earth's center, the difference is:

$$dHA = -d\alpha = 0^s.021(\rho\,\omega\cos\phi'/c)\cos HA/\cos\delta,$$

$$d\delta = 0''.320(\rho\,\omega\cos\phi'/c)\sin HA\,\text{sen}\,\delta$$

No sensible error is made if the equatorial radius of the Earth and the astronomical latitude are used in these formulae. Notice that when the star is on the meridian, the effect on declination is zero, while the displacement towards east causes a later transit. The delay is equal to:

$$\Delta t = 0^s.021\cos\phi/\cos\delta$$

operationally indistinguishable from a collimation error of the telescope.

For a satellite such as the Hubble Space Telescope, orbiting at approximately 550 km in a plane inclined by 27° to the equator and with a period of 90^m, the diurnal velocity has the higher value of about 7 km/sec. Therefore, the effect of the orbital aberration must be computed with care.

7.6 Planetary Aberration

We have seen that the correction for annual aberration provides the geometric direction to the star at the time when the light reaches the observer; no allowance is made for the motion of the star in the long time interval between emission and reception. In the case of the bodies of the Solar System, whose orbits are known with high accuracy, we can explicitly take into account the finite time τ of propagation of the light from the body to the observer. The term *planetary aberration* rigorously means the sum of the annual aberration (independent of the distance and motion of the source, and thus affecting in the same manner the apparent positions of the stars surrounding the body) plus the finite light time (note that some authors actually mean only the second term).

Let us examine the planetary aberration with an approximate, iterative procedure, assuming the orbital elements of the body are known. Given the distance ρ of the planet from the Earth barycenter at the time t of observation, the light time $\tau_t = \rho/c$ is also known. Since τ_t seldom exceeds 60 min, during this time the body can be considered in uniform motion, and the first approximation to the position at the time the light left the object is given by:

$$\alpha_{t-\tau}^{(1)} = \alpha - \tau_t\frac{d\alpha}{dt}, \quad \delta_{t-\tau}^{(1)} = \delta - \tau_t\frac{d\delta}{dt}$$

where the derivatives are known from the orbital elements. With this revised position, we obtain a second estimate of the light time, and we iterate the procedure.

Notice that the observed positions (α, δ) also contain the aberrational terms $\Delta\alpha$, $\Delta\delta$ which are, however, the same affecting the surrounding stars. With the usually differential methods for determining the celestial coordinates, the aberration terms are therefore automatically eliminated, and the observations directly provide the topocentric positions. By knowing the geographic location of the observer, the topocentric positions can be reduced to geocentric positions.

Often this simple procedure is quite adequate, but in some cases greater precision is needed. The ephemerides of the Solar System bodies are usually given in the barycentric system of reference, with respect to which the Sun moves in a complex path with velocity of about 12 m/sec and an approximate period of 20 years. In its turn, the Earth moves around the barycenter of the Earth–Moon system with a velocity 0.0005 times that of revolution, giving rise to monthly terms of amplitude $0''.008$ that enter in the numerical value of Bessel's C and D constants.

The comparison of observations with theory is particularly complicated in the case of the Moon, because usually the published lunar positions are those apparent to the geocentric observer, not those geometrically correct. There is therefore, the definite danger of ignoring, or instead, taking into account several constant terms twice, as has indeed happened many times in the literature.

As already stated in several places, when the positional precision needs to be better than a few hundredths of an arcsec, e.g., in the calculation of the occultation of a star by a planet, the problem becomes quite intricate, and resorting to precise formulae is absolutely mandatory. The foundations indeed of Newtonian mechanics can become inadequate, as we will briefly see in the next section.

7.7   The Gravitational Deflection of Light

There is another effect due to the propagation of light that was not included in the pre-1984 formulae, namely the gravitational deflection of the light by the Sun. Such deflection was already foreseen by the Newtonian theory, but with a value twice as small as that calculated on the basis of general relativity (Einstein, 1915). Dyson, Eddington, and Davidson, taking advantage of the solar eclipse of May 1919, confirmed (although certainly not in a conclusive way) the correctness of Einstein's prediction. The gravitational deflection of light, together with the precession of the perihelion of Mercury and the gravitational red-shift of the spectral lines, since then has been one of the fundamental astronomical proofs of that theory.

Karl Schwarzschild (1916) introduced a typical radius associated with a spherical mass M, the so-called Schwarzschild's radius r_S given by:

$$r_S = 2\frac{GM}{c^2}$$

whose value is about 3.0 km for the Sun and 1.0 cm for the Earth. The influence of the mass of the Sun on a grazing light ray will make its path slightly concave towards the Sun; this effect is the manifestation of the curvature of space due to mass. Therefore, in first approximation, a star near the limb of the Sun will be seen by the terrestrial observer in a direction slightly displaced, radially outward, by the quantity:

$$\Delta\theta = (1 + \Gamma)\frac{r_S}{R_\odot} = (1 + \Gamma)\frac{r_S}{a_\odot \alpha_\odot}$$

where $a_\odot$ is the AU and $R_\odot$ and $\alpha_\odot$ are the linear and the angular radius of the Sun ($\approx 16'$), respectively. The constant Γ is equal to 0 in the Newtonian theory, and to 1 in general relativity; therefore, $\Delta\theta$ is $0''.88$ in the first case, $1''.76$ in the second. Notice that the deflection is independent of the wavelength; in particular, it is the same in the optical and radio domain, and in fact, the radio measurements are much more precise than the optical ones.

To understand why the ratio between the Newtonian and the general relativity values is precisely 2, we can resort to a reasoning due to Eddington (1920). Gravitation acts as a refracting medium, having index of refraction $n \approx 1 + 2m/r$, where m and r are a-dimensional mass and radius of the deflecting body. By solving the light path in this medium, it can be shown that the asymptotic total deflection of the ray passing at distance r is $4m/r$ (radians), while in the Newtonian theory it is only $2m/r$. The specification of asymptotic deflection, in this context means that both observer and source are considered at infinite distance from the Sun. The discussion is much more complicated if the source of light is not at infinity but inside the Solar System, for instance in the case of Mercury passing behind the Sun (the gravitational deflection being smaller than the asymptotic value).

In conclusion, the outward radial displacement ΔE of the apparent direction of the star decreases linearly with the angular distance from the center of the Sun, so that, after some manipulation, the general formula can be easily derived:

$$\Delta E = \frac{r_S}{a_\odot}\frac{1 + \cos E}{\sin E} = 0''.00407\frac{1}{\tan\dfrac{E}{2}} \tag{7.9}$$

(see Figure 7.7).

At actual grazing incidence and when the solar diameter is $0°.25$, the value of the displacement is $1''.866$. Notice that at $45°$ from the center of the Sun the displacement is still at the level of $0''.01$, and of $0''.004$ at $90°$. An appropriate projection of this angle on the equatorial system will permit the determination of the corrections to be applied to the apparent coordinates:

$$\cos E = \sin\delta\sin\delta_\odot + \cos\delta\cos\delta_\odot\cos(\alpha - \alpha_\odot)$$

$$\Delta\alpha = 0^s.000271\cos\delta_\odot\sin(\alpha - \alpha_\odot)/(1 - \cos E)\cos\delta$$

$$\Delta\delta = 0''.00407[\sin\delta\cos\delta_\odot\cos(\alpha - \alpha_\odot) - \cos\delta\sin\delta_\odot]/(1 - \cos E)$$

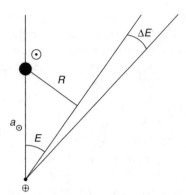

FIGURE 7.7
The gravitational deflection of light for a very distant source.

Since the displacement depends on the distance from the center of the Sun, the gravitational deflection amounts to a slight distortion of the celestial sphere, and also appears in differential measurements of high precision over large arcs. At a distance of 1 AU, the velocity V of a body in circular orbit around the mass M is:

$$V^2 = 2\frac{GM}{a_\odot}$$

so that for a generic distance $R < a_\odot$:

$$\frac{r_S}{R} \geq \frac{V^2}{c^2}$$

In other words, the gravitational deflection is as large, or larger, than the second-order terms of the annual aberration.

What has been said for the Sun could be repeated for Jupiter or Saturn or any other planet. In future projects aiming to achieve positional precision much better than $0''.001$, the gravitational field of all planets of the Solar System, will have to be computed (see Notes). For this level of precision, allowance must be made for the finite distance of the observer from the deflecting mass, and also for its position on the surface of the Earth (where the gravitational deflection can never exceed $0''.0003$).

A relativistic treatment can be found, for instance in Green (1985). Beyond the complexity of the formulae, it is shown there that the effect of aberration, light deflection, and parallax can still be considered separately, even if a priori space and time cannot be separated in general relativity. A more recent treatment, especially connected with the VLBI data, is found in Walter and Sovers (2000). At the level of precision already achieved by the VLBI, and that expected from future space missions such as GAIA, the passage of the decades may make the galactic rotation measurable, as aberration of very distant extragalactic sources. The distance indeed between the Sun and the galactic center is around 8 kpc, the rotational velocity is approximately

TABLE 7.1

Values of Γ

Method	Γ
Optical solar eclipses	0.90 ± 0.22
Radar echoes from Venus and Mercury	1.03 ± 0.06
Radar echoes from Voyager 2	1.00 ± 0.06
Signal from radio pulsar 1937 + 21	1.03 ± 0.05
Solar occultations of radio sources by VLBI	0.9996 ± 0.0017

250 km/sec; therefore the change in direction of this velocity produces a maximum aberrational displacement of a few microarsec per year that accumulates in time, and will eventually become detectable.

For completeness, it must be pointed out that the space-time curvature due to gravity also affects the times of a signal received by two antennae of the VLBI network or sent from Earth to a planet or to a spacecraft and relayed back to Earth. The delay caused by the gravitational field is usually referred to as Shapiro delay. Therefore, there are several ways to measure the parameter Γ. Table 7.1 gives a compilation of a few results.

More recently, such measurement has been carried out taking advantage of the passage of the Cassini spacecraft beyond the Sun on its way to Saturn. The results have validated the general relativity to an accuracy of 1 part over 10^5, a remarkable result indeed (see Bertotti et al., 2003).

Notes

- We have tried to avoid the common misconception that aberration depends on the relative velocity between star and observer. What matters in astronomical observations are the changes in the velocity vector of the observer with respect to the star.

- An excellent account of the traps incurred by many text books can be found in:
  Liebscher, D.-E., Brosche, P., 1998, Aberration and relativity, *Astronomische Nachrichten*, **5**, pp. 309–318.

- The following books and articles are worth reading, in addition to those quoted in the Bibliography.
  Eddington, A., 1920, *Space, Time and Gravitation*, Cambridge University Press, Cambridge.
  Stumpff, P., *A Series of Papers on Astronomy and Astrophysics*, published from 1979 to 1985.
  Hellings, R.W., 1986, Relativistic effects in astronomical time measurements, *Astronomical Journal*, **91**, p. 650.

Kovalevsky, J., Brumberg, V.A. Eds., 1986, I.A.U. Symposium 114, *Relativity in Celestial Mechanics and Astronomy.*

Bertotti, B., Iess, L., Tortora, P., 2003, A test of general relativity using radio links with the Cassini spacecraft, *Nature*, **425**, p. 374.

Will, C.M., 2003, Propagation speed of gravity and the relativistic time delay, *The Astrophysical Journal*, **590**, pp. 683–690.

- Solar eclipses in the visible:
  Dyson, F.W., Eddington, A., Davidson, C., 1920, A determination of the deflection of the light by the Sun's gravitational field, from observations made at the total solar eclipse of May 29, 1919, *Philosophical Transactions of the Royal Society of London A*, **220**, p. 291.
  Jones, B.F., 1976, Gravitational deflection of light: solar eclipse of 30 June 1973, *The Astronomical Journal*, **81**, p. 455.

- Solar radio occultations:
  Fomalont, E.B., Shramek, S.A., 1975, A confirmation of Einstein general theory of relativity by measuring the bending of microwave radiation in the gravitational field of the Sun, *The Astrophysical Journal*, **199**, p. 749.

- A pictorial representation of the gravitational bending of light by the planets is given, in the frame of the GAIA Project, in the ESA's Report to the 34th COSPAR Meeting, SP-1259, p. 96, 2002.

- Aberration will also be present in laser ranging to satellites with retro-reflectors. Therefore, the angle of the retro-reflectors must be slightly changed with respect to a stationary corner cube to achieve identical directions between incoming and outgoing laser rays (see for instance Minato, A., Sugimoto, N., Sasano, Y., 1992, *Applied Optics*, **31**, pp. 6015–6020).

- Aberration can still produce basic doubts. Two recent papers can be quoted in this respect: Ragazzoni, R., On the existence of transverse relatavistic aberrations in moving mirror, *Experimental Astronomy*, 7(3), pp. 209–219.
  Wing, W.H., 2003, On the aberration of light from a moving retroreflector, *Optic Communications*, **220**, pp. 1–6.

8

The Parallax

The phenomenon of the parallax is due to the finite distance of the planet or of the star from the observer. Two observers located in two different positions, or the same observer moving from one place to another because of the diurnal rotation or annual revolution, will see the object in two different locations on the celestial sphere. Of all the phenomena described so far altering the apparent direction of the source, the parallax is the first to give direct information on the distance, and therefore on the nature, of the heavenly body, and not only on the motions of the observer or on fundamental properties of light and of space-time. Furthermore, the knowledge of the distance between the observers provides a link between the terrestrial and the cosmic distance scales. The determination of the parallaxes is therefore one of the most fundamental astronomical measurements.

8.1 The Trigonometric Parallax

Given two observers, O and C, at a distance R from each other, and an object X at distance d' from O and d from C, observer O will see the object in direction z' and observer C in direction $z = z' - p$ with respect to the baseline OC (see Figure 8.1).

The angle p is called the parallax of X with respect to the baseline OC. The following exact relations hold:

$$CX' = d \cos z = R + d' \cos z', \quad XX' = d \sin z = d' \sin z',$$

$$CX = d = d' \cos p + R \cos z, \quad CC' = d \sin p = R \sin z',$$

$$C'X = d \cos p = d' + R \cos z', \quad OO' = d' \sin p = R \sin z,$$

$$\tan p = \frac{R}{d'} \frac{\sin z'}{1 + \dfrac{R}{d'} \cos z'} = \frac{d}{d'} \frac{R}{d'} \frac{\sin z}{1 + \dfrac{R}{d'} \cos z'}$$

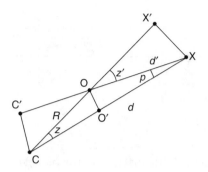

FIGURE 8.1
The geometric parallax.

We adapt now these generic relationships, useful in terrestrial triangulations, or when observing an artificial satellite, to the astronomical case. In the following sections we will distinguish between objects in the Solar System, where the Earth's radius is sufficiently large to provide a useful baseline (*diurnal parallax*), and stars for which this radius is exceedingly small, and the baseline is provided by the radius of the annual revolution around the Sun (*annual parallax*).

8.2   The Diurnal Parallax

When determining the celestial coordinates of an object of the Solar System (for instance a planet, but the same will hold for a comet or an asteroid), it will be necessary to take into account the location of the observer on the terrestrial surface, namely his topography, in order to translate these coordinates to an ideal geocentric observer. This process will therefore transform topocentric into geocentric coordinates. Let C be the center of the Earth, O the generic observer on the surface, at a distance $R = \rho a_\oplus$ from C ($\rho \approx 1$), X the planet distant d from C and d' from O (see Figure 8.2), the angle $p = X\hat{C}O$ is the instantaneous parallax of X. Notice that in a generic observation, the plane through X, O, and C will not coincide with the meridian of O.

In Figure 8.2, line OZ is the astronomical vertical, line CZ' the geocentric vertical; the observations provide the astronomical zenith distance z_0, that can be transformed into the distance z' from the geocentric vertical by means of the deviation of the vertical $v = \phi' - \phi$ given in Equation 2.4:

$$z' = z_0 - (\phi' - \phi) = z_0 - v$$

The angle p is then given by:

$$\sin p = \rho \frac{a_\oplus}{d} \sin z' \tag{8.1}$$

and finally, if the parallax p can be determined by the observations (see later) or if the distance d is known, the geocentric zenith distance z can be

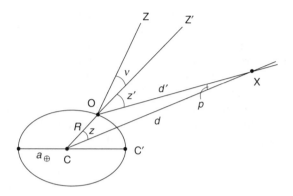

FIGURE 8.2
The instantaneous diurnal parallax. The ellipticity of the Earth is greatly exaggerated. The plane COX does not necessarily coincide with the meridian of O.

determined by:

$$z = z' - p \tag{8.2}$$

Notice that the topocentric zenith distance is larger than the geocentric one, and that the change in line of sight from the topocentric to the geocentric observer takes place in the plane OCX. This plane essentially coincides with the vertical plane of X as seen by O (disregarding the small difference between astronomical and geodetic vertical, and therefore the influence of the diurnal parallax is practically to increase the zenith distance (or to decrease the elevation above the horizon) measured by the topocentric observer with respect to the geocentric one (hence the other designation, *parallax of elevation*, although rigorously the circle through X'Z' is not exactly the vertical one). In particular, the observations in meridian give directly the geocentric right ascension, and the entire diurnal parallax is reflected in a variation of declination. As already stated, if the observations provide p, we can derive d and z from Equation 8.1 and Equation 8.2. If, instead, we know the geocentric zenith distance z and the distance d by some almanac, we can derive the topocentric parallax p by developing $\sin z' = \sin(z + p)$:

$$\tan p = \rho \frac{a_\oplus}{d} \frac{\sin z}{\left(1 - \rho \frac{a_\oplus}{d} \cos z\right)} \tag{8.3}$$

The angle p is variable with the rotation of the Earth, even disregarding the motion of the planet. Therefore, it has been agreed to name *horizontal equatorial parallax* π the angle under which X sees perpendicularly the Earth radius $a_\oplus$. In other words, π is the maximum value of p at any given distance of X from C, but π itself will vary according to the relative positions of X and C in their heliocentric orbits.

The horizontal parallax of the Moon, $\pi_{\leftmoon}$, varies between $54'$ and $61'$ due to the strong eccentricity of its geocentric orbit (notice that these values are about twice the apparent diameter of the lunar disk, therefore the occurrence of a lunar occultation depends very critically on the position of the observer). All other celestial bodies are much farther than the Moon, apart from some occasional objects than can pass inside its orbit, and their diurnal parallax can vary by great amounts. For instance, $\pi(\text{Venus})$ varies between $5''$ and $34''$. Therefore, essentially, in all cases (except artificial satellites) we are justified to assume $d \gg a_{\oplus}$ and π very small, so that:

$$d = \frac{a_{\oplus}}{\sin \pi} \approx \frac{a_{\oplus}}{\pi}, \quad \sin \pi \approx \pi = \frac{a_{\oplus}}{d}$$

For the Sun, averaging over the slight ellipticity of the orbit, the distance d is the astronomical unit $a_{\odot}$ ($a_{\odot} = 1\ \text{AU} \approx 1.49 \times 10^8\ \text{km}$), the horizontal parallax is essentially constant:

$$d_{\odot} = a_{\odot} \approx \frac{a_{\oplus}}{\pi_{\odot}}$$

Expressing the distance d to a generic body in AU, and the parallax in seconds of arc:

$$\pi'' = \left(\frac{1}{d(\text{AU})}\right)\pi''_{\odot}, \quad \pi''_{\odot} \approx 8''.8 \tag{8.4}$$

When the body is on the meridian:

$$\sin p = \rho \sin \pi \sin z' \approx \rho\pi \sin z' \approx p,$$

$$\tan p = \rho \sin \pi \frac{\sin z}{1 - \rho \sin \pi \cos z} \approx \rho\pi \frac{\sin z}{1 - \rho\pi \cos z} \approx p$$

because the angles p, π are always so small that no sensible error is made in using their arc instead of their sine or tangent; using arcsecs and AUs as units, we have:

$$p'' = \rho \frac{\pi_{\odot}}{d} \sin z' \approx 8.8'' \frac{\rho}{d(\text{AU})} \sin z' \tag{8.5}$$

Suppose now we observe the same planet by two stations having the same longitude, at the same instant and in meridian (meridian observations give directly geocentric right ascensions, the diurnal parallax produces only a variation of declination). It is then easy to find the horizontal equatorial parallax of the planet (this is an idealized case, but of great historical significance). From $p = \rho\pi \operatorname{sen} z' = z' - z$ written for the two observers 1 and 2, we derive:

$$z'_1 = z_{01} - v_1, \quad z'_2 = z_{02} - v_2, \quad \pi = \frac{(z'_1 + z'_2) - (z_1 + z_2)}{\rho_1 \sin z'_1 + \rho_2 \sin z'_2}$$

In these relations, the topocentric zenith distances are given by the observations, while the sum of the geocentric zenith distances is equal to the difference of the two astronomical latitudes: $z_1 + z_2 = \phi_1 - \phi_2$. To maximize this factor, the two observers should preferably be on opposite sides of the equator. In more realistic cases, appropriate corrections must be applied for the difference in longitudes, also taking into account the orbital motion of the planet. To minimize the effects of refraction, the measurements should be taken with respect to a set of nearby stars. Let us consider now the generic case of a body observed outside the meridian, with the simplifying assumption of astronomical vertical coinciding with the geodetic one. The geocentric zenith Z' is on the meridian at the small distance v from the astronomical one Z, and closer to the horizon (see Figure 8.3).

X' will be the observed topocentric position of the body, X the geocentric one on the great circle through X' and Z'. The angle $Z'\hat{P}X' = Z\hat{P}X'$ is the topocentric hour angle $HA' = HA(X')$, the arc $X_1'X'$ is the topocentric declination $\delta' = \delta(X')$. The corresponding geocentric quantities are $Z'\hat{P}X = Z\hat{P}X = HA(X)$, arc $X_1X = \delta(X)$. Given that:

$$PZ' = 90° - \phi', \quad Z'X' = z', \quad Z'X = z, \quad XX' = p, \quad z' = z + p$$

(as shown in Chapter 2), we see that the diurnal parallax augments the zenith distance along the vertical circle, and leaves the azimuth essentially constant (hence the other designation *parallax of altitude*, although the circle through $X'Z'$ is not rigorously the vertical one). Now call A the angle in Z', which is common to the spherical triangles $PZ'X'$ and $PZ'X$. Letting $\Delta HA = HA' - HA$, after some manipulation we find:

$$\frac{\cos \phi' \sin p}{\sin z \sin(z + p)} = \frac{\sin A \sin \Delta HA}{\sin HA \sin(HA + \Delta HA)} \tag{8.6}$$

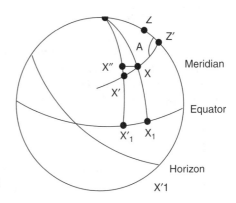

FIGURE 8.3
Effects of the diurnal parallax on the celestial coordinates.

Taking into account that:

$$\sin p = \rho \frac{a_\oplus}{d} \sin(z + p), \quad \sin A \sin z = \cos \delta \sin HA$$

we finally get the rigorous formulae:

$$\begin{cases} \tan \Delta HA = -\tan \Delta \alpha = -\rho \dfrac{a_\oplus}{d} \cos \phi' \dfrac{\sin HA}{\rho \dfrac{a_\oplus}{d} \cos \phi' \cos HA - \cos \delta} \\[3ex] \tan \delta' = \dfrac{\tan \delta - \rho \dfrac{a_\oplus}{d} \sin \phi' \sec \delta \cos \Delta HA}{1 - \rho \dfrac{a_\oplus}{d} \cos \phi' \sec \delta \cos HA} \\[3ex] \sin p = \sin \pi \dfrac{\sin HA' \cos \delta'}{\sin HA \cos \delta} \end{cases} \qquad (8.7)$$

Let us introduce the auxiliary quantity q:

$$q = \cot \phi' \frac{\cos[(HA + HA')/2]}{\cos[(HA - HA')/2]}$$

by which:

$$\tan \Delta \delta = \frac{\rho \dfrac{a_\oplus}{d} \sin \phi'(q \sin \delta - \cos \delta)}{1 - \rho \dfrac{a_\oplus}{d} \sin \phi'(q \cos \delta + \sin \delta)}$$

where $\Delta \delta = \delta' - \delta$. The ratio between topocentric and geocentric distances is expressed as:

$$\frac{d'}{d} = \cos(\delta' - \delta) - \rho \frac{a_\oplus}{a} \sin \phi'(q \cos \delta' + \sin \delta')$$

These rigorous formulae seldom need to be applied. For the occasional asteroids coming inside the orbit of the Moon, or even for meteors in the upper atmosphere, radar echoes can often be obtained that provide high-precision distances and velocities. However, in these cases it is more efficient to use Cartesian coordinates, and apply the refined methods used for tracking satellites (see Notes).

For the usually much more distant celestial objects, the quantity $\rho a_\oplus / d = \rho \pi$ is so small that Equation 8.7 becomes:

$$\begin{cases} -\Delta HA = \Delta \alpha = -\rho \dfrac{a_\oplus}{d} \cos \phi' \dfrac{\sin HA}{\cos \delta} = -\rho \pi \cos \phi' \, \mathrm{cosec}\, z \sec \delta \\[2ex] \Delta \delta = -\rho \dfrac{a_\oplus}{d} (\sin \phi' \cos \delta - \cos \phi' \sin \delta \cos HA) \\[1ex] \quad = -\rho \pi (\sin \phi' \, \mathrm{cosec}\, z \sec \delta - \tan \delta \cot z) \end{cases} \qquad (8.8)$$

Recalling Equation 8.5, another way of expressing the same quantities is:

$$
\begin{cases}
-\Delta HA'' = \Delta \alpha'' = -\rho \dfrac{\pi_\odot{}''}{d(\text{AU})} \cos \phi' \dfrac{\sin HA}{\cos \delta} \\[3mm]
\Delta \delta'' = -\rho \dfrac{\pi_\odot{}''}{d(\text{AU})} (\sin \phi' \cos \delta - \cos \phi' \sin \delta \cos HA)
\end{cases}
\tag{8.9}
$$

No sensible error is made if the observed quantities HA', δ' are substituted to the geocentric values in the right side of these equations. Notice, however, that the symmetry between geocentric and topocentric coordinates in Equation 8.9 is illusory: indeed, the knowledge of the geocentric coordinates depends on the previous knowledge of the geocentric distance. For a new comet or asteroid, d is unknown, so that only the trigonometric expressions (also named *parallactic factors*) can be computed. These equations provide a second method for obtaining distances (*diurnal method*), this time using observations taken at the same observatory but at two different hour angles (ideally, 12 h apart), provided the parallax of the Sun is known.

If we could follow the apparent path of the comet or asteroid with respect to the distant stars for a full Earth's revolution, we would observe a circle (projected as an ellipse) trailed by the revolutions around the Sun of the Earth and of the body itself. This kind of observation is indeed possible from the Hubble Space Telescope, whose orbital semiperiod is approximately 45 min; an asteroid will therefore leave, on a sufficiently long exposure, an appreciably curved trail (see Notes).

The diurnal parallax also has an interesting consequence on the apparent diameters. Let s be the diameter in km of a planet (treated here for simplicity as a disk), S its apparent geocentric angular radius when at geocentric distance d, and S', d' the corresponding topocentric values; the following relations will hold:

$$
\sin S = \frac{s}{d}, \quad \sin S' = \frac{s}{d'} = \frac{s}{d}\frac{d}{d'} = \sin S \frac{\sin z'}{\sin z}
$$

The angles are usually small quantities, so that:

$$
S' = S \frac{\sin z'}{\sin z}
\tag{8.10}
$$

From the Earth's surface, the effect for the Sun never exceeds $0.1''$, but it is noticeable for the Moon, whose geocentric apparent radius is given by $S_{\leftmoon} = 0.27245 \pi_{\leftmoon}$, with a maximum variation of approximately $40''$. Therefore, this change in lunar diameter with the location of the observer must be taken into account when calculating precise circumstances of eclipses and occultations. Another effect is usually considered for the Moon, namely the slight variation of the rising and setting instants. The diurnal parallax retards the rise, and anticipates the setting (contrary to the effect of the atmospheric refraction; notice that for the Moon the diurnal parallax can be larger than the refraction at the horizon, which amounts to some $35'$, see Chapter 11), by:

$$\Delta HA = -\frac{1^m 28^s}{\sqrt{\cos(\phi - \delta_\text{☽}) \cos(\phi + \delta_\text{☽})}} \tag{8.11}$$

8.3 Solar and Lunar Parallaxes

The direct determination of the solar parallax $\pi_\odot$ poses great difficulties, because of both its small amount and the lack of precise reference points on the surface. Copernicus and Tycho Brahe, for instance, adopted a value of 3 arcminutes! Soon, it was found advantageous to make recourse to indirect methods, of which we provide two examples:

1. Suppose an accurate table of geocentric positions (namely the ephemerides) of a given planet is available, and compute:

$$\alpha'(t_2) - \alpha'(t_1) = \alpha(t_2) - \alpha(t_1) + \pi_\odot[f_\alpha(t_2) - f_\alpha(t_1)]$$

$$\delta'(t_2) - \delta'(t_1) = \delta(t_2) - \delta(t_1) + \pi_\odot[f_\delta(t_2) - f_\delta(t_1)]$$

   where the f's are the parallactic factors in Equation 8.9. From a set of well-measured topocentric positions the unknown $\pi_\odot$ can be derived. Indeed, it is not even necessary to know the geocentric positions if advantage is taken of the moments when the planet appears stationary on the celestial sphere. Furthermore, if the measurements are made at the same time from two different locations, the unknown parallax of the planet is also found. With this method, applied by Picard (1672) in Paris ($\phi = +49°50'$) and by Lacaille in Cayenne ($\phi = +4°57'$) to planet Mars, a difference of $15''$ in the two topocentric declinations was found, corresponding to an equatorial parallax of $25.5''$. From this value, Cassini was able to derive the first reliable value of $\pi_\odot$, namely $9.5''$, a value he could essentially confirm measuring from Paris the right ascensions of the planet. (It is worth recalling that during the same expedition, Lacaille found that his pendulum was slower than in Paris, loosing some 2^m in 24^h, the first experimental proof of the nonsphericity of the Earth.) A variant of this method was used by Encke (1835), by a new reduction of the data taken on the rare occasion of the transit of Venus on the solar disk; his value of $\pi_\odot$ was, however, decidedly lower than the true one.

2. Consider Kepler's third law (see Chapter 12) for a given planet, in its slightly approximate expression $P^2 = (4\pi/GM_\odot)a^3$, P being the sidereal period, a the semi-major axis and $M_\odot$ the mass of the Sun. Comparing this expression with the corresponding one for

the Earth:

$$P^2/P_\oplus^2 = a^3/a_\oplus^3$$

we can derive only the relative dimensions of the orbits. At least one absolute determination in kilometers is needed to fix the scale of the Solar System. To reach this goal, it is advantageous to observe from several locations an asteroid (having a star-like image) whose orbit takes it as close as possible to the Earth. Several asteroids were tried, the one giving the best results being 433 Eros, whose perihelion is at 1.13 AU. Eros came at the opposition in 1900 to 1901, and again in 1930 to 1931. The first event, during which the minimum distance was 0.32 AU, was observed visually and gave the value $\pi_\odot = (8.806 \pm 0.004)''$. In the second opposition, the small planet came much closer, 0.17 AU, and the observations took advantage of photography; the derived value was $\pi_\odot = (8.790 \pm 0.001)''$. Eros was observed again in 1975, but this time with radar echoes, a technique that had been successfully applied in 1959 and 1961 to the inferior transits of Venus; these radar data also gave the first determination of its dimensions, approximately $16 \times 35 \text{ km}^2$. Finally, Eros was reached in 2001 by the spacecraft NEAR, who crashed on its surface at the end of a very successful mission.

Table 8.1 provides a compilation of historical values of $\pi_\odot$; notice that the variations from time to time usually greatly exceeded the quoted probable errors, an indication of the presence of systematic errors in several methods.

It is worth recalling that other indirect methods of finding the solar parallax were the radial velocities of ecliptic stars, and the aberration of the Medicean moons of Jupiter (see also Chapter 7 and Chapter 9).

On closer examination, the methods based on Kepler's third law (therefore also the Eros observations) must provide the mass of the Moon in addition to the solar parallax, because it is the Earth–Moon barycenter to follow an

TABLE 8.1

Values of the Solar Parallax

Period	$\pi_\odot$	Method
1801–1833	9.0″	Parallax of Mars, transits of Venus in 1761–1769
1834–1869	8.5776″	New discussion by Encke of previous Venus transits data
1870–1881	8.95″	(LeVerrier) From the lunar orbit (solar parallactic inequality)
1882–1900	8.848″	Best value according to Newcomb
1901	8.80″	Official value agreed in 1896
1930–1931	8.790″	Spencer–Jones, from Eros opposition
1964 (IAU)	8.79405″	Radar, corresponding to 149.60003×10^{11} m
1976 (IAU)	8.794148″	Radar, corresponding to 149.59787×10^{11} m

(almost) Keplerian orbit around the Sun. As also discussed in Chapter 6, the geocentric observer has a monthly inequality with respect to the dynamical ecliptic that can reach $6.44''$ when the Moon is in quadrature. This method of obtaining the mass of the Moon was used in addition to those expounded in Chapter 6.

Regarding the lunar parallax, of great historical significance is the method followed by Hipparchus. During a lunar eclipse, S, E, L are the centers of Sun, Earth, and Moon, respectively. The shadow of the Earth is a cone of vertex E_1 and height HK when intercepted by the Moon in L; as seen from E, the angular semiaperture of the shadow LH is $\psi/2$, while the angular solar radius is $\alpha/2$. From simple geometry we derive:

$$\psi/2 = \pi_\odot - \alpha_\odot/2 + \pi_{\leftmoon} \approx -\alpha_\odot/2 + \pi_{\leftmoon}$$

The angle $\psi/2 (\approx 2.5\alpha_\odot/2)$ can be determined by the instants of immersion and emersion of the Moon in the shadow. Although the method cannot be very precise, Hipparchus arrived at the value $3489''$, fairly close to the true value of $3422''$.

Finally, a direct and precise measurement was obtained by Lalande in 1752 from Berlin in Germany and Lacaille from the Cape in South Africa, taking advantage of the coincidence in longitude and strong difference in latitude.

In today's practice, the instantaneous distance and the rotation of the figure can be obtained with excellent precision (around 1 cm) by laser echoes from the retro-reflectors that were placed on its surface by American and Soviet missions in the 1970s (see for instance the Project Apollo, Notes).

8.4   The Annual Parallax

Annual parallaxes are the fundamental method for the direct determination of the stellar distances.

Figure 8.4 shows the Earth in two diametrically opposed positions along its orbit, which at the moment we take as circular with radius $a_\odot$ ($= 1$ AU). E_1 and E_2 are the two geocentric observers, S the ideal heliocentric observer,

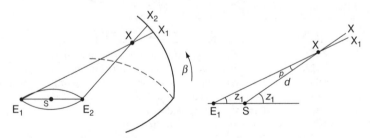

FIGURE 8.4
The annual parallax.

and X a nearby star, which will be seen from the Earth as projected in X_1 and X_2 on the celestial sphere (in other words, with respect to the background of the distant stars). Calling ES the direction Earth–Sun, z the angle of the line of sight with that direction from S and z' from E, d the heliocentric distance of X, we have:

$$a_\odot \sin z' = d \sin(z - z') = d \sin p, \quad \sin \pi = \frac{a_\odot}{d}, \quad \sin z' = \frac{\sin(z' - z)}{\sin \pi}$$

The angle π, under which the AU is seen perpendicularly from the star, namely the maximum value of p, is called the *annual parallax* of X. Until now, the star Proxima Centauri (in the triple system of α Cen) has the largest observed value, $\pi = 0.76''$; therefore, no sensible error is made by using arcs instead of their sine or tangent, and therefore:

$$z' = z - \pi \sin z' \approx z - \pi \sin z, \quad \pi \approx \frac{a_\odot}{d} \tag{8.12}$$

The finite distance has the effect that the geocentric observer sees the star closer to the Sun by the slight amount $(z' - z)$; on the celestial sphere this apparent movement occurs along the great circle passing through X and S. During the year, S moves along the ecliptic, and the locus X' of apparent geocentric positions will be an ellipse centered on the heliocentric position X. Notice that this ellipse has nothing to do with the ellipticity of the Earth's orbit (that we have assumed circular), it is simply the projection effect depending on the ecliptic latitude β of X; the slight ellipticity of the orbit ($e = 0.0167$) will introduce small modifications, which presently we will neglect. Notice also that the annual parallax could be determined, in principle, by measuring the variation of the angular distance of the nearby star from the Sun during the year but, in practice, it will be much easier to determine its relative positions in respect to the background of the other stars.

Let us put these qualitative conclusions in a more quantitative way, by reasoning in ecliptic coordinates (see Figure 8.5).

If (λ, β) are the heliocentric coordinates of X, the geocentric coordinates will differ by very small amounts that we can treat as differentials,

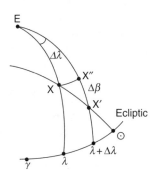

FIGURE 8.5
The annual parallax in ecliptic coordinates.

$(\lambda + \Delta\lambda, \beta + \Delta\beta)$, in other words, the small triangle $XX''X'$ can be considered as a plane triangle with infinitesimal sides:

$$X''X = \Delta\lambda \cos \beta, \quad X''X' = \Delta\beta, \quad XX' = z - z' = \pi \sin z$$

Calling $\lambda_\odot$ the longitude of the sun, after some calculations (similar to those made for the annual aberration), we get:

$$\Delta\lambda = \pi\frac{\sin(\lambda_\odot - \lambda)}{\cos \beta}, \quad \Delta\beta = -\pi \sin \beta \cos(\lambda_\odot - \lambda)$$

or else:

$$\cos^2\beta\left(\frac{\Delta\lambda}{\pi}\right)^2 + \left(\frac{\Delta\beta}{\pi \sin \beta}\right)^2 = 1 \tag{8.13}$$

However, $\cos \beta\Delta\lambda$ is the movement parallel to the ecliptic, and $\Delta\beta$ is the one perpendicular to it; the annual locus is therefore an ellipse, with semimajor axis π and semiminor axis $\pi \sin \beta$. For an ecliptic star this locus degenerates in a straight segment, for a star seen toward the ecliptic pole the locus is a circle. The star passes through the semimajor axis when its longitude is $\pm 90°$ from that of the Sun, for which reason there are two preferred dates in the year for any particular region of sky. If, in addition, observations in meridian are preferred, parallax work is better done at dusk and dawn. Notice also that the ellipse of parallax anticipates by three months that of annual aberration, and of course its dimensions are proportional to the distance and not fixed as those of aberration.

In equatorial coordinates, this ellipse will be rotated by a certain angle, and it is not difficult to prove the following relations:

$$\begin{cases} \Delta\alpha \cos \delta = \pi(\cos \varepsilon \cos \alpha \sin \lambda_\odot - \sin \alpha \cos \lambda_\odot) \\ \qquad = \pi(Y_\odot\cos \alpha - X_\odot\sin \alpha) \\ \Delta\delta = \pi(\sin \varepsilon \cos \delta \sin \lambda_\odot - \cos \alpha \sin \delta \cos \lambda_\odot - \cos \varepsilon \sin \delta \sin \alpha \sin \lambda_\odot) \\ \qquad = \pi(Z_\odot\cos \delta - X_\odot\cos \alpha \sin \delta - Y_\odot\sin \alpha \sin \delta) \tag{8.14} \end{cases}$$

where $(X_\odot, Y_\odot, Z_\odot)$ are the geocentric equatorial coordinates of the Sun. These corrections $\Delta\alpha, \Delta\delta$ must be added with their signs to the heliocentric coordinates, subtracted from the geocentric ones.

Using the AU as baseline, it is possible to institute the fundamental unit of astronomical distances, namely, the parsec: one parsec (pc) is the distance from where the AU subtends perpendicularly an angle of $1''$. Therefore:

$$1 \text{ pc} = 206264.8 \text{ AU} = 3.09 \times 10^{13} \text{ km}$$

where the last conversion factor is derived from the solar parallax. Any revision of this latter value will not change the distance as expressed in parsecs. On the practical side, this remark is not so important today, because the precision with which the solar parallax is known is much higher than the precision of stellar distances; it is useful to remember it, however, considering how the cosmic distances ladder has been connected to laboratory units.

A secondary unit of distance is the light-year, corresponding to the distance traveled in vacuum by the light in 1 year; it is easily found that 1 pc ≈ 3.26 light-years.

The measurement of the minute stellar parallaxes has always been a difficult task, quite often plagued by systematic errors. Part of the difficulty resides in the proper motion of the nearby stars, which trails the parallax ellipse, as we will discuss in the next chapter. Until recently, the precision could rarely be claimed to be better than $0.01''$; if we restrict our considerations to stars having a relative error $\sigma(\pi)/\pi < 1/3$, we then find an astrometric horizon around 30 pc. The astrometric satellite Hipparcos has enlarged this horizon by ten times, but still the volume of the Milky Way directly accessible to trigonometric distance determination is a very slight fraction of the total. The interest of having future satellites with a thousand times better precision is then readily understood.

The above treatment of stellar parallaxes is approximate for several reasons:

- The orbit of the Earth is slightly elliptical.
- The heliocentric observer should be substituted by the barycentric observer. The distance between the two never exceeds 0.01 AU (two solar radii), with a complex behavior in time according to the longitude of Jupiter, Saturn, and the other planets, as shown in Figure 8.6.
- The relativistic deflection of light affects the apparent directions at the level of $0.001''$.
- The geocentric observer should be replaced by the Earth–Moon barycentric observer; the effect is smaller than the diurnal aberration on that particular star, because the barycenter is inside the body of the Earth.

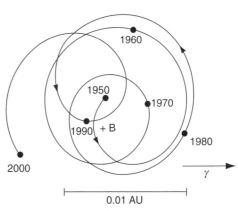

FIGURE 8.6
The ecliptic projection of the center of the Sun with respect to the barycenter B of the Solar System from 1950 to 2000.

The differences between our approximate treatment and a more rigorous one are of the order of $(a_\odot/d)^2$ from the trigonometric point of view. Furthermore, a correct treatment in the frame of general relativity would introduce terms depending on the radial velocity V_r and the proper motion μ of the star, because parallaxes cannot be totally separated from aberration and velocities.

8.5   Secular and Dynamical Parallaxes

The previous discussion has shown that, in general, stellar distances will not be measurable by direct trigonometric means. Indirect criteria must be found, quite often having only a statistical validity in the sense of not providing the value for a particular star, but only for a group or an association. In later chapters, we will examine some photometric and spectroscopic distance indicators; here, we extend the concept of trigono-metric parallax to encompass the secular movement of the Solar System, and the consequences of Kepler's third law to binary stars.

Secular parallaxes: The Solar System moves in respect to the ensemble of the nearby stars with a velocity of approximately 20 km/sec in direction of the constellation Lyra (a point named *apex* of solar motion). A hypothetic observer, at rest in the frame of this group of nearby stars, would observe the Sun in rectilinear motion toward this apex, and the planets of the Solar System describing open orbits around it. The distance covered by the traveling Sun is approximately 4 AU per year. This baseline is four times that of the annual parallaxes: we shall call *secular parallax H* of the star X the angle under which this baseline is seen perpendicularly:

$$H \approx 4\pi \tag{8.15}$$

However, this is a secular, not a periodic, motion, and it is observed entangled with the proper motion of that particular star. It might be, however, a useful distance indicator for a group of stars having the same distance, as we will also discuss in the next chapter.

Dynamical parallaxes: Let A, B be the two components of a binary star, having, respectively, M_A and M_B masses, and orbital period P. Kepler's third law states that:

$$P^2 = \frac{4\pi}{G} \frac{a^3}{(M_A + M_B)} \tag{8.16}$$

In astronomical units (M in solar masses, P in years):

$$M = M_A + M_B = \frac{a^3}{\pi^3} \frac{1}{P^2} \tag{8.17}$$

(pay attention to $\pi = 3.1415...$ in Equation 8.16, and $\pi =$ parallax in Equation 8.17). In Equation 8.17, a and π are both in arcsec.

Masses in general are unknown, but putting $M = 2$, an acceptable value for π is obtained, because of the small weight of M on the error. The method can be refined by using the luminosities of the two stars and an appropriate mass-luminosity function, calibrated on well-known systems, but usually the uncertainties on a and P dominate the error on π.

Parallaxes from radial velocity and angular expansion: If an object has an opaque surface and is in contraction or expansion, from the measurement of the angular variation of its dimensions (arcsec/y), and of the radial velocity (km/sec, by the spectroscopic Doppler effect), it is possible to derive its distance. We quote this method here as an example of a nontrigonometric mean for finding distances. We will discuss it again in Chapter 9.

Notes

- Aristarchus of Samus (310–230 B.C.) attempted a determination of the ratio between the lunar and solar distances by measuring the angles Moon, Sun, Earth at the first or last quarter. At quadratures, the angle at the Moon must be 90°, the angle at the Earth can be measured, hence the third angle can be calculated; from the lunar parallax, the solar parallax could then be derived. The measurements of Aristarchus were fairly crude; he produced an angle of 87° instead of 89°51′, nevertheless, for the first time it could be concluded that the Sun must be much more distant (and consequently much larger) than the Moon.

- The first annual parallax was determined by Bessel (1838) for the star 61 Cyg, suggested by Piazzi as likely to be near to the Sun because of its very large proper motion of 5″/year. Bessel obtained the value $\pi = 0.30″$ by using a so-called heliograph, namely a telescope with a split objective. Almost contemporarily, Struve derived the parallax of Vega ($\pi = 0.12″$), and Henderson at the Cape that of the triple star α Cen. The third star, Proxima Cen, is some 2° away from A, B, and it holds the record of being the closest star, with $\pi = 0.76″$. For a recent readable account of the historical developments, see the book *Parallax*, by Hirshfeld (Freeman and Co., 2001).

- For the details of the NEAR mission to 433 Eros see http://near.jhuapl.edu/

- For a discussion of asteroids trails in HST images, see the paper by Evans et al., 1998, *Icarus*, **131**, pp. 261–282. An image can be found in http://oposite.stsci.edu/pubinfo/pr/1998/08

- Since the publication of the Hipparcos and Tycho Catalog (1997), doubts have been raised about the reliability of part of the data, most

noticeably those related to the distance determination of some open clusters. For instance, Makarov pointed out the possibility of systematic errors in the parallaxes given by the Hipparcos Catalog, see Makarov, V.V., 2002, *The Astronomical Journal*, **124**, p. 3299 and *Ibidem*, **126**, p. 2048. These and other doubts motivated a rereduction of the Hipparcos measurements, which was initiated in 1998 and is currently in progress. An extensive discussion on the defects in the raw data and the provisions introduced in the new reduction can be found in: van Leeuwen, F., 2005, Rights and wrongs of the Hipparcos data, *Astronomy and Astrophysics*, **439**, p. 805.
van Leeuwen, F., Fantino, E., 2005, A new reduction of the raw Hipparcos data, *Astronomy and Astrophysics*, **439**, p. 791.

- As examples of Doppler parallaxes applied to distant objects see: Panagia, N., Gilmozzi, R., Macchetto, D., Adorf, H.M., Kirschner, R.P., 1991, Properties of the SN 1987A ring and the distance to the LMC, *The Astrophysical Journal*, **380**, p. L23.
  Schmidt, B.P., Kirschner, R.P., Eastman, R.G., 1992, Expanding photospheres of type II SNs and the extragalactic distance scale, *The Astrophysical Journal*, **395**, p. 366.

- The Project Apollo to determine high-precision distances to the Moon by laser echoes with the 3.5 m Apache Point Telescope is described in http://physics.ucsd.edu/~tmurphy/apollo/apollo.html

Exercises

1. Describe the diurnal parallax in terms of Cartesian coordinates. In the geocentric reference system centered in C, let

$$\mathbf{r}(a_\oplus \sin \pi \cos \alpha \cos \delta, \; a_\oplus \sin \pi \sin \alpha \cos \delta, \; a_\oplus \sin \pi \sin \delta)$$

be the vector to the object P, and $\mathbf{R}(\rho a_\oplus \cos \phi' \cos CO,$ $\rho a_\oplus \cos \phi' \sin CO, \; \rho a_\oplus \sin \phi')$ the vector to the observer O, at any given instant ST of sidereal time. The topocentric vector position of P is $\mathbf{r}' = \mathbf{r} - \mathbf{R}$, with components:

$$x' = r' \cos \alpha' \cos \delta' = a_\oplus \sin \pi \cos \alpha \cos \delta - \rho a_\oplus \cos \phi' \cos OC$$

and corresponding expressions for y' and z'. The topocentric equatorial coordinates can be computed in the usual way:

$$\alpha' = \arctan \frac{y'}{x'}, \quad \delta' = \arctan \frac{z'}{\sqrt{x'^2 + y'^2}}$$

with due consideration of the quadrant of α' (add 12^h if $x' < 0$).
    A worked example of the inverse procedure for a low Earth orbiting satellite can be found in the book by Green (1985).

2. Derive the effect of the diurnal parallax on the ecliptic coordinates. Determine the ecliptic coordinates of the geocentric zenith $(\lambda_{Z'}, \beta_{Z'})$ at the instant of observation, and from Equation 8.9 compute:

$$
\begin{cases}
\Delta\lambda'' = -\rho \dfrac{\pi_\odot''}{d(\text{AU})} \sec\beta \sin(\lambda_{Z'} - \lambda) \cos\beta_{Z'} \\[4mm]
\Delta\beta'' = -\rho \dfrac{\pi_\odot''}{d(\text{AU})} (\sin\beta_{Z'}\cos\beta - \cos\beta_{Z'}\sin\beta \cos(\lambda_{Z'} - \lambda))
\end{cases}
$$

3. Determine the position of the Solar System barycenter for the year 2000. Table 8.2 shows the heliocentric ecliptic position of Earth, Jupiter, and Saturn with respect to the ecliptic and mean equinox at J2000.0 every two months since December 27, 1999 UT1 = 0^h, until December 21, 2000. We wish to compute the approximate position of the barycenter during that period, discussing the influence of the other planets on the values found.

TABLE 8.2

Heliocentric Ecliptic Positions of Earth, Jupiter, and Saturn with Respect to the Mean Ecliptic and Equinox of J2000.0

Date (UT1 = 0^h)	Longitude	Latitude	Distance (UA)
Earth			
1999 Dec 27	94°46′25″.4	− 0°00′00″.6	0.983441058
2000 Feb 25	155 41 26.6	− 0 00 00.5	0.989699191
2000 Apr 25	215 05 48.3	+ 0 00 00.2	1.006041897
2000 Jun 24	272 47 36.7	+ 0 00 00.9	1.016436859
2000 Aug 23	330 10 09.3	+ 0 00 00.5	1.011210382
2000 Oct 22	28 53 16.4	− 0 00 00.6	0.995157692
2000 Dec 21	89 25 07.2	− 0 00 01.1	0.983751820
Jupiter			
1999 Dec 27	35°47′36″.0	− 1°10′46″.2	4.964672267
2000 Feb 25	41 15 16.0	− 1 07 15.9	4.973298467
2000 Apr 25	46 41 39.7	− 1 03 10.1	4.983834442
2000 Jun 24	52 06 32.9	− 0 58 31.5	4.996175289
2000 Aug 23	57 29 42.4	− 0 53 23.3	5.010203092
2000 Oct 22	62 50 56.6	− 0 47 49.0	5.025787816
2000 Dec 21	68 10 05.3	− 0 41 52.0	5.042780745
Saturn			
1999 Dec 27	45°31′24″.7	− 2°18′23″.1	9.185031429
2000 Feb 25	47 41 40.9	− 2 16 11.0	9.172285980
2000 Apr 25	49 52 18.1	− 2 13 46.7	9.159972260
2000 Jun 24	52 03 15.5	− 2 11 10.3	9.148103239
2000 Aug 23	54 14 32.5	− 2 08 22.2	9.136696298
2000 Oct 22	56 26 08.3	− 2 05 22.3	9.125772846
2000 Dec 21	58 38 02.3	− 2 02 10.9	9.115348720

TABLE 8.3

Barycentric Coordinates (J2000.0) of the Center of the Sun from December 1999 to January 2003, Given by MICA

Date (UT1 = 0^h)	X(UA)	Y(UA)	Z(UA)
1999 Dec 27	− 0.007168681	− 0.002605028	− 0.000904252
2000 Feb 25	− 0.006833660	− 0.002999961	− 0.001082030
2000 Apr 25	− 0.006467923	− 0.003369512	− 0.001249870
2000 Jun 24	− 0.006069952	− 0.003709978	− 0.001406332
2000 Aug 23	− 0.005645329	− 0.004017280	− 0.001549221
2000 Oct 22	− 0.005198099	− 0.004292663	− 0.001678861
2000 Dec 21	− 0.004726104	− 0.004534023	− 0.001794670
2001 Jan 20	− 0.004481906	− 0.004639521	− 0.001846341
2002 Jun 14	− 0.000169340	− 0.004961843	− 0.002100710
2002 Jul 14	+0.000069611	− 0.004897186	− 0.002079631
2002 Aug 13	+0.000304080	− 0.004825351	− 0.002055347
2003 Jan 10	+0.001422324	− 0.004353087	− 0.001884571

Table 8.2 shows that during that period, the longitude of Jupiter gradually reached and overtook that of Saturn. From the orbital elements of the two planets, we could notice that the distance of Saturn from the Sun was in that period decidedly smaller than the semimajor axis of its orbit. From these data, and from the masses of the planets, we can calculate the (x,y,z) coordinates in the ecliptic heliocentric system, and then by rotation the equatorial ones (X,Y,Z).

To verify your calculations, consider Table 8.3, which contains the values given by Multiyear Interactive Computer Almanac (MICA, U.S. Naval Observatory, 1990–2005) taking into account all other planets. The table has been extended until January 2003 in order to show that X changes sign with the passage of Jupiter in the second quadrant.

9

Radial Velocities and Proper Motions

The motions of the stars with respect to the observer reveal themselves in two different ways, namely, as radial velocities V_r along the line of sight and as angular tangential motions on the celestial sphere. The first effect is detected and measured spectroscopically thanks to the Doppler effect, directly in km/s or other convenient units. The second one, named proper motion μ, is seen as a variation in time of the position of the star on the celestial sphere, and is expressed as angular velocity per unit time, for instance milliarcsec/yr or arcsec/century. It is much easier to measure proper motions than parallaxes and radial velocities because the displacements, and therefore the precision of the measurements, increase with the passage of time. Historically, it was Halley in 1718 who discovered the first proper motion, namely that of Arcturus (α Boo), by comparing his own coordinates with those given by Hipparchus two millennia earlier: after allowance had been made for precession, the two determinations differed by more than one degree, too large to be attributed to errors. Accurate radial velocities could instead be measured only in the second half of the 19th century, after Christian Doppler discovered the effect that bears his name. The intrinsic precision of the radial velocities does not depend on the distance of the star, but only on its apparent brightness.

From a formal point of view, if $r(t_0)$ is the heliocentric (or better barycentric) position vector of a star at a given initial epoch t_0, and $r_0 = 1/\sin \pi_0 \approx 1/\pi_0$ its distance in terms of the trigonometric parallax π_0, the instantaneous velocity vector $V(t_0)$ could be easily derived:

$$\mathbf{r}(t_0) = \begin{bmatrix} r_0 \cos \alpha_0 \cos \delta_0 \\ r_0 \sin \alpha_0 \cos \delta_0 \\ r_0 \sin \delta_0 \end{bmatrix},$$

$$\mathbf{V}(t_0) = \dot{\mathbf{r}}(t_0) = \begin{bmatrix} -\sin \alpha_0 \cos \delta_0 & -\cos \alpha_0 \sin \delta_0 & \cos \alpha_0 \cos \delta_0 \\ \cos \alpha_0 \cos \delta_0 & -\sin \alpha_0 \sin \delta_0 & \sin \alpha_0 \cos \delta_0 \\ 0 & \cos \delta_0 & \sin \delta_0 \end{bmatrix} \begin{bmatrix} r_0 \mu_{\alpha 0} \\ r_0 \mu_{\delta 0} \\ \dot{r}_0 \end{bmatrix}$$

where $\mu_{\alpha 0}$, $\mu_{\delta 0}$ are the proper motions in right ascension and declination, and $\dot{r}_0 = V_r(t_0)$ is the radial velocity. The position of the star at a following date T, but referring to the same epoch, would then be derived from:

$$\mathbf{r}(T) = \mathbf{r}(t_0) + (T - t_0)\dot{\mathbf{r}}_0$$

However, the knowledge of the necessary quantities is usually incomplete, so that in the following the radial velocities and the proper motions will be considered separately. It also clear that the treatment of the system of proper motions and velocities is fundamentally linked to that of precession, a great complication, indeed, for transferring catalogs based on stellar observations from one epoch to another, if the highest precision is to be maintained.

9.1 Radial Velocities

The radial velocity is measured through the Doppler effect, namely through the variation in wavelength λ of the radiation, caused by the relative motion V_r along the line of sight; the velocities of the planets or of the stars of the Milky Way are usually so small in comparison with the velocity of the light c that no sensible error is made in using the prerelativistic formula:

$$V_r = c\frac{\lambda_O - \lambda_S}{\lambda_S} = c\frac{\Delta\lambda}{\lambda} = cz, \quad z = \frac{\Delta\lambda}{\lambda} = \frac{V_r}{c}, \quad \frac{\lambda_O}{\lambda_S} = 1 + \frac{V_r}{c} = 1 + z \quad (9.1)$$

where λ_O is the wavelength measured by the observer, and λ_S is that measured in the rest-frame of the source. The use of the letter z to indicate $\Delta\lambda/\lambda$ is fairly widespread. Notice that the radial velocity can be positive ($z > 0$, namely the wavelength is *red-shifted*), or negative ($z < 0$, namely the wavelength is *blue-shifted*).

When the velocities exceed say $0.01c$, then special relativity must be taken into account through Lorentz transformations, as already seen in Chapter 7. It is more advantageous here to work in terms of frequency ν than of λ (in vacuum, $\nu = c/\lambda$, $d\nu = -cd\lambda/\lambda^2$); the classical formula would be: $1 + z = \nu_S/\nu_O$. To derive the relativistic formula, let O be an observer in uniform motion with velocity $\mathbf{V}$ with respect to an inertial observer S, and let τ be the time measured by O (proper time) and t the time measured by S. Between the two times the following relation applies:

$$d\tau = \sqrt{1 - \frac{V^2}{c^2}}\,dt < dt \quad (9.2)$$

(slowing down of the moving clock). As a consequence, the frequency of the light measured by O, and the frequency of the light measured by S are related by:

$$\nu_O = \nu_S \sqrt{1 - \frac{V^2}{c^2}} \left(1 - \frac{1}{c}\mathbf{V}\cdot\mathbf{n}\right)^{-1} \tag{9.3}$$

where n is the normal to the wavefront. Therefore special relativity foresees a transverse Doppler effect (when $\mathbf{n}$ is perpendicular to $\mathbf{V}$), which is not present in the prerelativistic formula. In other words, the whole velocity vector, and not only its radial component, enters in the observed frequency displacement.

After some simple manipulation of Equation 9.3 we obtain the series expansion:

$$z = \frac{\Delta\lambda}{\lambda} = -1 + \sqrt{\frac{(1+V)/c}{1-V)/c}} \approx \frac{V}{c} + \frac{1}{2}\left(\frac{V}{c}\right)^2 + \cdots \tag{9.4}$$

which shows that actually the Doppler effect is not symmetric in the sign of $\mathbf{V}$. See the Exercises for a fuller discussion.

Inside the Solar System, velocities of a few tens of km/s prevail, with the notable exception of comets and asteroids skimming the surface of the Sun, whose heliocentric velocity can exceed 700 km/s. The classical formula is therefore adequate for most applications; however, another factor must be carefully taken into account, that of the extreme precision (say ± 1 mm/s) with which the velocity of a spacecraft, possibly orbiting a planet, can be measured. The consequence is that for accurate navigation inside the Solar System, general relativity must be taken into account in expressing the metric of space-time.

The radial velocities of the normal stars of the Galaxy rarely reach 500 km/s; among the nearest stars, velocities higher than 50 km/s are seldom encountered, with notable exceptions such as Barnard's star, which has a radial velocity of -108 km/s with respect to the Sun; this group of nearby high velocity stars is extremely important for the comprehension of the overall kinematics and dynamics of the Milky Way (see e.g., Mihalas and Binney (1981)). For the great majority of stars, a precision of at best ± 100 m/s is reached; only in favorable cases and with refined techniques, e.g., in searches for extra-solar planets by means of radial velocity variations, ± 3 m/s are achieved, the limiting factors being on one side the technical limitations and on the other the turbulent structure of the stellar spectral lines themselves. Therefore the classical formula is usually adequate. There are cases, however, where special relativity must be applied, e.g., for the variable star SS433, or for the expanding gaseous envelopes of explosive variables (novas, supernovas), where velocities of tens of thousands km/s are encountered.

For galaxies, beyond a certain distance roughly coinciding with that of the cluster of galaxies in Virgo, whose velocity is approximately $+1000$ km/s, only positive velocities are encountered (for closer galaxies, negative velocities can be found, e.g., for M31 in Andromeda). The spectral lines of the distant galaxies and of other objects of cosmological significance, e.g., the Quasi stellar objects (also named quasars or QSOs), are always red-shifted, with an amount z_c increasing with the distance, as was discovered by Hubble around 1930. This observational effect is at the basis of all cosmology, and is referred to as *expansion of the Universe*. Values of z_c around 10 have already been measured, with an immense displacement of the wavelength of the spectral lines (for instance, if $z = 6$, the ultraviolet spectral line Lyman-α of hydrogen, having $\lambda = 1216 \, \text{Å}$ in the laboratory, is observed in the near infrared at $\lambda = 8512 \, \text{Å}$). To derive the radial velocity from such high redshifts, the relativistic formula must be employed; however, in cosmology the simple connection between z and velocity breaks down, because the expansion of the universe is an expansion of the metric itself, and it does not reflect the motion of the source in a fixed coordinate frame. Nor is it possible to reason in terms of three spatial coordinates and one time coordinate. For these reasons, the radial velocities derived from Equation 9.4 are better called indicative velocities.

General relativity predicts another redshift for the light emitted by atoms in a gravitational field, e.g., on the surface of a star. From a spherical source of mass M and radius R, this gravitational redshift is expressed by:

$$\frac{\lambda}{\lambda_0} = \sqrt{1 - 2\frac{GM}{c^2 R}} = \sqrt{1 - \frac{r_S}{R}}$$

where G is the gravitational constant, and r_S the already introduced Schwarzschild's radius (see Chapter 7). If the radius $R \gg r_S$:

$$z_g = -1 + \frac{1}{\sqrt{1 - \frac{2GM}{c^2 R}}} \approx \frac{GM}{c^2 R} = \frac{r_S}{2R} \qquad (9.5)$$

On the Sun, the effect amounts to 0.64 km/s. On the white dwarf Sirius B, having mass approximately equal to the Sun's, but radius only 80% of that of the Earth, the effect is correspondingly much larger. The measurement of the gravitational redshifts from white dwarfs therefore constitutes another test of the correctness of general relativity. In the limit of R shrinking to r_S (namely, at the surface of a black hole) the light will be infinitely reddened.

If a given object displays the kinematical, cosmological, and gravitational redshifts, the combination of the different effects is given by:

$$1 + z_{\text{tot}} = (1 + z_V)(1 + z_c)(1 + z_g), \quad z_{\text{tot}} = z_V + z_c + z_g \qquad (9.6)$$

where the last approximate equality holds true only for small redshifts.

When measuring radial velocities, the observations must be corrected for the annual and diurnal motions of the observer. The necessary formulae are easily determined if the wanted precision is around 100 m/s. The heliocentric velocity of the Earth $V_\oplus$ varies between 29.3 km/s at aphelion and 30.3 km/s at perihelion; its projection towards a direction of ecliptic coordinates (λ, β) is:

$$V_\oplus = -Kc \cos \beta \, [\sin(\lambda - \lambda_\odot) + e \sin(\lambda_\Pi - \lambda)] \qquad (9.7)$$

(see Chapter 7), where $Kc = 29.79$ km/s, $\lambda_\odot$ is the longitude of the Sun in that particular date, λ_Π is the longitude of the perigee, and e the eccentricity of Earth's orbit. Notice that the term in eccentricity, amounting at most to 0.50 km/s, is essentially constant for any given star, with a secular variation due to the very small changes in λ_Π and e.

Another way of expressing this correction is by making use of equatorial coordinates of the line of sight and of the Cartesian components of the Earth's velocity, given in AU/day for each day by the *Astronomical Almanac*:

$$V_\oplus = 1731.5 \, (\dot{X} \cos \alpha \cos \delta + \dot{Y} \sin \alpha \cos \delta + \dot{Z} \sin \delta) \; \text{km/s} \qquad (9.8)$$

Regarding the diurnal rotation, the velocity of the observer at the equator is approximately 0.465 km/s: therefore, for a generic geocentric latitude ϕ' the projection of this velocity on the line of sight to a star of declination δ and hour angle HA will be:

$$V_{\text{rot}} = 0.465 \cos \phi' \cos \delta \sin HA \; \text{km/s} \qquad (9.9)$$

with the slight complication that V_{rot} can appreciably change on very long exposures. Furthermore, for better precision one should take into account the elevation of the observer. This is particularly true for the Hubble Space Telescope, for which a complete knowledge of the geocentric velocity vector is required. Finally, the wanted heliocentric radial velocity is obtained by:

$$\mathbf{V} = \mathbf{V}_r - \mathbf{V}_\oplus - \mathbf{V}_{\text{rot}}$$

Should one need a precision better than 100 m/s (very rarely achievable, as already commented), then more accurate formulae must be employed; at this level of precision, one should refer the velocities to the Earth–Moon and Solar System barycentric observers, the relative velocities with respect to the geocentric and the heliocentric observers being both of approximately 12 m/s.

9.2 Proper Motions

Let us consider first the proper motion of the nearer stars. Be S the heliocentric observer, and X a generic star at a distance d with heliocentric velocity $\mathbf{V}$, at a certain date (see Figure 9.1). Owing to the enormous distances, for many decades or centuries the velocities can be considered as rectilinear and uniform, apart from very few exceptions such as Barnard's star. Expressing the modules in km/s, and indicating with n the number of seconds in one year, after 1 year the star will be seen in X', having traveled a course of Vn km along a rectilinear path forming an unknown angle θ with

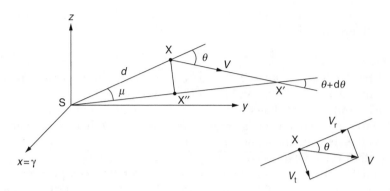

FIGURE 9.1

Heliocentric velocity and proper motion of star X. Notice that when the star reaches X', the angle between the line of sight and V will change by $d\theta = -\mu$.

axis SX. On the plane tangent to the celestial sphere, the star will appear to have moved by the small angle:

$$\mu = n\frac{V}{d}\sin\theta \text{ (rad)} \tag{9.10}$$

The component of $\mathbf{V}$ perpendicular to the line of sight, $\mathbf{V}_t = \mathbf{V}\sin\theta$, is called transverse or tangential velocity. The corresponding apparent angular velocity μ is said to be proper motion of the star X, and is commonly measured in arcsec/yr, or arcsec/century. By using the parallax π in arcsec instead of the distance d in kilometers, and taking into account the appropriate conversion factors (namely: 1 km/s = 0.21095 AU/yr, 1 AU/yr = 4.74045 km/s), we derive the following values of the modules:

$$V_t = V\sin\theta = 4.740\frac{\mu}{\pi} \text{ km/s}, \quad \mu = \frac{\pi}{4.740}V_t \text{ arcsec/yr}$$

Barnard's star has a proper motion as high as $10''$/yr; very few stars have μ larger than $2''$/yr. Obviously, the nearer stars have, in general, the larger proper motions, but the vice versa is not true; many nearby stars have small motions because of the orientation of their velocity vectors. For the nearer stars, the annual ellipse of parallax is trailed by the proper motion, as evidenced in Figure 9.2.

If the radial velocity of star X can be measured from the Doppler effect:

$$V_r = V\cos\theta = c\frac{\Delta\lambda}{\lambda} = 4.740\frac{\mu}{\pi}\cot\theta \text{ km/s} \tag{9.11}$$

(the nonrelativistic formula being certainly valid), then the full velocity vector of the star can be reconstructed. However, as we have already noted, for the great majority of cases the parallaxes are not available.

It is convenient to consider the proper motion as a vector $\boldsymbol{\mu}(\mu, q)$ on the plane tangent to the celestial sphere, with modulus μ expressed in angular units (i.e., arcsec/yr) along the great circle XX'', and direction expressed by

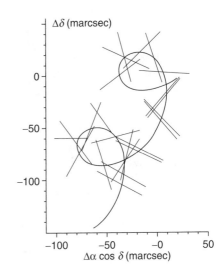

FIGURE 9.2
The proper motion of $\mu = 0''.03/\text{yr}$ trails the annual parallax ellipse of a nearby star having $\pi = 0''.08$: the segments give the instantaneous positions and associated errors, the continuous line is the best fitting path. (Adapted from Hipparcos satellite data.)

a position angle q measured from the north through the east ($0° \leq q < 360°$). Alternatively, the two equatorial components (μ_α, μ_δ) can be given; attention must be paid to the units of μ_α, because the angular distance between two successive positions of the star must be measured along the great circle passing through XX″, but often μ_α is derived by the difference between two successive right ascensions and expressed in seconds of time.

Therefore:

$$\mu_\alpha(^s/\text{y}) = \frac{1}{15}\mu(''/\text{y})\sin q \sec \delta, \quad \mu_\delta(''/\text{y}) = \mu(''/\text{y})\cos q \qquad (9.12)$$

$$\mu_\alpha(''/\text{y}) = 15\mu(^s/\text{y})\cos \delta, \quad \tan q = \frac{\mu_\alpha(''/\text{y})}{\mu_\delta(''/\text{y})}$$

In the same manner, the tangential velocity components are derived as:

$$V_{t\alpha} = 4.740\frac{\mu_\alpha(''/\text{y})}{\pi}, \qquad V_{t\delta} = 4.740\frac{\mu_\delta(''/\text{y})}{\pi} \qquad (\text{km/s})$$

From the previous relationships, we also derive:

$$\mu = \sqrt{\mu_\alpha^2 + \mu_\delta^2} = \frac{\pi}{4.740}\sqrt{t_\alpha^2 + t_\delta^2}, \quad \tan q = \frac{\mu_\alpha}{\mu_\delta} = \frac{t_\alpha}{t_\delta}$$

namely, the position angle of the proper motions is the same as that of the tangential velocity, and is independent of the parallax of the star. This property, which applies also to the distribution function of the proper motions in a given small area of the sky, has been used to study the stellar kinematics in several investigations where only the position angles were known. We will discuss this argument further in the following sections. Again, with reference to Figure 9.3, from the spherical triangle XPX′ we notice that:

$$\cos \delta \sin q = \cos \delta'' \sin q''$$

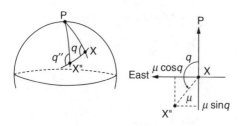

FIGURE 9.3

The proper motion resolved in two equatorial components.

Namely, that the quantity $\cos \delta \sin q$ is conserved during the movement of the star, implying that:

$$\frac{d}{dt}(\cos \delta \sin q) = 0, \quad \dot{q} = \tan q \tan \delta \, \dot{\delta} \qquad (9.13)$$

a property that will be utilized in the following.

9.3 Variation of the Equatorial Coordinates

In order to find the effect of the proper motion on the celestial coordinates of a given star X, a first-order approximation will be sufficient for short time intervals; Δt years after the epoch t_0, the mean coordinates will be:

$$\alpha(t_0 + \Delta t) = \alpha(t_0) + \mu_\alpha \Delta t + (m + n \sin \alpha \tan \delta)\Delta t,$$
$$\delta(t_0 + \Delta t) = \delta(t_0) + \mu_\delta \Delta t + (n \cos \alpha)\Delta t \qquad (9.14)$$

a formula which includes the general precession (do not confuse in this section the constant of precession n with the number of seconds in 1 year), with due attention to the units (the sign of Δt can obviously be reversed). However, to be rigorous we must note that (μ_α, μ_δ) do vary in time even if the velocity is rectilinear and constant. This happens for two distinct reasons: on one hand, the reference system rotates because of precession; on the other, the changing perspective alters the apparent length of equal arcs, as shown (with great exaggeration) in Figure 9.4.

Therefore, consider the successive terms in 9.14:

$$\alpha(t_0 + \Delta t) = \alpha(t_0) + \mu_\alpha \Delta t + (m + n \sin \alpha \tan \delta)\Delta t + \frac{1}{2}\dot{\mu}_\alpha(\Delta t)^2 \qquad (9.15)$$

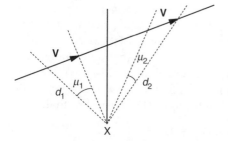

FIGURE 9.4

Perspective acceleration of a star X in uniform motion. The proper motion changes with time even if vector **V** stays constant. The perspective effect is obviously present also in the radial velocity.

$$\delta(t_0 + \Delta t) = \delta(t_0) + \mu_\delta \Delta t + (n \cos \alpha)\Delta t + \frac{1}{2}\dot{\mu}_\delta(\Delta t)^2 \tag{9.16}$$

The time derivatives of (μ_α, μ_δ) are composed of two terms: a first term due to precession, independent from radial velocity and distance, and clearly the dominating one; a second term, due to the variation of the projection of the velocity on the line of sight, which can be written down explicitly only if parallax and radial velocity are both known.

The components due to precession, that clearly modify only the direction of $\boldsymbol{\mu}$, but not its modulus $(d\mu/dt = 0)$, are found in a somewhat elaborate manner which is not given here (see also Exercises in Chapter 6):

$$\dot{\mu}_{\alpha,p} = \frac{n}{R''}(\mu_\alpha \cos \alpha \tan \delta + \mu_\delta \sin \alpha \sec^2 \delta) \ (\text{arcsec}/\text{y}^2)$$

$$\dot{\mu}_{\delta,p} = -\frac{n}{R''}\mu_\alpha \sin \alpha \ (\text{arcsec}/\text{y}^2) \tag{9.17}$$

where n is the general precession in declination (arcsec/yr), μ_α is in s/yr and μ_δ in arcsec/yr.

Regarding the second, intrinsic, term, let us take the time derivatives of Equation 9.12, with the caution of remembering that all derivatives must be in circular units (for instance, $\dot{\delta} = \mu_\delta/R''$). Then, recall Equation 9.13 and that:

$$\dot{\mu} = \frac{V}{4.740}[\dot{\pi} \sin q + \pi \cos q\dot{q}], \quad \dot{\pi} = -\frac{\pi^2 V_r}{4.740} \tag{9.18}$$

After some manipulation, it can be shown that these intrinsic components are:

$$\dot{\mu}_{\alpha,i} = \frac{1}{R''}(2\mu_\alpha\mu_\delta\tan \delta - 0.422V_r\pi\mu_\alpha),$$

$$\dot{\mu}_{\delta,i} = -\frac{1}{R''}(\mu_\alpha^2\sin \delta \cos \delta + 0.422V_r\pi\mu_\delta)$$

As before μ_α is in s/yr, μ_δ in arcsec/yr, π in arcsec and V_r in km/s (remember that the required quantities for the determination of the derivatives are known for a very limited number of stars). Therefore, for long time intervals we must include in Equation 9.15 and Equation 9.16 the terms $1/2(\dot{\mu}_{\alpha,p} + \dot{\mu}_{\alpha,i})T^2$, $1/2(\dot{\mu}_{\delta,p} + \dot{\mu}_{\delta,i})T^2$. Recourse to even more complicated procedures (for instance, considering another term in the series expansion) is seldom necessary, unless very long time spans are considered, or utmost precision is sought.

The operations for transferring the coordinates from one epoch to another can be performed in three different ways: transformation by components, transformation by time series, and rigorous matrix rotations. Whatever the method, it is necessary to apply the proper motions corrections before precessing.

9.4   Interplay between Proper Motions and Precession Constants

The proper motion of a star can be derived from the difference of its equatorial coordinates at two successive dates, provided the effects of precession and nutation are first removed; namely, the two coordinate sets must be reported to the same equinox before taking the difference. Therefore, the uncertainties in the precessional constants will enter into the uncertainty of the proper motion. Such uncertainty is not so important for a single star, but instead for the system of proper motions. For instance, the FK5 could be affected by a spurious rotation at the level of about $0''.15$/century. This seemingly small systematic error enters into the knowledge of the overall field of motions and forces of the Milky Way, as will be discussed in a later paragraph. Looking at the problem from the other side, a reasonable model of the distribution of proper motions can lead to an improved determination of the precessional constants. It is worth, therefore, looking into the connection between proper motions and precession with more insight. For simplicity, let us limit the discussion to the first-order formulae 9.14, which can be rewritten as:

$$\begin{cases} \mu_\alpha = \dot{\alpha} - (m + n \sin \alpha \tan \delta) \\ \mu_\delta = \dot{\delta} - n \cos \alpha \end{cases} \tag{9.19}$$

where (see Chapter 5) $m = \psi \cos \varepsilon - g$, $n = \psi \sin \varepsilon$, ψ being the lunisolar and g the planetary precession. To be precise, the value of ψ derived by astrometric means contains, in addition to the true lunisolar term ψ_0, the very small geodesic precession ψ_g, $\psi = \psi_0 + \psi_g$ (see Notes, Chapter 5). This geodesic precession derives from general relativity, which foresees a very slow rotation of the frame connected with the Earth orbiting around the barycenter of the Solar System. According to deSitter and Brouwer (1938), and to Lieske et al. (1977), its value is:

$$\psi_g = \frac{3}{2} K^2 (1 - e)n \approx -1''.92/\text{century}$$

where K is the aberration constant in radians, e the eccentricity of the Earth's orbit and n its mean motion in arcsec/century. Note the negative sign; the geodesic precession is a motion of γ on the ecliptic opposite to the lunisolar one. We discuss here Equation 9.19 in order to show how the determination of the proper motions of a large number of stars well-distributed over the celestial sphere and in the volume containing the Sun, can lead to the determination of the precessional constants and how the errors on the former propagate on them. This method of determining the precessional constants is therefore based on the stellar kinematics; it has been discussed, for instance by Fricke (1977). From the following paragraphs, it will become clearer that the system of proper motions contains the reflex of the motion of the Sun with respect to the nearer stars (the so-called local standard of rest, LSR), and the effect of the overall galactic rotation and of its variation with distance and direction (namely the presence of the A and B constants in Oort's and Lindblad's theory of galactic rotation), in addition to any spurious undetected systematic error.

Efforts must therefore be made to obtain a system of proper motions as precise as possible and free from systematic effect. The satellite Hipparcos could not produce a major improvement, because its operational life was too short. An alternative method is to derive proper motions with respect to a nonrotating background of fixed objects, such as the distant quasars; the reference system ICRF is by definition devoid of rotation, so many efforts are currently being made to reference the proper motions to it. See, for instance the paper by Charlot et al. (1995), who found a correction of (-3.00 ± 0.20) mas/yr to the IAU (1977) lunisolar constant, by combining data provided by VLBI and lunar laser ranging.

A further difficulty is associated with the task of transforming the proper motions from the pre to the post 1984 reference frames, e.g., from FK4 to FK5 or vice versa. The reasons for caution, and the fairly complex procedure needed to accomplish this seemingly simple task, are explained in great detail by Hohenkerk et al. (1992), and are reproduced in the *Astronomical Almanac*.

9.5 Astrometric Radial Velocities

The precision of Hipparcos has allowed astrometric radial velocities to be derived without passing for the intermediary of the spectroscopy, at a precision level of 0.3 to 1 km/s; future projects, such as Gaia will undoubtedly provide even better precisions for a large sample of stars. The advantage of astrometry is that the velocities are not affected by the systematic errors of spectroscopy (e.g., convective motions in the star's atmosphere, transverse Doppler effect, or gravitational redshift). This idea had already been put forward, in 1901 by Seeliger and by Schlesinger in 1917, but only recently has been brought to fruition (see Dravins et al., Notes). With reference to Figure 9.4, we notice that the distance to the star changes with the passage of time, so that the radial velocity can also be expressed by:

$$V_r = -a_\odot \frac{\dot{\pi}}{\pi^2}$$

where $a_\odot$ is the astronomical unit; the difficulty is to have a reliable determination of $\dot{\pi}$, that even for Barnard's amounts to only $+34\ \mu$ arcsec/yr. However, at the same time:

$$V_r = -a_\odot \frac{\dot{\mu}}{2\pi\mu}$$

A third effect of the passage of time is visible in the relative distance between two stars having parallel and equal velocities but slightly different positions, e.g., two stars belonging to the same cluster; the cluster will appear to contract or to expand with the same amount of the changing parallax.

Although the astrometric method to determine radial velocities is attractive and promising, it nevertheless also has some systematic biases that must be accurately checked, in particular, gravitational perturbations due to nearby stars (it is well-known that the majority of stars are members of binary systems) or to the overall gravitational field of the Milky Way.

9.6 Apex of Stellar Motions and Group Parallaxes

Let us consider, in Figure 9.5, the heliocentric equatorial reference system, and let $(\dot{x}, \dot{y}, \dot{z})$ be the velocity components of star X. These Cartesian components can be expressed in terms of $(V_{t\alpha}, V_{t\delta}, V_r)$ by a suitable rotation of coordinates. Table 9.1 shows the rotation matrix that transforms $(V_{t\alpha}, V_{t\delta}, V_r)$ into $(\dot{x}, \dot{y}, \dot{z})$ and vice versa; for instance:

$$\dot{x} = -V_{t\alpha} \sin \alpha - V_{t\delta} \cos \alpha \sin \delta + V_r \cos \alpha \cos \delta,$$

$$V_{t\alpha} = -\dot{x} \sin \alpha + \dot{y} \cos \alpha$$

and similar for other components.

Recalling the expression of the tangential velocity V_t, we derive the Cartesian equatorial components by the observable quantities:

$$
\begin{cases}
\dot{x} = -\dfrac{4.740}{\pi}(\mu_\alpha \sin \alpha + \mu_\delta \cos \alpha \sin \delta) + V_r \cos \alpha \cos \delta \\[2mm]
\dot{y} = +\dfrac{4.740}{\pi}(\mu_\alpha \cos \alpha - \mu_\delta \sin \alpha \sin \delta) + V_r \sin \alpha \cos \delta \\[2mm]
\dot{z} = +\dfrac{4.740}{\pi}\mu_\delta \cos \delta + V_r \sin \delta
\end{cases}
\qquad (9.20)
$$

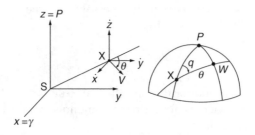

FIGURE 9.5

On the celestial sphere, the direction of V corresponds to point W (apex of the stellar motion), and the great circle XW to the plane passing through SX and containing the vector **V**.

TABLE 9.1

Rotation Matrix between $(V_r, V_{t\alpha}, V_{t\delta})$ and $(\dot{x}, \dot{y}, \dot{z})$

	$V_{t\alpha}$	$V_{t\delta}$	V_r
$\dot{x}$	$-\sin\alpha$	$-\cos\alpha\sin\delta$	$\cos\alpha\cos\delta$
$\dot{y}$	$\cos\alpha$	$-\sin\alpha\sin\delta$	$\sin\alpha\cos\delta$
$\dot{z}$	0	$\cos\delta$	$\sin\delta$

From the knowledge of the entire velocity vector **V**, the direction of the motion of the star in the heliocentric reference system $S(x, y, z)$ can be determined (see again Figure 9.5).

On the celestial sphere, point X corresponds to direction SX, and point W to direction XX′ of Figure 9.1, so that the great circle XW represents the plane SXX′. The point W is called apex of the stellar motion, and has equatorial coordinates given by:

$$
\begin{cases}
\sin\delta_W = \cos\theta\sin\delta + \sin\theta\cos\delta\cos q \\
\cos(\alpha_W - \alpha)\cos\delta_W = \cos\theta\cos\delta - \sin\theta\sin\delta\cos q \\
\sin(\alpha_W - \alpha)\cos\delta_W = \sin\theta\sin q
\end{cases}
\tag{9.21}
$$

where (α, δ) are the initial coordinates of the star. The angular distance between X and W is given by:

$$
\cos\theta = \sin\delta_W\sin\delta + \cos\delta_W\cos\delta\cos(\alpha_W - \alpha)
$$

so that the equatorial Cartesian components of vector **V** can be expressed in a second way:

$$
\begin{cases}
\dot{x} = V\cos XW = V\cos\delta_W\cos\alpha_W \\
\dot{y} = V\cos YW = V\cos\delta_W\sin\alpha_W \\
\dot{z} = V\cos ZW = V\sin\delta_W
\end{cases}
\tag{9.22}
$$

In general, however, the parallax is unknown so that the observations do not provide the length of the arc θ. Nevertheless, the statistical study of the stellar motions in certain areas of the sky has shown the existence of stars having motions converging towards the same apex: therefore, these stars seem to belong to a group having a common motion, *as if* their space velocities were parallel. We could therefore speak of a *co-moving group*, or even of a *stellar current*. However, having convergent motions is not *per se* a proof of parallel velocities, unless the parallaxes are known to be identical, or unless there are reasons to believe that the stars are at the same distance. In such instances (a strong case is when the stars are in the same open cluster, Hyades, Pleiades, etc.), we can obtain the parallax of the group by the determination of the convergent point and of the radial velocities of the stars; a hidden assumption is that the diameter of the cluster do not vary in time

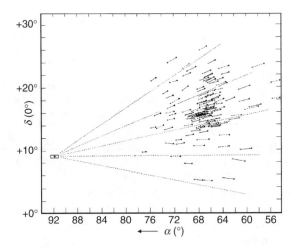

FIGURE 9.6
The convergence of the proper motions of the Hyades stars. Notice the great angular distance between the cluster and its convergent point.

(see Hanson, 1975). We have indeed:

$$V \cos \theta = V[\sin \delta_W \sin \delta + \cos \delta_W \cos \delta \cos(\alpha_W - \alpha)] = V_r, \quad \pi = \frac{4.740\mu}{V_r \tan \theta}$$

Let us consider in particular the very nearby open cluster of the Hyades, whose distance is around 45 pc, and whose extension in space is approximately 10 pc. The convergence of the proper motions of the stars to a common apex was ascertained long ago (see Figure 9.6), but the data obtained by the Hipparcos satellite has provided us with a much improved knowledge (see Perryman et al., 1998). This is one of the few cases where the tridimensional structure can be obtained directly from the observations, and therefore the Hyades play a central role in the calibration of several relationships, e.g., the Hertzsprung–Russell diagram (see Chapter 16).

9.7 The Peculiar Motion of the Sun

In this section we will complete the discussion of the secular parallaxes initiated in Chapter 8. The observed proper motions contain the reflex of the motion of the Sun, and therefore from the ensemble of the motions of the nearer stars we can construct a velocity reference system (not a spatial one!) to which the motion of the Sun can be referred. In 1783, Herschel had already laid down the foundations of this method, utilizing only the position angles of 12 stars. Indeed, q is independent from the parallax, and it coincides with the position angle of the tangential velocity. Herschel's method can be easily visualized: draw on the celestial sphere the great circles defined by the proper motions, and consider the semicircle oriented as the motion itself. All these semicircles will intersect, within the errors, in a point (more realistically, in a small area) which is the antapex of the solar motion; the modulus of the velocity of the Sun will remain undetermined by this method.

Let us now examine the case of an assembly of N nearby stars, for which the four quantities (π, μ, q, V_r), and therefore the heliocentric velocity vectors $\mathbf{V}(\dot{x}_i, \dot{y}_i, \dot{z}_i)$, $(i = 1, ..., N)$, are known. Let us calculate the average value, change its sign and define this quantity as the *peculiar motion* of the Sun with respect to the given ensemble:

$$\dot{x}_\odot = -\langle \dot{x} \rangle = -\frac{1}{N} \sum_1^N \dot{x}_i, \quad \dot{y}_\odot = -\langle \dot{y} \rangle = -\frac{1}{N} \sum_1^N \dot{y}_i,$$

$$\dot{z}_\odot = -\langle \dot{z} \rangle = -\frac{1}{N} \sum_1^N \dot{z}_i$$

The modulus $s_\odot$ and apex $(\alpha_\odot, \delta_\odot)$ will be derived in the usual manner:

$$s_\odot = \sqrt{\dot{x}_\odot^2 + \dot{y}_\odot^2 + \dot{z}_\odot^2}, \quad \tan \alpha_\odot = \frac{\dot{y}_\odot}{\dot{x}_\odot}, \quad \sin \delta_\odot = -\frac{\dot{z}_\odot}{s_\odot}$$

The early observations provided a velocity of approximately 20 km/s, in direction $\mathbf{W}_\odot(\alpha_\odot, \delta_\odot) \approx (18^h + 30°)$, not far from Vega.

We stress that this value depends on the particular set of stars used to define it. Ideally, if we could take into account all nearby stars, we would obtain a LSR, a velocity reference system of great interest for the study of the velocity field of objects in the Milky Way. Therefore, it is useful to transform the coordinate system from the equatorial to the galactic one (see Chapter 3). Let us denote with (u_i, v_i, w_i) the three velocity components derived by appropriate rotation of $(\dot{x}_i, \dot{y}_i, \dot{z}_i)$, with axis $\mathbf{u}$ directed toward the galactic center, axis $\mathbf{v}$ at 90° in the galactic plane (toward the constellation of Cygnus), and axis $\mathbf{w}$ toward the galactic pole. Note that this LSR has a purely kinematics significance, the masses of the stars not having been taken into account, and does not possess a precise origin in space. We can assume that the Sun (or better, the barycenter of the Solar System) is passing through the LSR origin at the present time.

Table 9.2 and Table 9.3 contain three determinations of the LSR deriving from three different sets of stars chosen to define it; the precise meaning of the spectral types is given in Chapter 16 and Chapter 17. The designation of *typical* is for the value determined using all available stars in the catalogs, it is the value to be used in the determination of secular parallaxes; *basic* refers to the nearer stars; *fundamental*, with respect to the hypothetical LSR in circular orbit around the center of the galaxy, with zero velocity dispersion. The noticeable differences in the various columns have a profound significance in the frame of all theories on the origin and evolution of the Milky Way.

TABLE 9.2

Solar Motion in Galactic Coordinates, Resulting from Three Different
Choices of the Defining Stars

	$u_\odot$	$v_\odot$	$w_\odot$	$s_\odot$	$l_\odot$	$b_\odot$	$\alpha_\odot$	$\delta_\odot$
Typical	− 10.4	+ 14.8	+ 7.3	20.0	56°	23°	$18^h.0$	+ 30°
Basic	− 9.0	+ 11.0	+ 6.0	15.0	51°	23°	$17^h.9$	+ 26°
Fundamental	− 9.0	+ 12.0	+ 7.0	16.5	51°	23°	$17^h.8$	+ 23°

TABLE 9.3

Solar Motion Referred to Different Classes of Stars

Spectral Types	$u_\odot$	$v_\odot$	$w_\odot$	σ_u	σ_v	σ_w	σ^2
Supergiants O-B5	− 9.0	+ 13.4	+ 3.7	12	11	9	364
A Giants	− 13.4	+ 11.6	+ 10.3	22	13	9	734
M Giants	− 4.5	+ 18.3	+ 6.2	31	23	16	786
Subgiants	− 8.0	+ 28.0	+ 8.0	48	23	16	3875
GO-V	− 14.5	+ 21.1	+ 6.4	26	18	20	1400
White dwarfs	− 6.0	+ 37.0	+ 8.0	50	33	25	5614
Planetary nebulae	− 8.0	+ 29.0	+ 8.0	45	35	20	3650

We mention only that after the initial hypothesis put forward by Kapteyn of
the evidence for two star currents, Schwarzschild and Charlier (1907, 1915)
proposed that the distribution function of the velocities is a three-dimensional
Gaussian function having different dispersions along the three axes.
Moreover, the dispersion depends on the particular class of stars. The tables
show that the dispersion is always smaller in the direction perpendicular to
the plane, and larger towards the galactic center (the dispersion ellipsoid
indeed points to the galactic center). Other relationships in the tables were
noted long ago, for instance a linear augmentation of σ^2 with v (Stromgren,
1930, a dependence called *asymmetric drift*), so that one could extrapolate the
solar motion to the hypothetical population having $\sigma^2 = 0$.

For a fuller analysis of this rich and evolving field, reference is made to
several texts on the dynamics of the Milky Way (see Mihalas and Binney
(1981), and Binney and Merrifield (1998)).

9.8   Secular and Statistical Parallaxes

The knowledge of the solar motion ($s_\odot$, $W_\odot$) allows definition of two distance
indicators whose statistical validity can be extended to several hundred
parsecs, namely the *secular* and the *statistical parallaxes*. Suppose we have a
set of stars all at the same distance from the Sun and well distributed over a
large area of sky (ideally, over the entire celestial sphere); furthermore, let

their velocities and their intrinsic proper motions be randomly distributed. Following a method first indicated by Kapteyn, the individual proper motions can be resolved in two orthogonal components, one along the great circle passing for the star and the solar apex, called component *upsilon* (v, μ_v), and one perpendicular to it, called component *tau* (τ, μ_τ):

$$\mu_v = 15\mu_\alpha \cos \delta \sin \psi - \mu_\delta \cos \psi, \quad \mu_\tau = 15\mu_\alpha \cos \delta \cos \psi - \mu_\delta \sin \psi$$

where both components are in arcsec/yr, and ψ is the position angle of the solar apex $W_\odot$ with respect to star X (see Figure 9.7). The same decomposition can be made for the velocity vectors.

Obviously, the solar motion is present only in component v, being easily demonstrated that for each star:

$$\mu_{v,i} - \frac{\pi s_\odot \sin \psi_i}{4.740} = \frac{\pi V_{v,i}}{4.740}$$

By averaging over all stars, the right side will vanish; however, before averaging it is advisable to multiply both sides for $\sin \psi_i$, in order to properly apply the least squares method. Thus finally, the secular parallax H is obtained as:

$$H = \langle \pi \rangle = \frac{4.740}{s_\odot \sum \sin^2 \psi_i} \sum \mu_{v,i} \sin \psi_i$$

The analysis of component τ can be made by adding the knowledge of the radial velocities, which contain the reflex of the solar velocity, with the additional hypothesis that the individual stars have a small random isotropic velocity component superimposed to that of the entire group. Since the average of the peculiar τ velocities and proper motions is zero, it is necessary to consider their absolute values before averaging. Then the statistical parallax is:

$$\langle \pi \rangle = \frac{4.740 \sum |\mu_{\tau,i}|}{\sum |V_{r,i} + s_\odot \cos \psi_i|}$$

The methods of secular and statistical parallaxes have been applied to stars as distant as 500 pc, but with great uncertainties. The apparently simple

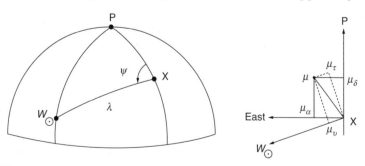

FIGURE 9.7
Components v, τ of the stellar motions. $W_\odot$ = apex of the solar motion.

process of averaging is indeed affected by the selection of the stars and other subtle choices; even the assumptions of the method can be questioned. The astrometric satellite Hipparcos has extended the trigonometric parallaxes to a point where those two statistical indicators are less useful than in the past.

9.9 Differential Rotation of the Galaxy and Oort's Constants

If the distance from the Sun increases beyond a few hundred parsecs, it is not legitimate to ignore the effects of the differential rotation of the Milky Way on radial and tangential velocities. Let us assume, in a very simplified model, that the stellar system is a thin disk, and that each star, including the Sun, performs a coplanar circular orbit with angular velocity $\omega(R)$ around the galactic center GC distant R:

$$\mathbf{V}(R) = \omega(R)\mathbf{R}$$

With these assumptions, peculiar motions are not allowed: the stars perform circular orbits around the galactic center GC with an angular velocity $\omega(R)$, which is an unknown function of the distance (if $\omega(R) = $ const, the rotation would be a rigid one). Observations of sufficiently distant stars can reveal the behavior of $\omega(R)$. In other words, we should be able to detect the differential effects of the rotation both on the radial velocities and on the proper motions. We have already highlighted in Chapter 3 that the sense of rotation of the Milky Way is clockwise if seen from the North pole of the heliocentric galactic coordinates, so that the galactic longitudes l increase contrary to the rotation, and that the distance of the Sun from the GC is approximately $R_\odot = 8.5$ kpc. In the process of analyzing radial velocities and proper motions we are led, following Oort, to introduce the two constants A and B, thus defined:

$$A = -\frac{1}{2}R_\odot\left(\frac{d\omega}{dR}\right)_\odot, \quad B = A - \omega_\odot$$

which have dimensions $(\text{time})^{-1}$, in practical units km/s·kpc.

Let us calculate the heliocentric radial velocity of a star having galactic coordinates (l, b) and distant R from C, r from the Sun. Since only circular velocities are allowed, the value will be:

$$V_r = R_\odot(\omega - \omega_\odot)\cos b \sin l \approx R_\odot\left(\frac{d\varpi}{dR}\right)(R - R_\odot)\cos b \sin l$$

In the usual observations in the visible band we will be limited to $r \ll R$, because of the strong obscuration on the galactic plane (this assumption is not justified for radio observations), so that after a little algebra relative to the triangle Sun–star–GC, we obtain:

$$V_r \approx Ar \cos^2 b \sin 2l$$

Therefore, Oort's constant A measures the local shear, i.e., the radial velocity gradient, in the orbital motions of stars in the solar neighborhood. Current values of A are approximately 14.5 ± 1.5 km/s·kpc (as already mentioned, if the Milky Way rotated as a rigid disk, then $\omega = $ const, and $A = 0$).

Regarding the proper motions, let us decompose each vector μ in its components along l and along b. After some passages, we obtain:

$$4.740\mu_l = A \cos 2l + B, \quad 4.740\mu_b = -A \sin 2l \cos b \sin b$$

As the quantity $VR = \omega R^2$ is the specific angular momentum of the material in the galactic disk, Oort's constant B measures the angular momentum gradient. Notice that the effect of the galactic rotation on the proper motions is independent from the distance of the star. Averaging the first relation over all longitudes, an overall rotation of the entire celestial sphere of $(B/4.740)$ mas/yr is predicted. Since the current values of B are around -12 ± 3 km/s·kpc, this systematic rotation is at the level of $0''.2$/century, namely of the same amount of the uncertainty of the rotation of the FK5. These considerations are a further proof of the importance of all systematic effects in the proper motions, and of the superiority of any reference system based on very distant, nonrotating objects, such as the ICRF provided by the VLBI.

From the values of A and B, we can derive the speed and the period of rotation of the Milky Way at the distance of the Sun:

$$V(R_\odot) = \omega(R_\odot)R_\odot = (A - B)R_\odot \approx 220 \text{ km/s},$$

$$P(R_\odot) = 2\pi/\omega(R_\odot) = 2.5 \times 10^8 \text{ yr}$$

Notes

- Stumpff has called attention to the rigorous definition of the heliocentric motions of the stars in a truly inertial system, see in Stumpff, P., 1985, *Astronomy and Astrophysics*, **144**, p. 232 and in Stumpff, P., 1986, *Relativity in Celestial Mechanics and Astronomy*, IAU Symposium, **114**. Fortunately, the differences with the loose definitions given here do not amount to more than a few mas after several decades. See also in van de Kamp, P., 1977, *Vistas in Astronomy*, **20**, p. 501 and **21**, p. 289.

- Ives and Stilwell, 1938, *Journal of the Optical Society of America*, **28**, p. 215 were among the first to determine with precision the validity of Equation 9.3, by shooting hydrogen atoms down a vacuum tube, with velocities (relative to the laboratory) of about 1×10^3 km/sec. As the hydrogen atoms were in flight, they emitted light in all directions. Looking into the end of the tube, with the atoms coming toward them, Ives and Stilwell measured spectral lines in the light coming forward from the hydrogen, Doppler-shifted towards the blue. Placing a mirror at the opposite end of the tube, behind the hydrogen atoms, they could look at the same light from behind, i.e., red-shifted. The measurements were in favor of the relativistic expression, although Ives and Stilwell themselves did not trust the theory!

- A discussion of the transverse Doppler effect in the cosmological situation is given, for instance by Dishon, G., Weber, T.A., 1977, Redshifts and superluminal velocities of expansion, *The Astrophysical Journal*, **212**, p. 31, whose cosmological conclusion however must be treated with great caution. Also, the previously quoted papers by Stumpff deal with relativistic velocities and superluminal expansions.

- The geodesic precession is discussed by de Sitter, W., Brouwer D., 1938, On the system of astronomical constants, *Bulletin of the Astronomical Institutes of the Netherlands*, **8**, pp. 213–231. The value quoted here is that adopted by Lieske, J.H., Lederle, T., Fricke, W., Morando, B., 1977, Expression for the precession quantities based upon the IAU (1976) system of astronomical constants, *Astronomy and Astrophysics*, **58**, pp. 1–16.

  For the connection between proper motions and precessional constants see:

  Fricke, W., 1977, Basic material for the determination of precession and galactic rotation, and a review of methods and results, *Veröffentlichungen Astronomisches*, **28**, Rechen-Institut, Heidelberg.

  Charlot, P., Sovers, O.J., Williams, J.G., Newhall, X.X., 1995, Precession and nutation from joint analysis of radio interferometric and lunar laser ranging observations, *The Astronomical Journal*, **109**, pp. 418–427.

  Hohenkerk, C.Y., Yallop, B.D., Smith, C.A., Sinclair, A.T., 1992, *The Explanatory Supplement to the Astronomical Almanac*, Chapter 3.

- The astrometric radial velocities are discussed in detail by Dravins, Lindegren and Madsen in three papers appearing in Dravins, D., Lindegren, L., Madsen, S., 1999, *Astronomy and Astrophysics*, **348**, p. 1048; Dravins, D., Lindegren, L., Madsen, S., 2000, *Astronomy and Astrophysics*, **356**, p. 1119; Dravins, D., Lindegren, L., Madsen, S., 2002, *Astronomy and Astrophysics*, **281**, p. 446. The papers also contain many references to previous literature, in particular to the paper by Schlesinger, A., 1917, *The Astronomical Journal*, **30**, p. 137.

- Eichorn, H., 1981, *The Astronomical Journal*, **86**, p. 915, proposed the opposite method, of deriving parallaxes from radial velocities and proper motions extended over 100 years.

- For the moving cluster method, see Hanson, R.B., 1975, *Astronomical Journal*, **80**, p. 379, and also in E. Hesser, Ed., 1980, *Star Clusters*, IAU Symposium 85.
- A detailed account of the results obtained by the satellite Hipparcos on the Hyades cluster is given in the paper: Perryman, M.A.C., Brown, A.G.A., Lebreton, Y., Gomez, A., Turon, C., Cayrel de Strobel, G., Mermilliod, J.C., Robichon, N., Kovalevsky, J., Crifo, F., 1998, The Hyades: distance, structure, dynamics, and age, *Astronomy and Astrophysics*, **331**, pp. 81–120.
- For Oort's constants *A* and *B* see the classic paper by Oort, J., 1965, *Stellar Dynamics*, Stars and Stellar Systems Vol. 5, University of Chicago Press, Chicago.

Exercises

The relativistic Doppler effect has been given in Equation 9.3 as:

$$\nu_O = \nu_S \frac{\sqrt{1 - \dfrac{V^2}{c^2}}}{1 - \dfrac{V \cos \theta}{c}}$$

where ν_O, ν_S are the frequencies of the light in the reference frame of the observer O and of the source S, V is the relative velocity, and θ is the angle between the line of sight and the normal to the wavefront. Now using the wavelengths, letting $\beta = V/c$, and $\alpha = \pi - \theta$ the angle between the line of sight and the velocity vector of S we have:

$$\frac{\lambda_O}{\lambda_S} = -1 + \frac{1 + \beta \cos \alpha}{\sqrt{1 - \beta^2}}$$

or else:

$$1 + z = 1 + \frac{\lambda_O - \lambda_S}{\lambda_S} = \frac{1 + \beta \cos \alpha}{\sqrt{1 - \beta^2}} \qquad (9.23)$$

which is the expression we wish to discuss.

Let V be all along the line of sight, $V = V_r$. If $\alpha = 0°$ (source receding):

$$1 + z = 1 + \frac{\lambda_O - \lambda_S}{\lambda_S} = \frac{1 + \beta}{\sqrt{1 - \beta^2}} = \sqrt{\frac{1 + \beta}{1 - \beta}} \approx 1 + \beta + \frac{1}{2}\beta^2 + \cdots$$

If $\alpha = 180°$ (source approaching):

$$1 + z = 1 + \frac{\lambda_O - \lambda}{\lambda} = \frac{1 - \beta}{\sqrt{1 - \beta^2}} = \sqrt{\frac{1 - \beta}{1 + \beta}} \approx 1 - \beta + \frac{1}{2}\beta^2 + \cdots$$

To the first order the classical Doppler effect is:

$$1 + z = 1 + \frac{\lambda_O - \lambda_S}{\lambda_S} \approx 1 \pm \beta$$

The second term however shows that the relativistic expression is not symmetric in the sign of V: the redshift of a receding source can become arbitrarily high (galaxies and quasars have been observed with $z \approx 10$), but if the source approaches the observer, its blueshift tends to the limit $z = -1$, as can be seen by discussing the cubic equation obtained with $\cos \alpha = 1$:

$$(\beta - 1)z^2 + 2(\beta - 1)z + 2\beta = 0$$

and taking β with the appropriate sign. For approaching velocities, the Doppler effect is always numerically very close to the classical one.

Let V now be all transverse, $V_r = 0$; the shift is always positive and approximately equal to:

$$1 + z \approx 1 + \frac{1}{2}\beta^2$$

(see Chapter 1), vanishingly small for small velocities.

In the general case, the calculations can be carried out for all values of α between 0 and 180°, maintaining β as a parameter, e.g., up to $\beta = 0.999999$. Notice that for angles greater than 90°, redshifts can be observed even with approaching sources! Indeed, the observed redshift depends on a combination of the radial and transverse ones, and the latter only produces positive values for z.

10

The Astronomical Times

In previous chapters, time was found necessary to describe properly the movements of the celestial sphere with respect to the meridian. Time also enters into the Newtonian dynamical theory, as the fundamental independent variable in differential equations. In this chapter, several operative definitions of time will be given, together with the transformations among them, in order to complete the notions already expounded in Chapter 4 by taking into account precession, nutation, and Earth's rotation variations. We will consider four different timescales: sidereal, solar, dynamic, and atomic, the three first scales being associated with astronomical observations. Furthermore, when general relativity matters, it will be necessary to distinguish between proper time and coordinate time, and time will become a component of the overall space-time geometry.

10.1 The Sidereal Time *ST*

By definition, the sidereal time (*ST*) is the hour angle of the vernal point γ, which is in diurnal motion with respect to the meridian because of the Earth's rotation. Correspondingly, the variation of $HA(\gamma)$ can be utilized to define a timescale in which the elementary unit of time is the interval between two successive passages of γ through the meridian. Such a time unit is called a sidereal day, which is divided into 24 hours of 3600 seconds of sidereal time. However, even if the irregularities of the Earth's rotation are disregarded, the sidereal time is only approximately a uniform one, because the position of γ is affected by nutation.

We have seen that the nutation is composed by the superposition of many different periodic terms, in particular by the one depending on the longitude of the node of the lunar orbit. Therefore, one has to distinguish between apparent and mean sidereal time: the difference $EE = \Delta\psi \cos \varepsilon$ (in the sense apparent *ST* minus mean *ST*), is called equation of the equinox (before 1960, *EE* was also called nutation in right ascension). The amount of *EE* is always between $\pm 1^s.179$, with a main periodicity of 18.6 yr. For instance, in 2001 *EE* was $-0^s.94$ at the beginning and $-1^s.02$ at the end of the year. *EE* became easier to measure around 1930, when the precision of clocks became better than 1 ms/d. The Mean Time is more uniform than the apparent one, but it is the latter that registers telescopic observations of high precision.

Sometimes, the sidereal day is confused with the interval of time between two culminations in the upper meridian of an equatorial star (ideally devoid of parallax and proper motion). Although very similar, this interval does not coincide with the sidereal day, first, because of the precession, which generates a difference of about $0^s.008$ per day (the stellar day being longer than the sidereal one), and second, because of the diurnal variation of the annual aberration. Therefore, this stellar day is never used in astronomy. It would however provide a measurement of the Earth's rotation period. The ratio between the mean sidereal day and the Earth's rotation period at the present epoch is 0.99999990; it varies very slowly, because of the varying precessional constant, by about one part over 6×10^{13} each century.

We open at this point a parenthesis about the many irregularities of the Earth's rotation, some having a short-term behavior and others a secular one. These irregularities appear in exactly the same manner in the observations of sidereal and of stellar times, thus that they cannot be detected by meridian observations unless very precise clocks, independent of astronomical observations, are available or unless reference is made to phenomena independent of the Earth's rotation, such as the ephemerides of planets. Therefore, the timescales based on the Earth's rotation are not fully adequate for dynamical purposes. This matter will be treated in more detail in the following sections.

10.2   The Solar Time $T_\odot$

A second timescale was defined in Chapter 4 by means of the hour angle of the Sun. The *solar day* is the interval of time between two consecutive upper culminations of the true Sun (we use in this context the adjective true to indicate the apparent Sun). The true solar time $T_\odot$ is therefore the hour angle of the apparent Sun, augmented by 12 h in order to comply with the contemporary convention that the day starts at midnight, not at noon: $T_\odot = HA_\odot + 12^h$. Furthermore, we recall that the largest nonuniformities of $T_\odot$ are eliminated through the introduction of the mean Sun $M_\odot$, which is in uniform motion on the equator, and whose hour angle defines the local Mean Time. In particular, the Universal Time is the hour angle of $M_\odot$ with respect to the meridian of Greenwich, plus the same 12 h:

$$UT = HA_{\text{Greenwich}}(M_\odot) + 12^h$$

Actually, the Mean Time is not a fully satisfactory concept, because it presupposes the availability of a strictly uniform timescale (we will introduce ephemeris time in a later section), and the absence of interpolation errors.

The instant of the passage of the true Sun in meridian is affected by nutation, by the two terms of the annual aberration (constant plus elliptical), and by the diurnal aberration. The duration of the solar day instead is independent from the constant term of the annual aberration and from the diurnal one. By definition, therefore, ST and UT, although different in rhythm and origin, have the same degree of uniformity, namely that of the terrestrial rotation.

Let us discuss the origins of the two times. According to Newcomb, the mean longitude of the nonaberrated fictitious Sun on the ecliptic $F_\odot$ at 12 h UT (noon) of January 1, 1900 had the value:

$$\lambda(F_\odot) = 280°40'56''.37 = 18^h42^m42^s.391$$

At the same instant, that was also the ST at Greenwich. Notice that the nonaberrated Sun, not the apparent one, which is $20''.45$ behind it, appears in this definition. After a whole Julian year of $365^j.25$, the value of ST increases by $86401^s.845$ (1 d in a tropical year, plus the difference corresponding to 0.0078 d), plus the minute acceleration of the precessional constants. Using the current values of the constants, and counting the time T in Julian centuries since January 1, 2000 at 12^h UT (namely the number of days since JD 2451545.0 divided by 36525), the complete expression of the mean ST at the midnight of Greenwich at any date T is:

$$ST_{\text{Greenwich}}(0^h\text{UT}) = 6^h41^m50^s.5481 + 8640184^s.812866T + 0^s.093104T^2$$
$$- 6^s.2 \times 10^{-6}T^3 \tag{10.1}$$

where the last two terms derive from the variation of the precessional constants. We will see in the following sections that the time T appearing in the previous expression should be shown in the scale UT1, not in the scale UT, but this distinction is unnecessary at the moment.

It is worth mentioning that while ST originates from the diurnal rotation, the mean solar time has a more hybrid nature, because the yearly revolution appears in definition in addition to the rotation. To remove this intrinsic source of ambiguity, an ephemeris time ET, based entirely on dynamical considerations, was introduced in 1960 and used until 1984, as detailed later.

10.3 The Year

The yearly revolution of the Earth provides the definition of a new timescale and of a new unit of time, namely the year. The year can be defined in several different ways.

10.3.1 Tropical Year

The tropical year is the interval of time between two consecutive passages of the Sun through the vernal point γ, or else the time needed for the right ascension of the Sun to increase by 360°. Its value, expressed in mean solar days (j), can be obtained with great precision simply by counting the number of days between two widely separated equinoxes. From a discussion of all the available data since antiquity, Newcomb derived the expression:

$$1 \text{ tropical year} = 365^j.24219879 - 0^j.00000614T \qquad (10.2)$$

T being the number of Julian centuries since mean noon at Greenwich on January 1st 1900.

Taking into account that the duration of the mean solar day has a minute change over the centuries, we should add to this definition that j is that measured at the same initial date. The second term in Equation 10.2 is due to the variation of the precessional constant in right ascension; it amounts to a diminution of the duration of the tropical year of approximately $0^s.53$ per century. Correspondingly, the mean motion of the Sun, which refers to the moving equinox, will also increase. The mean longitude of the fictitious Sun on the ecliptic at any date T is expressed by:

$$\lambda(F_\odot) = 280°40'56''.37 + 129602768''.13T + 1''.089T^2 - 20''.47 \quad (10.3)$$

where the last term distinguishes between the geometric and apparent (aberrated) Sun.

A more refined expression for the average duration of the tropical year, based on Laskar's orbital elements (1986) is:

$$1 \text{ tropical year} = 365^j.2421896698 - 0^j.0000065359T - 7.29 \times 10^{-10}T^2 \quad (10.4)$$

where T is now measured in centuries from the new fundamental epoch J2000.0, $T = (JD - 2451545.0)/36525$.

The actual duration of the tropical year can vary by several minutes around this mean value.

10.3.2 Besselian Year B, or Annus Fictus

As already stated in Chapter 4, the tropical year does not have a definite origin. Following Bessel, the origin of the year was defined as the instant when the apparent longitude of the fictitious Sun $\lambda(F_\odot)$, affected by aberration, and referring to the mean equinox of date, is exactly 280°. A year with this origin is called the Besselian year, and is indicated by the letter B before the numeral. The initial instant, designated by apposing .0 after the numeral (e.g., B1950.0), is always within 1 d of the midnight of

December 31. Any other instant in the course of the year is designated by an appropriate number of decimal digits, e.g., B1986.12345.

The next Besselian year will start at the next passage of the sun for the same longitude. The duration of the Besselian year is therefore essentially the same as the duration of the tropical year, apart from a minute secular acceleration coming from the different definition of $\lambda(F_\odot)$ and of $\alpha(M_\odot)$:

$$\text{duration Besselian Year} = \text{duration Tropical Year} - 0^s.148T' \quad (10.5)$$

being T' in tropical centuries since 1900.0. This different duration means an increasing difference in the origins; for instance, in year 2000 the Besselian year started about $7^s.4$ before the Julian one, a shift so small as to be irrelevant in almost all instances, e.g., in precessional calculations.

10.3.3 Sidereal Year

The sidereal year is the interval of time between two passages of the sun over an ecliptic star devoid of proper motion. Therefore, the sidereal year is longer than the tropical one by the amount of the precession of γ along the ecliptic, namely by approximately $(1296000'' - 50''.4)/1296000''$, corresponding to 20^m24^s, or else to 35,000 km along the orbit of the Earth. Therefore, the duration of the sidereal year is $365^j.25636$. This value is *not a measured one*, but is derived from the length of the tropical year. From it, we also get the mean sidereal solar motion:

$$n = 1296000''/365^j.25636 = 3548''.1928''/j \quad (10.6)$$

whose value is not affected by the secular variation of the precessional constant, and therefore has the same uniformity of the diurnal rotation.

10.3.4 Anomalistic Year

The anomalistic year is the interval of time between two consecutive passages of the Sun through the perigee. The direction of the major axis of Earth's orbit (or in other terms, the *line of the apses*) however is not fixed in the inertial space, it slowly precesses in the same direction of the yearly motion, by an amount of $11''.63/\text{yr}$ determined by the gravitational perturbations of the other planets. Therefore the longitude of the perigee, referred to as the moving equinox, increases by approximately $11''.63 + 50''.26 = 61''.89/\text{yr}$. The anomalistic year is longer than the previous ones, its average duration being approximately $365^j.25964$, with a secular acceleration of $0.263^s/\text{century}$. It is easily seen that perigee and equinox coincide every 21,000 yr: because the duration of the seasons depends on the distance between equinox and perigee, the consequence is their appreciable variation of duration, at a level of 1 h per century.

10.3.5 Draconic and Gaussian Years

We quote two more years: the draconic (or draconitic) and the Gaussian year.

The draconic year is the interval of time between two passages of the Sun through the ascending node of the lunar orbit. It is therefore connected with the occurrence of eclipses. Owing to the retrograde motion of the lunar nodes on the ecliptic, this year is the shortest, its duration being $346^j.6201$.

The Gaussian year derives from Kepler's third law, it is the period of revolution of a massless body in circular orbit around the Sun at the distance of 1 AU, whose value is $365^j.25690$. We shall discuss the Gaussian year again in Chapter 12.

10.4 The Dynamical Time

We discuss now the nonuniformities of sidereal and Universal times caused by the nonuniformities of the diurnal rotation. Let us set aside the acceleration due to the secular variation of the precessional constant, in order to concentrate our attention on the rotation itself. Kepler had already alluded to the possibility of a nonuniform rotation of the Earth, and Kant, in 1754, had the idea of a progressive slowing down due to the tides. The really sound experimental evidence of secular effects came however around 1870, when Newcomb compared the available observations with the very precise tables of lunar positions calculated by Hansen. The agreement in longitude was very good in the second part of the 1700s and in the first part of the 1800s; it gradually deteriorated before 1750 (the data being the instants of occultations measured in Paris) and after 1850. Newcomb had serious doubts however, that the slowing down of the rotation was the only reason for the disagreement. More data coming from different sources, such as the instants of ancient eclipses and better clocks available after 1930, were needed to solve the problem. We can nowadays distinguish three types of irregularities (see Figure 10.1):

- A secular slowing down of the rotation, amounting to a variation of the mean solar day by approximately 2 ms/century, partly but not totally explained by tidal dissipation of the rotational energy. Since this increase accumulates over the ages, the effect on phenomena that took place several millennia ago can amount to several hours. The records of eclipses in the distant past, which are available from the year 4000 B.C., are of considerable help in establishing this variation of the length of the day.

- Seasonal variations due to meteorological causes, of periodic nature, and amplitude of a few milliseconds.

- Irregular fluctuations of geophysical origin, implying a transfer of angular momentum between core and mantle.

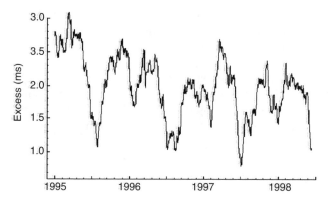

FIGURE 10.1
Fluctuation of the duration of the day, expressed as excess to 86,400 SI, from 1995 to 1998.
(Adapted from the IERS site.)

Table 10.1 provides a summary of the largest long-term variations. The mathematical expressions given there are simply convenient interpolation formulae, with no physical basis. Different authors give different formulae (see later for the meaning of TT and UT1).

Those variations, however small and difficult to measure, must also be reflected in the ephemerides of the planets and of the Moon (further discussed in Chapter 14). A second effect is expected for the Moon: the tides dissipate a fraction of the rotational energy of the Earth, therefore the Earth–Moon distance has to increase to keep the total angular momentum of the Earth–Moon system constant. By virtue of Kepler's third law, the mean motion of the Moon must decrease, and this secular acceleration is of opposite sign to that due to the lengthening of the day. This is the reason for the many debates on lunar accelerations during the 19th and 20th centuries. Today, the Earth–Moon distance can be continuously monitored by means of laser and radar echoes, with a precision (at a level of a few centimeters per year) that allows any variation to be ascertained and measured.

TABLE 10.1

Long-Term Variations of the Duration of the Day. T in Julian Centuries after 1800

From 390 B.C. to 948 A.D.	$\Delta j = 2.4$ msec/century
	$\Delta(TT - UT1) = 1360^s + 320T + 44.3T^2$
From 148 A.D. to 1600 A.D.	$\Delta j = 1.4$ msec/century
	$\Delta(TT - UT1) = 25.5T^2$
From 1600 A.D. to date	$\Delta j = (7^s.286 \pm 0^s.170) \times 10^{-6}(\text{year} - 1819)$
	≈ 0.7 msec/century
	$\Delta(TT - UT1) = 5^s.156 \pm 0^s.404$
	$\quad + (13.3066 \pm 0.3264)(T - 0.19 \pm 0.01)^2$

Following the previous considerations, we can discuss with greater insight the Mean Time, and in particular the Universal Time, arriving in successive steps to four realizations of it, UT0, UT1, UT2, and finally, UTC, which will be discussed in a later section.

We shall call UT0 the apparent Mean Time at Greenwich. UT0 cannot be used in high-precision works because of the polar motion. By measuring the local sidereal time and knowing the longitude and latitude of the observing station, the effects of the polar motion can be removed by a correction of the type:

$$UT1 = UT0 - (u_x \sin \Lambda + u_y \cos \Lambda) \tan \phi' \qquad (10.7)$$

where (u_x, u_y) are the coordinates of the pole in time units, and Λ, ϕ' the geocentric longitude and latitude. Therefore, UT1 is the observatory-independent time, the one that should enter in all previous formulae referring to UT. UT1 is determined by a posteriori analysis of the data coming from many stations, and then distributed by the IERS.

Although much more uniform than UT0, UT1 is still affected by the nonuniformities of the Earth's rotation. By removing the periodic components of these nonuniformities the third level UT2 is derived, which is however not used in astronomy.

Even UT2 is not entirely satisfactory for dynamical purposes because of the remaining irregularities. To realize a truly uniform time, one could resort to the mean Sun originally defined by Newcomb, both as origin and rhythm. According to the definition, the geometric mean longitude of the Sun is:

$$\lambda = 279°41'48''.04 + 129602768''.13T + 1''.089T^2 \qquad (10.8)$$

T being expressed in Julian centuries after 1900, Jan.0, 12^h UT.

The formula 10.8 could be used to indicate a uniform time, whose rhythm is given by the coefficient of T, and which has a small precessional acceleration. This time was called ephemeris time (ET). The second of ET is then defined by the number N of seconds in the tropical year 1900:

$$N = \frac{1296000 \times 35525 \times 86400}{129602768.13} = 31556925.9747$$

In other words, 1 second ET is the fraction $1/N$ of the length of the tropical year 1900. Indeed, at this stage we should redefine the initial epoch as 12^h ET, not UT. Note that ET is a dynamical solar time defined by the revolution of the Earth, and apparently independent from the rotation.

The body of knowledge accumulated since the original definition has proved that two different mean Suns have to be distinguished between: one whose right ascension increases uniformly with UT, and one whose right ascension increases uniformly with ET; only if the rotation of the Earth were strictly uniform would the two suns coincide in a single body.

In other words, in order to free ET, whose origin is purely the annual revolution, by the rotational slowing down of the Earth we must introduce a mobile Greenwich meridian (ephemeris meridian), in very slow motion toward east with respect to the conventional one. The hour angle of γ with respect to the ephemeris meridian is said to be ephemeris sidereal time. The two meridians were assumed to coincide in 1902, at the present epoch they are at about $2''$ from each other.

Furthermore, while UT can be obtained from meridian observations, ET must be derived from the longitude of the Sun; the Sun however moves too slowly along the ecliptic to provide a good clock, the Moon is much better for this purpose. Thus, ET can be better identified with the argument (dynamical time) that appears in the ephemerides of the Moon. However then, the knowledge of ET implies the reduction of a great amount of data, the difference UT − ET is known only a posteriori and is affected by the residual errors in the lunar ephemerides.

Finally, and perhaps more decisively, ET is still a prerelativistic concept; therefore its utilization in the almanacs, introduced in 1960, was discontinued in 1984. Nevertheless, it still retains some usefulness, as we will discuss in Section 10.5.

10.5 The Atomic Time

In the previous sections, the time has been defined, or mathematically derived, by the motion of heavenly bodies. Since 1955, a different physical time of very high regularity has been available, namely the international atomic time (TAI), which was officially adopted as standard time in 1972.

TAI is defined by the radiation coming from two hyperfine levels of the fundamental energy level of cesium, when the atom is far from magnetic fields and at sea level. The frequency of this resonant transition is 9 192 631 770 Hz, with a stability of about 2×10^{-13}. It defines the international second SI, which, always by definition, is equal to the second of ET. The duration of the mean solar day is then 86400 SI. In practice, some 200 stations well-distributed over the Earth keep the atomic time to within one nanosecond per day, and distribute it via radio and navigational systems (e.g., Loran-C, Omega, GPS).

Regarding the origin of TAI, by international agreement its zero point was at epoch 1958 January 1, at 0^h UT2. This decision implied an offset between ET and TAI: ET = TAI + $32^s.184$.

Adopting now TAI as the fundamental timescale, the quantity TAI + $32^s.184$ was said to be terrestrial dynamic time, TDT. Since 1986, TDT is the tabular argument of the ephemerides. TDT maintains the continuity with ET, but its realization no longer depends on observations of the Sun or the Moon, but on laboratory clocks. From 2001 onwards, by IAU decision, TDT has been renamed TT.

TAI (and as a consequence also TT) is certainly a very uniform time; nevertheless, according to general relativity, the frequency of any clock varies with the varying gravitational potential in which it is immersed. Therefore, TAI must be interpreted as a proper time. In the differential equations of mechanics it must be transformed to coordinate time, according to the position and velocity of the observer with respect to the barycenter of the Solar System. This ideal time of the inertial observer is said to be dynamical time of the barycenter (TDB). The TT − TDB difference is expressed by purely periodic terms; if precision of 1 ms is sufficient, the following expression can be used:

$$
\begin{aligned}
\text{TDB} = \text{TT} + 1.658 \times 10^{-3}(\sin E + 0.0368) + 2.03 \times 10^{-6}\cos \phi' \\
(\sin(\text{UT} + \Lambda) - \sin \Lambda) + \cdots
\end{aligned}
\tag{10.9}
$$

where E is the eccentric anomaly of the Sun, and ϕ', Λ are latitude and longitude of the clock. The difference TDB − TT never exceeds 0.0017 sec, so at this level of precision we could adopt a generic dynamical time TD.

In the end, however, what matters for astronomy (and also for navigation) is the true angle of rotation of the Earth, namely UT in its various realizations. Therefore, by international agreement, a time is broadcasted having the rhythm of TAI, but the origin always coincident, within 900 ms, with that of UT1. This hybrid time is called UTC (coordinated universal time). Since UT is not uniform, UTC cannot be a continuous function; according to the need, a leap second is added (or in theory subtracted, but since 1972 this never happened) at the beginning or at the midpoint of each particular year. No necessity for the leap second has occurred since January 1, 1999; until further notice: UTC − TAI = − 32 s (see Bulletin C of the IERS). UTC is therefore extremely practical and inexpensive, and sufficient for many astronomical purposes; however, at least in principle, it is not correct to measure the duration of an event by differencing UTC because of its discontinuities. The broadcasted radio-signals actually contain also the UTC-UT1 difference.

By means of UTC it is also possible to redefine precisely the ephemeris meridian longitude with respect to Greenwich:

$$
\Lambda(\text{ME}) = 1.002738(\text{TT} - \text{UT1})
$$

The adoption of the dynamical time suggests redefining the Julian day (JD) using TD and not UT. This JD is often indicated by JED, or JDE according to others.

The TD − UT difference has only been well-known only for the last decades. The *Astronomical Almanac* contains Table K8 giving its values (actually ET − UT before 1984) since 1620. The numbers are extremely

TABLE 10.2

Values of $\Delta T = ET - UT$ from 1890 to 1983, $\Delta T = TDT - UT$
from 1894 to 2001, $\Delta T = TT - UT$ from 2001

Year	ΔT (^{s})	Year	ΔT (^{s})
1620	+124	1950	+ 29.15
1630	+72	1960	+ 33.15
1700	+9	1970	+ 40.14
1750	+13	1975	+ 45.48
1800	+13.7	1980	+ 50.54
1850	+7.1	1985	+ 54.34
1885	−5.8	1990	+ 56.86
1900	−2.72	1995	+ 60.78
1902	−0.02	2000	+ 64

uncertain before 1890, being based on an adopted value of $-26''/cy^2$ for the tidal term in the mean motion of the Moon. Some values of ΔT are given in Table 10.2.

Notes

- After the great work of Huygens at the end of the seventeenth century, the precision of clocks reached a few tens of seconds per day; it gradually improved through several mechanical innovations, mostly in England (Harrison, Kendall) and France (Bertoud), until the master work of Shortt in 1920 (a combination of electrical and mechanical devices). The Shortt bipendulum could claim a precision of about 10 ms per day, and it provided decisive evidence of the nonuniformity of the Earth's rotation. Then, in 1925 piezoelectric technology became available, and finally atomic technology. The site http://tycho.usno.navy.mil/ contains a wealth of useful information on timescales and on historical and modern clocks.

- For the utilization of the pulsars as high-precision clocks see: Backer, C.D., Hellings, R.W., 1986, Pulsar timing and general relativity, *Annual Review of Astronomy and Astrophysics*, **24**, p. 537.

- For the definition of Universal Time: Aoki, S., Guinot, B., Kaplan, G.H., Kinoshita, H.H., McCarthy, D.D., Seidelman, P.K., 1982, The new definition of universal time, *Astronomy and Astrophysics*, **105**, p. 359.

- For the duration of the tropical year: Laskar, J., 1986, Secular terms of classical planetary theories using the results of general relativity, *Astronomy and Astrophysics*, **157**, pp. 59–70.

- For the utilization of historical eclipses to determine the slowing down of the Earth rotation, see:
  Stephenson, F.R., Morrison, L.V., 1984, Long term changes in the rotation of the earth: 700 B.C. to A.D. 1980, *Philosophical Transactions of the Royal Society of London, Series A*, **313**, p. 47; an updated version is given by Morrison, L.V., Stephenson, F.R., 1998, The sands of time and the Earth's rotation, *Astronomy and Geophysics*, 39, 5.8.
- The variations of Earth's oblateness have been discussed for instance by:
  Dickey, J.O., Marcus, S.L., deViron, O., Fukumori, I., 2002, Recent Earth oblateness variations: unraveling climate and postglacial rebound effects, *Science*, **298**, pp. 1975–1977.
- The web site http://www.iers.org/map/ gives the excess of the duration of the day to 86,400 sec, and angular velocity of the Earth's rotation, since 1623. As already stated in Chapter 6, in addition to the IERS web site (http://www.iers.org/), the reader is advised to check on the U.S. Naval Observatory site (http://www.usno.navy.mil/)

Exercise

Table 10.3 provides the dates of the perigee of the Sun from 1996 to 2006, as taken from Section A1 of the *Astronomical Almanac*.

The second row gives the different durations. Why only in 2004 is the duration reasonably close to the standard value of the anomalistic year?

Solution: It is simply a matter of definition. In Section A1 (*Phenomena*) the times are defined by the instant when the Earth–Sun difference is the minimum, namely when $\dot{R}(t) = 0$. These times do not agree with those corresponding to the passage through the perigee in the mean elliptical orbit, because of perturbations by the planets.

The same reasoning applies to the instances of the apogee.

TABLE 10.3

Dates of the Perigee of the Sun

	1996	1997	1998	1999	2000	2001	2002	2003	2004	2005	2006
January (UT)	4.29	2.00	4.88	3.54	3.13	4.37	2.58	4.21	4.75	2.04	4.62
Difference (ΔJD)		363.7	367.9	363.7	364.6	367.2	363.2	366.6	365.5	363.3	367.6

11

The Terrestrial Atmosphere

This chapter is devoted to the examination of the influence of the Earth's atmosphere on the apparent coordinates of the stars and on the shape of their images; the discussion will be limited essentially to the visual band. The effects of the atmosphere on astronomical photometry and spectro-photometry will be expounded in Chapter 16 and Chapter 17.

Tycho Brahe had already recognized the importance of the atmospheric refraction, and great contributions came later thanks to Cassini and Laplace. We will treat here only the case of sources at infinite distance from the observer; much more complex is the case of nearby sources embedded in the same refracting medium, which often produce quite spectacular phenomena such as the mirage and the *fata Morgana* (see for instance the books of Danjon (1980) and of Minnaert (1993)).

11.1 The Vertical Structure of the Atmosphere

Figure 11.1 provides a schematic representation of the vertical structure of the atmosphere; the visual band is affected essentially by the troposphere, namely by the air in the first 15 km or so of height, where approximately 90% of the total mass of the atmosphere is contained. The temperature profile in the troposphere is actually more complicated than shown in Figure 11.1. The height of the tropopause (a layer of almost constant temperature) from the ground ranges from 8 km at high latitudes to 18 km above the equator; it is also highest in summer and lowest in winter. The average temperature gradient (lapse rate) is approximately $-6°C/km$, but often, above a critical layer situated in the first few kilometers, the temperature gradient is inverted, with beneficial effects on astronomical observations thanks to the intrinsic stability of all layers with temperature inversion (such as the stratosphere and the thermosphere), essentially because convection cannot develop. This is the case, for instance, for the Observatory of the Roque de los Muchachos (Canary Islands, height 2400 m above sea level, a.s.l), where the

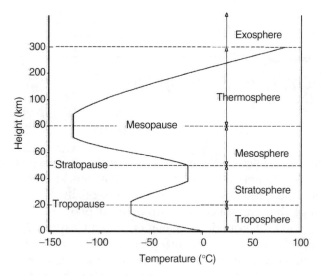

FIGURE 11.1
The indicative vertical structure of the Earth's atmosphere.

inversion layer is usually a few hundred meters below the telescopes at the top of the mountain (see Figure 11.2).

The chemical composition of the troposphere is mostly molecular nitrogen N_2 and molecular oxygen O_2 (approximately 3:4 and 1:4, respectively), with traces of the noble gas argon and of water vapor (the water

FIGURE 11.2
The summit of the Roque de los Muchachos (2400 m a.s.l) seen from the Teide mountain in Tenerife at sunset. The upper 400 m of the Roque stand above the inversion layer, materialized by the thick cloud layer at 2000 m height.

vapor concentration may be as high as 3% at the equator and decreases toward the poles). Above the tropopause, at higher heights in the stratosphere, the temperature rises considerably thanks to the solar UV absorption by the ozone (O_3) molecule with the process: UV photon $+ O_3 = O_2 + O +$ heat. The mesosphere ranges from 50 to 80 km; in this region, concentrations of O_3 and H_2O vapor are negligible, hence the temperature is lower than in the stratosphere. The chemical composition of the air becomes strongly height-dependent, with heavier gases stratified in the lower layers. It is in this region that meteors and spacecraft entering the atmosphere start to warm up. Following the smooth decrease in the mesosphere, the temperature rises again in the thermosphere, because the solar UV and x-rays and the energetic electrons from the magnetosphere can partly ionize the very thin gases of the thermosphere. The weakly ionized region which conducts electricity, and reflects the radio waves with frequencies less than 30 kHz, is called the ionosphere; it is divided into the regions D (60 to 90 km), E (90 to 140 km), and F (140 to 1000 km), according to the respective electron density profiles. Finally, above 1000 km, the gas composition is dominated by atomic hydrogen escaping the Earth's gravity, which is seen by satellites as a bright geocorona in the resonance line $Ly - \alpha$ at $\lambda = 1216$ Å.

11.2 The Refraction

It is well-known that the light propagates in a straight line in any medium of constant refraction index n, with a phase velocity given by $v = c/n = 1/(\varepsilon\mu)^{1/2}$, where ε is the dielectric constant and μ is the magnetic permeability of the medium, both quantities being wavelength dependent. The group velocity is $u = v - \lambda dv/d\lambda$. At the separation surface between two media of different refraction index (e.g., vacuum/air), the ray changes direction, so that the observer immersed in the second medium sees the light coming from an apparent direction different from the true one (see Figure 11.3).

The propagation in a medium of continuously varying index of refraction can be treated by successive recourse to Huygens' principle, but in the following sections a simpler treatment based on geometrical optics will suffice.

Consider again Figure 11.3, and suppose that the atmosphere can be treated as a succession of parallel planes (hypothesis of plane-parallel stratification), by virtue of its small vertical extension with respect to the

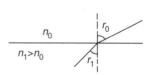

FIGURE 11.3
Vacuum-air refraction.

Earth's radius. According to Snell's laws, when the ray coming from the region of refraction index n_0 encounters the separation surface with a medium of refraction index $n_1 > n_0$, part of the energy will be reflected back to the same hemispace (to the left in the figure) with the same angle r_0 with respect to the normal. This reflected fraction will not be considered here, it only implies a dimming of the source, which will be examined in Chapter 16. The remaining part will be refracted, in the same plane defined by the incident ray and the normal, to an angle $r_1 < r_0$. In reality, a clear atmosphere without clouds does not have sharp separation surfaces. The refraction index gradually increases from 1 to a final value n_f near the ground, with typical scale lengths for significant changes of n much greater than the wavelength of light (as already mentioned, we limit our treatment to the visual band), so that the continuously varying direction can be considered as a series of finite steps in the plane passing through the vertical and the direction to the star (see Figure 11.4):

$$n_0 \sin r_0 = n_1 \sin r_1, \cdots, n_i \sin r_i = n_{i+1} \sin r_{i+1}, \cdots, n_{f-1} \sin r_{f-1} = n_f \sin r_f$$

where $n_{i+1} > n_i$, and $r_{i+1} < r_i$. By equating each term:

$$n_0 \sin r_0 = n_f \sin r_f \quad (n_0 = 1) \tag{11.1}$$

Hence, the important result: in a plane-parallel atmosphere the total angular deviation only depends on the refraction index close to the ground, independent of the exact law with which it varies along the path. The net effect is shown in Figure 11.5.

The star is seen with respect to the vertical in a direction z' smaller than the true direction z, namely closer to the local zenith, by an amount R which is the atmospheric refraction: $z' = z - R$. By virtue of 11.1, and for small

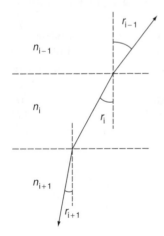

FIGURE 11.4
The successive refraction angles in a plane-parallel stratified atmosphere.

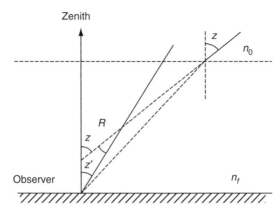

FIGURE 11.5
The total atmospheric refraction.

Rs (in practice, if $z < 45°$):

$$n_f \sin z' = \sin z = \sin(z' + R) = \sin z' \cos R + \cos z' \sin R$$

$$\approx \sin z' + R \cos z' \tag{11.2}$$

or else:

$$R = (n_f - 1) \tan z' \tag{11.3}$$

In the visual band, for average values of temperature and pressure ($T = 273$ K, $P = 760$ mmHg), $n_f \approx 1.00029$, so that in round numbers $R(15°) \approx 16''$, $R(45°) \approx 60''$: a very large effect indeed.

For zenith distances larger than $45°$, the path of the ray inside the atmosphere is so long that the curvature of the Earth cannot be ignored, and the mathematical treatment becomes more intricate, even restricting it to successive refractions in the same plane and with n decreasing outwards with continuity.

With reference to Figure 11.6, let dn be an infinitesimal variation of the refraction index in passing through the surface AA' separating layer M from layer M':

$$n \sin r = n' \sin r' = (n + dn) \sin(r - dr), \quad dr = dR = \frac{dn}{n} \tan r$$

and integrating from $n = 1$ to $n = n_f$:

$$R = \int_1^{n_f} \tan r \, \frac{dn}{n} = \int_0^{\ln n_f} \tan r \, d(\ln n) \tag{11.4}$$

Let us call d, d' the distances from the center of the Earth of point P, P'; then: $d' \sin r'' = d \sin r'$, and therefore:

$$dn \sin r = d'n' \sin r'' = \text{const} \tag{11.5}$$

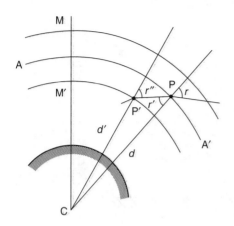

FIGURE 11.6
The spherical shell model of the atmosphere.
C is the center of the Earth.

In other words, the product of the distance of a particular point P from the center of the Earth times the refraction index at that point times the sine of the incidence angle is constant along the path of the ray. The radius of the Earth being $a_\oplus$, by the invariant relation Equation 11.5, we have on the ground:

$$dn \sin r = a_\oplus n_f \sin z', \quad \tan r = \frac{a_\oplus n_f \sin z'}{\sqrt{d^2 n^2 - a_\oplus n_f^2 \sin^2 z'}}$$

$$R = a_\oplus n_f \sin z' \int_1^{n_f} \frac{dn}{n \sqrt{d^2 n^2 - a_\oplus^2 n_f^2 \sin^2 z'}} \tag{11.6}$$

which is an integral of fundamental importance in several studies of ray propagation inside an atmosphere.

Here, we simplify the problem arriving at an expression which is valid down to 75°, and independent of the relationship between the height and density of the air. Let us introduce a new parameter s by letting: $d = a_\oplus + H = a_\oplus(1 + s)$.

The parameter s is in any case a small quantity. Ignoring s^2 and higher powers, the integral in Equation 11.6 can be split into two parts, one independent and the other dependent on s:

$$R = \int_1^{n_f} \frac{n_f \sin z' dn}{n \sqrt{n^2 - n_f^2 \sin^2 z'}} - n_f \sin z' \int_1^{n_f} \frac{s n \, dn}{\sqrt{(n^2 - n_f^2 \sin^2 z')^3}} = R_1 + R_2$$

The primitive of the integrand function in R_1 is $-\arcsin(n_f \sin z'/n)$, so that after several passages the first integral provides:

$$\tan \frac{R_1}{2} = \frac{(n_f - 1)}{(n_f + 1)} tg\left(z' + \frac{R_1}{2}\right) = \frac{2n_f(n_f - 1)}{(n_f + 1)^2 - (n_f^2 - 1)\tan^2 z'} \tan z'$$

For angles z' smaller than approximately $75°$, the term in $(n_f^2 - 1)\tan^2 z'$ can be ignored; furthermore n_f is approximately $= 1$, so that R_1 is:

$$\frac{R_1}{2} \approx \frac{2n_f(n_f - 1)}{(n_f + 1)^2} \tan z' \approx \frac{(n_f - 1)}{2} \tan z'$$

Namely, the same as in the elementary theory.

For the second term R_2, again down to $z' = 75°$ we can approximately put $n = n_f = 1$, so that:

$$R_2 = \frac{\sin z'}{\cos^3 z'} \int_1^{n_f} s dn = (\tan z' + \tan^3 z') \int_1^{n_f} s dn$$

$$= (\tan z' + \tan^3 z') \frac{1}{a_\oplus} \int_1^{n_f} h dn \tag{11.7}$$

The last integral can be solved by parts:

$$\int_1^{n_f} s dn = \frac{1}{a_\oplus} \int_1^{n_f} H dn = \frac{1}{a_\oplus} [Hn]_1^{n_f} - \frac{1}{a_\oplus} \int_{H_1}^0 n dH$$

where H_1 is the value of H for which $n = 1$. Recalling that $n = n_f$ corresponds to $H = 0$, we conclude that:

$$R_2 = (\tan z' + \tan^3 z') \int_0^{H_1} (n - 1) dH$$

As shown in Section 11.5, the term $(n - 1)$ is nearly proportional to the density of the air, and therefore the above integral is simply proportional to the total mass of the air column from the ground to the summit of the atmosphere, quite independent of the exact law between height and density. Therefore, we can conclude that:

$$R_2 = (\tan z' + \tan^3 z') l \rho_0$$

l being the length of an hypothetical air column of constant density having the same temperature and pressure observed on the ground.

Collating the two results, we have finally:

$$R = A \tan z' + B \tan^3 z' = (n_f - 1) \left[\left(1 - \frac{l}{a_\oplus} \right) \tan z' - \frac{l}{a_\oplus} \tan^3 z' \right] \tag{11.8}$$

where:

$$A = (n_f - 1) + B, \quad B = -\frac{l}{a_\oplus}(n_f - 1)$$

Typical values of l at sea level are around 8 km; in addition to the temperature, the actual value depends on the elevation and latitude (namely on the gravity) of the site; typical values for B are therefore around $-0''.07$.

11.3 Effects of Refraction on the Apparent Coordinates

We have seen that the main effect of the refraction is to move the star closer to the zenith in the vertical plane, thus raising its elevation h but leaving essentially unchanged its azimuth A. From Figure 11.7 we get:

$$\begin{cases} -\Delta HA \cos \delta = (\alpha' - \alpha) \cos \delta = R \sin q \\ \Delta \delta = \delta' - \delta = R \cos q \end{cases} \tag{11.9}$$

where the parallactic angle q can be found by one of the two formulae (see Chapter 3):

$$\cos \delta \cos q = \sin \varphi \cos h + \cos \varphi \sin h \cos A$$

$$\sin A \sin h = \cos HA \sin q + \sin HA \cos q \sin \delta$$

In meridian, the refraction is all in declination, and this is true also for the Sun at true noon.

For zenith distance not greater than approximately 45°, after several passages we finally get:

$$\begin{cases} \Delta \alpha = (n_f - 1) \dfrac{\sec^2 \delta \sin HA}{\cos HA + \tan \varphi \tan \delta} \\[4mm] \Delta \delta = (n_f - 1) \dfrac{\tan \varphi - \tan \delta \cos HA}{\cos HA + \tan \varphi \tan \delta} \end{cases} \tag{11.10}$$

By means of formulae 11.10 we can derive the true (or the apparent, according to the sign) topocentric positions. Obviously, no such correction is necessary for a telescope in outer space.

11.4 The Chromatic Refraction of the Atmosphere

The refraction index n depends on the wavelength, diminishing from blue to red, and the same will be true for the refraction angle R: the image on the

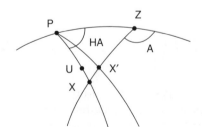

FIGURE 11.7
The effect of the refraction on the equatorial coordinates. $XX' = R = \Delta h$, $PXX' = PXZ = q$, $ZX = z$, $PX = 90 - \delta$, $XU = \Delta \delta$.

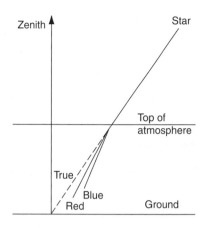

FIGURE 11.8
The atmospheric chromatic refraction.

ground of the star is therefore a succession of monochromatic points aligned along the vertical circle; the blue ray will be below the red one, and thus the blue star appears above the red one (see Figure 11.8). Therefore, the atmosphere behaves similar to a prism producing a short spectrum in the vertical plane, whose length increases with the zenith distance, reaching several arc seconds at low elevations.

The relationships $n(\lambda)$ can be expressed by the so-called Cauchy's formula:

$$n(\lambda) = 0.00028\left(1 + \frac{0.00566}{\lambda^2} + \frac{0.000047}{\lambda^4}\right) \qquad (\lambda \text{ in micrometers}) \quad (11.11)$$

corresponding to a variation of approximately 2% over the visible range. Table 11.1 gives representative values of the refraction at a site such as the Roque (at 2400 m a.s.l) in several photometric bands from the blue to the near infrared.

At sea level and lower elevations the spectrum can become several arcsec long. Therefore, systematic effects due to chromatic dispersion of the atmosphere must be feared on broad-band astrometry, if the color of the object under investigation is very different from that of the reference stars (e.g., when measuring the coordinates of a red asteroid against a net of blue stars, or of a very blue quasar against redder stars). The chromatic refraction

TABLE 11.1

Chromatic Refraction of the Atmosphere at 45° Elevation on a High Mountain

Band	Spectral Range (nm)	Refraction (milliarcsec)
B	391–489	730
V	505–595	290
R	590–810	350
I	780–1020	190

also introduces severe complications in accurate spectrophotometric works if the light of star enters the spectrograph through a slit not aligned in the vertical plane. Indeed, it is always advisable to align the slit of the spectrograph in the vertical plane, as we will also discuss in Chapter 16 and Chapter 17. To minimize these adverse effects, resort can be made to appropriate (and complex) optical devices called atmospheric dispersion correctors.

11.5 Relationships between Refraction Index, Pressure and Temperature

At any given wavelength λ, the refraction index depends on the density ρ according to Gladstone–Dale's law:

$$n - 1 = k\rho \qquad (11.12)$$

With the hypothesis that the atmosphere behaves as a perfect gas of pressure P, temperature T, and molecular weight μ, the density is given by:

$$\rho = \frac{\mu P}{RT} \qquad (11.13)$$

where R is the gas universal constant, so that:

$$n - 1 = k' \frac{P}{T} \qquad (11.14)$$

By referring to normal values of temperature and pressure P_0, T_0 (e.g., 760 mmHg, 0 °C) one has:

$$\frac{n - 1}{n_0 - 1} = \frac{P}{P_0} \frac{T_0}{T}, \quad n - 1 \approx 78.7 \times 10^{-6} \frac{P}{T} \qquad (P \text{ in mmHg, } T \text{ in K})$$

Therefore in the visible and for small refraction angles:

$$R \approx 60''.4 \frac{(P/760)}{(T/273)} \tan z$$

Calling H the height over the ground, from Equation 11.14 we have:

$$dn = k' \frac{1}{T} \left(dP - \frac{P}{T} dT \right),$$

$$\frac{dn}{dH} = k' \left(\frac{1}{T} \frac{dP}{dH} - \frac{P}{T^2} \frac{dT}{dH} \right) = k' \frac{P}{T^2} \left(\frac{T}{P} \frac{dP}{dH} - \frac{dT}{dH} \right)$$

The variation of pressure with the height is equal to the weight of the air in the elementary volume having unitary base and height dH,

$dP = -g\rho\,dq$, so that:

$$\frac{dn}{dq} = k'\mu\frac{P}{T^2}\left(\frac{-g}{R} - \frac{dT}{\mu dq}\right) \tag{11.15}$$

where the constant g/R equals approximately 3.4 K/km, and is called the adiabatic lapse. From Equation 11.15 we reach the conclusion that the variations of the refraction index depend on the vertical gradients of the temperature. A practical consequence is that all effort must be made to control and minimize those gradients over the accessible volume of the telescope enclosure.

11.6 Scintillation and Seeing

The unavoidable thermal turbulence of the atmosphere above the observatory will therefore produce unpredictable (although small) variations of the direction of arrival of the rays from a given star, or, from a different point of view, an erratic distortion of the wavefront. To the unaided human eye, the most impressive phenomena are the scintillation and the color variation of the star (in popular terms, the stars "twinkle"). Indeed, the small aperture of the pupil accepts only a small proportion of the wavefront, so that at a certain instant all rays converge and the star will appear brighter, at other instants all rays diverge and the star will dim; the colors will also change for the same reason of small spatial sampling of the wavefront. If we look at a planet of nonnegligible angular size (e.g., Jupiter) instead of a point-like star, all points of the planetary disk will scintillate with different phases, and the total effect will essentially vanish (planets do not twinkle).

On the focal plane of a telescope, the position and shape of the stellar image will rapidly vary, with characteristics that depend on the size of the aperture; these phenomena are collectively called stellar seeing. While intensity scintillations originate from wavefront curvature, angular displacements affecting image structure and position are traced to wavefront tilt. A complete discussion should take into account, in addition to the size of the telescope, also the wavelength and the temporal scale over which the seeing is sampled. A crucial parameter is the ratio between the typical timescale of the passage of a wavefront disuniformity over the aperture, which ranges from a few thousandths to a few seconds, and the response time of the particular detector (human eye, photographic plate, CCD, etc.). While the human eye has response times of a few tenths of a second and does not integrate the signal, receivers such as CCDs integrate the faint stellar light for seconds, minutes, and even hours.

As a consequence of its fast response time the eye can sometimes discern details invisible on long exposures (typical is the case of visual binary stars) and, in general, rapid detectors such as photon counters are required to remove the detrimental effects of seeing and to exploit the full optical quality of the telescope.

Instead, at the end of a long exposure, the light from the star will fill a circle whose diameter is called the seeing diameter. Inside this circle, the light distribution can be approximated by a Gaussian, or by a sum of two Gaussians, or by an empirically determined function with no simple mathematical expression. Therefore, a convenient parameter to characterize the curve is its full width half maximum (FWHM), namely the width of the curve (usually in arcsec) at half its maximum value. Figure 11.9 provides an example in the visual band obtained from years of measurements at the European Southern Observatory (ESO) of Cerro Paranal, in the Chilean Atacama desert.

The fraction of the stellar energy contained in any circular radius from the center at a given seeing FWHM depends on the actual shape of the curve. Figure 11.10 provides an indicative estimate.

The fairly complex statistical theory of the air turbulence is associated with names such as Kolmogorov and Tatarski, who provided the theoretical foundations of all methods employed to contrast the detrimental effects of the seeing on the image quality. The theoretical complexity is reflected by the very sophisticated technologies implemented on all large modern telescopes. These instruments are known collectively as adaptive optics devices. We quote some of the figures of merit employed by these techniques:

- The coherence length on the plane perpendicular to the propagation vector, namely the distance over the wavefront necessary to

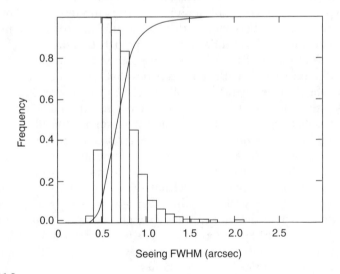

FIGURE 11.9

The seeing distribution at Cerro Paranal (2400 m a.s.l) in the visual band. The ordinate is the frequency of a given seeing bin in the histogram. The curve is the cumulative distribution function. The median FWHM is $0''.67$. (Courtesy of ESO.)

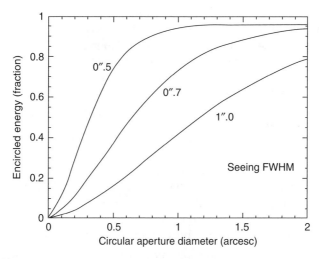

FIGURE 11.10

Indicative value of the fraction of the stellar energy contained inside a circle of a given radius on the focal plane, according to three different values of the seeing FWHM.

change the phase by more than 1 radian. Such distance is known as Fried's parameter r_0, which is also, roughly speaking, the diameter of a telescope whose diffraction figure has an angular extension equal to the seeing:

$$\theta_{\text{seeing}} \approx \lambda/r_0 \qquad (11.16)$$

where the constant is very close to 1. The analogy between diffraction and seeing is not entirely correct, because the extension of the diffraction figure increases in proportion to λ, while r_0 increases with $\lambda^{6/5}$. In the visual band, $r_0 = 10$ cm corresponds to a seeing diameter of $1''$. The best astronomical sites are characterized by seeing FWHMs ranging between $0''.3$ (a very rare figure indeed, as shown by Figure 11.9) and $2''.0$, in the visual band. In the near infrared, the seeing diameters are correspondingly smaller but, in general, they remain always larger than the diffraction figure. Therefore, without adaptive optics devices, it is generally true that ground telescopes are seeing, and not diffraction limited, at least in the usual astronomical observations.

- The correlation time τ_0, which is approximately the dimension of the average air bubble divided by the velocity of the wind. Typical values of τ_0 are a few milliseconds. As r_0, also τ_0 increases with $\lambda^{6/5}$.

- The isoplanatic angle θ_0: if the atmosphere were a single thin layer of air, any ray would suffer the same deflection, independently of the arrival direction; the distortion would therefore be the same all over the field of view (isoplanatism). However, this is not the case, and already with two layers isoplanatism would not occur. In practice, in the visual band the field is isoplanatic over no more than a few seconds of arc. As a consequence, the adaptive optics device can correct the seeing figure of the star on the optical axis of the telescope and its immediate surroundings only. In space, the entire field of the telescope is obviously isoplanatic.

Notes

- Regarding the behavior of the air density with height, Cassini had thought of a constant density, Bouguer had made the assumption of a power-law: $(n/n_f)^k = a_\oplus/d$, where k is an appropriate index, while Mayer assumed a linear decrease $\rho(H) = \rho_0(1 - H/H_1)$.

- The refraction alters the apparent position of the Pole, and also the apparent speed of rotation of the celestial sphere. This effect cannot be ignored in telescopes with a large field of view (FoV), such as the Schmidt, in equatorial mounting. It is advisable to introduce, by some mechanical devices, the possibility of altering the inclination of the polar axis with respect to the horizontal plane according to the elevation of the FoV. Nevertheless, the differential refraction over a FoV of several degrees can be noticeable.

- A spectacular visualization of differential refraction is provided by photographs taken on board the International Space Station, see e.g., in:
  http://science.nasa.gov/ppod/y2003/11apr_mysterymeteors.htm

- Another differential effect is the contraction of apparent diameters, essentially all in the vertical plane. If s is the true diameter of say a planet, from 11.3 we get: $s' = s - dR = s - (n_f - 1)s/\cos^2 z'$. Even at $z \approx 80°$ the effect is small, but it grows rapidly, and at the horizon it becomes about $6'$ for the Sun and the Moon. This shortening of the vertical axis is therefore well measurable (see an image in http://www.sundog.clara.co.uk/atoptics/sunflat.htm). The subjective judgment can however be deceiving: we tend to judge the Moon (or the Sun) much larger and flattened at the horizon than at the zenith. Several explanations have been put forward for this so-called Moon's illusion; see Minnaert, M.G.J., 1993, *Light and Colors in the Outdoors*, Springer-Verlag, New York Inc. Two interesting web sites are: http://griffithobs.org/IPSMoonIllus.html, and http://www.thursdaysclassroom.com/15jun00/ponzo.html

- In principle, the refraction also alters the instant of the eclipses and of the occultations. Owing to refraction, the observer aligned with the Sun and Moon at a certain instant t is not the ground observer in P, but one slightly higher in P'. Chauvenet gave a formula, valid down to $z \approx 70°$, for the distance H of P' from P: $1 + h/a = n \sin (z - R)/\sin z$, where a is the distance of P from the center of the Earth. Typical values of H are around 20 m.

- Several important papers can be quoted on the astronomical refraction:

  - Owens, J.C., 1967, Optical refractive index of air: dependence on pressure, temperature, and composition, *Applied Optics*, **6**, p. 51.

  - Garfinkel, B., 1967, Astronomical refraction in a polytropic atmosphere, *Astronomical Journal*, **50**, p. 235.

  - Saastamoinen, J., 1979, On the calculation of refraction in model atmospheres, *Refractional Influences in Astrometry and Geodesy: Proceedings from IAU Symposium no. 89*, E. Tengstrom, G. Teleki, I. Ohlsson, Ed., D. Reidel, Dordrecht, p. 73.

  - Kristensen, L.K., 1998, Astronomical refraction and airmass, *Astronomische Nachrichten*, **319**, pp. 193–198, also gives interesting historical notes.

  - Auer, L.H., Myles Standish E., 2000, *The Astronomical Journal*, **119**, pp. 2472–2484, give a computational method to solve 11.4 for all zenith angles.

- The scintillation is treated in great detail in the following papers: Dravins, D., Lindegren, L., Mezey, E., Young, A.T., 1997, *Atmospheric Intensity Scintillation of Stars, I. Statistical Distributions and Temporal Properties*, The Publications of the Astronomical Society of the Pacific, Vol. 109, p. 173.
  II. Dependence on Optical Wavelength, ibidem, Vol. 109, p. 725.
  III. Effects for Different Telescope Apertures, ibidem, Vol. 110, p. 610.
  Regarding Active and Adaptive Optics, see: http://www.eso.org/projects/aot/introduction.html, and references quoted there, and: Vernet, E., Ragazzoni, R., Esposito, S., Hubin, N., 2002, *Beyond Conventional Adaptive Optics: A Conference Devoted to the Development of Adaptive Optics for Extremely Large Telescopes*, European Southern Observatory Conference and Workshop Proceedings, Vol. 58, ISBN 3923524617. In general, all web sites of the largest observatories carry a description of their active and adaptive optics systems. Two precious booklets from SPIE Press (2004) are: Andrews, L.C., 2004, *Field Guide to Atmospheric Optics*, SPIE FG02 and Tyson, R.K., Frazier, B.W., 2004, *Field Guide to Adaptive Optics*, SPIE FG03.

- In addition to our atmosphere, all transparent media along the light path can affect the propagation of the light, as distant they might be from the observer. An interesting discussion is given by Moniez, M., 2003, Does transparent hidden matter generate optical scintillation? *Astronomy and Astrophysics*, **412**, p. 105.

- In the last decades, radio astronomy has become progressively more and more important for fundamental astrometric work, as we have discussed in the context of the VLBI and of the ICRF. The ionosphere and the troposphere affect the propagation of the radio waves in a variable way, following the variation of the electron and water vapor content, and as a function of the wavelength. Two other media affect the propagation of the radio waves, namely the solar corona and the ionized interstellar medium. The discussion of these effects is outside the scope of this book.

- Useful web sites on the atmosphere are:
  http://www.atmos.ucla.edu/web/
  http://www.ngdc.noaa.gov/stp/IONO/ionohome.html
  http://atschool.eduweb.co.uk/kingworc/departments/
  geography/nottingham/atmosphere/pages/atmosphere.html
  The American Geophysical Union (AGU, http://www.agu.org/) published in 1995 a special report on the water vapor in the climate system.

12

The Two-Body Problem

In this chapter we consider the dynamics of two bodies under their mutual gravitational attraction, a fundamental problem in many astronomical situations, from the Earth, Moon, and Sun-planet systems, to binary stars and pairs of galaxies. The two bodies will be considered of very small extent with respect to their mutual distance, in order to regard them as point-like masses (particles). The present approach is based on Newtonian dynamics, with the underlying assumptions of Galilean relativity (space is homogeneous and isotropic, time is uniform, all inertial system are equivalent with respect to the dynamical properties), but occasionally we will introduce some modifications owing to general relativity. One of the basic disagreements with Newtonian dynamics indeed originated from astronomical evidence, namely the secular advance of Mercury's perihelion: once allowance was made for all known planetary perturbations, it was realized that a fraction of the value observed, namely $43''$/century out of the total $550''$/century, could not be explained. The correct value was derived from general relativity, as described in Chapter 14.

Some of the results obtained for the gravitational interaction can be applied to the electrostatic force between two charged particles (although the electrostatic force can also be repulsive), because of the common line of action and dependence of the intensity from the inverse square of the distance.

12.1 The Barycentric Treatment

Let P_1 and P_2 be two particles of masses m_1 and m_2, respectively. Given an arbitrary but inertial frame of reference Oxy, let $\mathbf{r}_1$, $\mathbf{r}_2$, $\mathbf{V}_1$, $\mathbf{V}_2$ be the position and velocity vectors of the two particles in it. The gravitational force between them is directed along the joining line, with an intensity depending only on their relative distance:

$$\mathbf{F} = F(r)\frac{\mathbf{r}}{r}, \quad r = \mathbf{r}_2 - \mathbf{r}_1 \tag{12.1}$$

Furthermore, the force can be derived from a potential function $-U(r) = -U(|r_2 - r_1|)$:

$$\mathbf{F} = -\frac{dU}{dr}\frac{\mathbf{r}}{r}, \quad F_x = -\frac{\partial U}{\partial x}, \quad F_y = -\frac{\partial U}{\partial y}, \quad F_z = -\frac{\partial U}{\partial z} \qquad (12.2)$$

The gravitational force is then a particular case of central forces, and the potential belongs to the class of conservative fields. Notice the minus sign in the definition of the potential; the function $U(r)$ is then the potential energy of the system. While $\mathbf{F}$ is a vector, the potential $-U(r)$ is a scalar quantity.

The movements $r_1(t)$, $r_2(t)$ of the two particles can be derived from the fundamental Newtonian equations:

$$m_1 \frac{d^2\mathbf{r}_1}{dt^2} = m_1\ddot{\mathbf{r}}_1 = \mathbf{F}_1 = -\mathbf{F}, \quad m_2\frac{d^2\mathbf{r}_2}{dt^2} = m_2\ddot{\mathbf{r}}_2 = \mathbf{F}_2 = \mathbf{F} \qquad (12.3)$$

It is convenient to translate the origin of the inertial system in the barycenter B, a point always on the line joining the two particles. With respect to B, the distances of the two particles satisfy the relationships:

$$m_1\mathbf{r}_1 + m_2\mathbf{r}_2 = 0, \quad \frac{r_1}{r_2} = \frac{m_2}{m_1}, \quad \mathbf{r}_1 = -\mathbf{r}\frac{m_2}{m_1 + m_2},$$

$$\mathbf{r}_2 = \mathbf{r}\frac{m_1}{m_1 + m_2} \qquad (12.4)$$

Therefore, the barycenter is closer to the heavier of the two masses.

The Lagrange function L of the system, namely the difference between the kinetic energy T and the potential energy U, is:

$$L = T - U = \frac{1}{2}m_1\dot{\mathbf{r}}_1^2 + \frac{1}{2}m_2\dot{\mathbf{r}}_2^2 - U(r)$$

As is well known, the Lagrange function L satisfies the conditions:

$$\frac{d}{dt}\frac{\partial L}{\partial \mathbf{V}} = \frac{\partial L}{\partial \mathbf{r}} = \frac{dU}{dr}, \quad \frac{d}{dt}\frac{\partial L}{\partial \dot{r}_i} - \frac{\partial L}{\partial r_i} = 0 \quad (i = 1, 2, 3)$$

In the barycentric system, from Equation 12.4 we derive:

$$\dot{\mathbf{r}} = \mathbf{V} = \dot{\mathbf{r}}_2 - \dot{\mathbf{r}}_1, \quad T = \frac{1}{2}(m_1 V_1^2 + m_2 V_2^2) = \frac{1}{2}\frac{m_1 m_2}{m_1 + m_2}V^2 = \frac{1}{2}mV^2$$

where:

$$m = \frac{m_1 m_2}{m_1 + m_2} = \frac{m_2}{1 + m_2/m_1}, \quad \frac{1}{m} = \frac{1}{m_1} + \frac{1}{m_2}$$

is the so-called reduced mass of the two particles. If m_1 is the heavier of the two:

$$\frac{1}{2}m_2 \leq m \leq m_2$$

Therefore, in the barycentric system the Lagrange function formally coincides with that of a single particle of mass m in movement in an external field $U(r)$, which is symmetric with respect to the fixed origin of coordinates. If we determine the movement $\mathbf{r}(t)$ of this fictitious particle, then by Equation 12.4 we can obtain separately $\mathbf{r}_1(t)$ and $\mathbf{r}_2(t)$. As is well-known, the equations of movement are three second-order differential equations (or six first-order equations in the Hamiltonian notation). Therefore, six initial arbitrary constants are needed in order to specify fully the movement of the particles, for instance position and velocity at a given time. In astronomical applications, those initial constants are the six orbital elements, which will be introduced in Chapter 13. Furthermore, the masses of the two particles (or of one of them) are usually unknown, at least initially.

In addition to the reduced mass, it is convenient to consider the total mass M of the two particles: $M = m_1 + m_2$. Again, if m_1 is the heavier of the two, then:

$$M = m_1 + m_2 = m_1\left(1 + \frac{m_2}{m_1}\right), \quad m_1 \leq M \leq 2m_1$$

In particular, for the case Sun planet:

$$m_1(= M_\odot) \gg m_2(= m_{\text{planet}}), \quad m \approx m_2(= m_{\text{planet}}), \quad M \approx m_1(= M_\odot) \quad (12.5)$$

(the same result also applies to a planet-comet system, or to the proton–electron pair). The largest deviation from Equation 12.5 is for Jupiter, because $m_{\text{Jupiter}} \approx 0.001 M_\odot$.

Owing to the properties of the force field (which is central and conservative), the isotropy of space imposes the conservation of the angular momentum $\mathbf{K}$ of the fictitious particle, namely of the vector product:

$$\mathbf{K} = \mathbf{r} \times m\mathbf{V} = K\frac{\mathbf{r}}{r} = K\mathbf{k} \quad (12.6)$$

Therefore, the vector $\mathbf{K}$ must be constant, both in direction and modulus. The consequence of the constant direction is immediate: being the angular momentum perpendicular to the plane of $\mathbf{r}$ and $\mathbf{V}$, the movement must be confined at all times in a plane perpendicular to $\mathbf{K}$: the central movement takes place always on such plane (the orbital plane), whatever the mathematical expression of $U(r)$, even for a repulsive force (e.g., proton–proton).

To examine the consequences of the constancy of the modulus of $\mathbf{K}$, let us define on the orbital plane a polar system (r, φ) with origin in the barycenter B and arbitrary fundamental direction. The vectors velocity $\mathbf{V}$ and acceleration $\mathbf{a}$ can be decomposed in two components, one along the radial direction and one along the perpendicular to the radius vector. If $\mathbf{i}, \mathbf{j}$ are the unitary vectors along those directions, then:

$$\mathbf{V} = \dot{r}\mathbf{i} + r\dot{\varphi}\mathbf{j}, \quad \mathbf{a} = (\ddot{r} - r\dot{\varphi}^2)\mathbf{i} + \frac{1}{r}\frac{d}{dt}(r^2\dot{\varphi})\mathbf{j}$$

The modulus of the angular momentum and the kinetic energy are:

$$K = mr^2\dot{\varphi}, \quad T = \frac{1}{2}m(\dot{r}^2 + r^2\dot{\varphi}^2) \tag{12.7}$$

Notice that the variable φ does not appear explicitly in K, T, and L. The variable φ is then said to be a cyclic variable, and the associated generalized momentum $K = mr^2\dot{\varphi}$ constitutes a so-called integral of motion. This result too is entirely independent of the actual mathematical expression of $U(r)$. There is a very useful geometrical interpretation of such constancy of K; let:

$$dA = \frac{1}{2}r \cdot r\,d\varphi \tag{12.8}$$

be the elementary area swept by the radius vector in the time dt; the areal velocity dA/dt is then constant, a result know as Kepler's second law in the gravitational case, but which is valid in any central and conservative force field. In the astronomical notation, it is customary to introduce the area constant C, (several authors would prefer $c = C/2$) defined by:

$$\frac{C}{2} = \frac{dA}{dt} = \frac{1}{2}r \cdot r\dot{\varphi} = \frac{K}{2m}, \quad K = Cm \tag{12.9}$$

Then:

$$\dot{r} = \frac{dr}{d\varphi}\dot{\varphi} = \frac{C}{r^2}\frac{dr}{d\varphi} = -C\frac{d(1/r)}{d\varphi}, \quad \ddot{r} = -\frac{C^2}{r^2}\frac{d^2(1/r)}{d\varphi^2}$$

The transverse acceleration is always nil, as it should in a central field, while the radial one is given by Binet's formula:

$$a_r = -\frac{C^2}{r^2}\left[\frac{1}{r} + \frac{d^2(1/r)}{d\varphi^2}\right] \tag{12.10}$$

The kinetic energy is given by:

$$T = \frac{1}{2}m\dot{r}^2 + \frac{K^2}{2mr^2} \tag{12.11}$$

The total energy (another integral of motion, which has its root deep in the uniformity of time) is:

$$E = T + U = \frac{1}{2}m[\dot{r}^2 + r^2\dot{\varphi}^2] + U(r) \tag{12.12}$$

Or else:

$$dt = \pm\sqrt{\frac{2(E-U)}{m} - \frac{K^2}{m^2r^2}}\,dt, \quad d\varphi = \frac{K}{mr^2}dt \tag{12.13}$$

Relation 12.13 contains two differential equations in the variables (r, φ, t): by eliminating t the equation of the trajectory $r(\varphi)$ in its plane is obtained, by eliminating φ the equation of the movement $r(t)$, and finally, by eliminating r the equation of the anomaly $\varphi(t)$. From the constancy of the sign of $d\varphi/dt$, we conclude that the function $\varphi(t)$ is a monotonic one: the movement cannot reverse direction, whatever the function $U(r)$. Furthermore, the radial movement takes place as if the effective potential was the sum of that due to the external force plus a centrifuge potential:

$$U_{\text{eff}} = U(r) + \frac{K^2}{2mr^2}$$

If $K \neq 0$, the particle cannot reach the center (the two particles cannot collide), unless $U(r)$ approaches zero faster than $1/r^2$ or as U_0/r^2 provided $U_0 > K/2m$.

For those values of r for which $U_{\text{eff}}(r) = E$, it is also $dr/dt = 0$: namely in the points of minimum or maximum $r(t)$, the distance inverts its behavior. However, even in those points $d\varphi/dt$ must be different to zero, and maintain the same sign.

Notice that, in general, there is no insurance that r varies between a minimum and a maximum value (r_{min} and r_{max}): there is no a priori reason why the orbit should be confined to a ring in its plane. There are cases, indeed, when r_{max} can become arbitrarily large: the two particles approach a minimum distance, and then separate again.

Furthermore, even a confined orbit will not necessarily be a closed one. The general case is that of a succession of arches going from r_{min} to r_{max} and vice versa, for instance in a "rosette" configuration as shown in Figure 12.1.

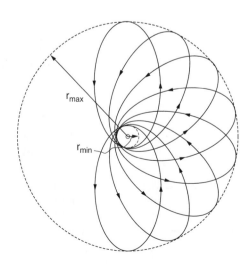

FIGURE 12.1
A so-called "rosette" movement.

After an infinitely long time, the particle will pass through every point in the ring. The figure also shows that the line joining r_{min} to r_{max} (the so-called line of apses) rotates in the plane in the same direction of the movement.

The orbit is necessarily a closed one in two important cases, namely when:

$$U(r) = U_0 r^2, \quad U(r) = -\frac{U_0}{r}$$

The first case is that of the so-called spatial oscillator, and will not be considered further. The second case covers the gravitational forces; the minus sign has been introduced in order to have $U_0 > 0$ (attractive force).

12.2 The Gravitational Attraction

Let us now restrict the discussion to the gravitational attraction:

$$U(r) = -\frac{U_0}{r}, \quad F(r) = U_0/r^2, \quad U_0 = Gm_1 m_2$$

where G is Newton's gravitational constant. From Equation 12.13 we obtain:

$$\varphi = \varphi_0 + \arccos \frac{K/r - mU_0/K}{\sqrt{2mE + m^2 U_0^2/K^2}} \tag{12.14}$$

By introducing the two positive quantities p and e defined by:

$$p = \frac{K^2}{mU_0} \text{ (parameter)}, \quad e = \sqrt{1 + \frac{2EK^2}{mU_0^2}} \text{ (eccentricity)}$$

we finally obtain the trajectory $r(\varphi)$ as:

$$\frac{1}{r} = \frac{1}{p}[1 + e\cos(\varphi - \varphi_0)] \tag{12.15}$$

The generic orbit is therefore a conic curve having its focus in the center of the coordinates. The sign of total energy E determines the value of the eccentricity e:

if $E < 0$ then $0 \le e < 1$ (the orbit is an ellipse)

if $E = 0$ then $e = 1$ (a limiting case, the orbit is a parabola)

if $E > 0$ then $e > 1$ (the orbit is an hyperbole)

It is convenient to identify the origin of the angles φ with the point of minimum distance (pericenter, or perihelion, or periastron) which is located on the line passing also for the point of maximum distance (apocenter, aphelion, aphastron). The angle φ is then called the true anomaly, and is generally indicated with the letter v; the true anomaly increases in the same direction as the movement of the particle in its orbit. As already said, the line joining pericentre and apocentre is called the *line of apses*.

The case $E < 0$ ($0 \le e < 1$) applies to the case of two bodies gravitationally bound together. In the case of the Solar System, remembering condition 12.5 on the relative masses, we see that the results obtained with the general treatment are the rigorous expression of Kepler's first and second laws: the planets move along ellipses having the Sun in one of the two foci, with constant areal velocity. Even substituting the Sun with the barycenter of the Solar System, these two laws are an approximation to the reality. There are, indeed, several planets (their perturbations to the two-body solution will be briefly examined in Chapter 14), and finally the Galilean relativity must be replaced by the general relativity. Nevertheless, they constitute an accurate description of the observations, and an extremely useful departure point for further theoretical developments.

Let us examine again the eccentricity e as function of the energy E: the ellipse becomes a circle ($e = 0$) if:

$$\frac{2EK^2}{mU_0^2} = -1, \quad E = -\frac{mU_0^2}{2K^2}$$

which is the minimum possible value for E. For larger values of E, the semimajor and semiminor axes of the ellipse are given respectively by:

$$a = \frac{p}{1 - e^2} = \frac{U_0}{2|E|}, \quad b = \frac{p}{\sqrt{1 - e^2}} = \frac{K}{\sqrt{2m|E|}}$$

Therefore, the value of the semimajor axis depends on the total energy but not on the angular momentum of the particle, as instead the semiminor does.

The apocentre is given by $r_{max} = p/(1 - e) = a(1 + e)$, the pericentre by $r_{min} = p/(1 + e) = a(1 - e)$.

Let us derive the period of revolution P on the ellipse, whose area is $A = \pi ab$. From Kepler's second law, $2mA = PK$, and therefore:

$$P = 2\pi a^{3/2} \sqrt{\frac{m}{U_0}} = \pi U_0 \sqrt{\frac{m}{2|E|^3}} = 2\pi \frac{a^{3/2}}{\sqrt{GM}} \tag{12.16}$$

In conclusion, the period depends only on the total energy, but not from the eccentricity. Its square is proportional to the third power of the semimajor axis and inversely proportional to the sum of the masses of the two bodies. Therefore the usual statement of Kepler's third law, that the square of the periods are proportional to the cubes of the semimajor axes, is incorrect, in the sense that the constant of proportionality actually depends on the mass of the particular planet (as already said the largest deviation is for Jupiter, whose mass is 0.001 the mass of the Sun). This remark must be taken into full account when discussing binary stars or pairs of galaxies. Note also that the measure of P and a produces the product $G \cdot M$, not separately G and M (for instance, the product $GM_\oplus$ is equal to 398601.8 km^3/s). In other words, the high precision of the astronomical data does not transfer on the precise determination of G. Indeed, the value of G is the least precise of all fundamental constants, the fifth decimal digit already being uncertain.

We do not derive explicitly the equations for $E \geq 0$; if the two bodies are not gravitationally bound, the fictitious body approaches the center of force coming from and returning to infinity, with a velocity at infinity which is greater than zero (or equal to zero for the parabolic orbit). The minimum distance from the focus is $r_{min} = p/(1+e)$, which becomes $= p/2$ for the parabola.

12.3 The Relative Movement

According to the general results obtained in the previous sections, the movements of the two particles in the barycentric system are obtained by solving the differential equation:

$$m\frac{d^2\mathbf{r}}{dt^2} = m\ddot{\mathbf{r}} = \mathbf{F} = Gm\frac{\mathbf{r}}{r^3}$$

and then separately:

$$\ddot{\mathbf{r}}_1 = -Gm_2\frac{\mathbf{r}}{r^3} = -Gm_2\frac{m_1^3}{(m_1+m_2)^3}\frac{\mathbf{r}_1}{r_1^3}, \quad \ddot{\mathbf{r}}_2 = Gm_1\frac{\mathbf{r}}{r^3} = -Gm_1\frac{m_2^3}{(m_1+m_2)^3}\frac{\mathbf{r}_2}{r_2^3}$$

It is like having in B a particle of fictitious mass m_f:

$$m_f = \frac{m_1^3}{(m_1+m_2)^3}$$

attracting a particle of mass m_2 at distance r_2, and vice versa when the role of the two masses is inverted. However, in astronomical applications the barycentric system is not always convenient, because we have no a priori knowledge of the two masses (although in the Solar System the barycentre closely coincides with the center of the Sun), nor can we determine its position by observations. Let us consider the specific example of a visual binary star. From the observational point of view, it is more immediate to refer the positions not to the invisible barycenter B, but to a nonrotating reference system centered in the heavier object P_1 (in the case of binary stars, the heavier star is usually also the brighter), and describe the relative motion of P_2 with respect to P_1. To be sure, P_1 itself will be accelerated toward P_2, and therefore this reference system will simply be parallel to the inertial barycentric one. Thus, this translation of origin introduces a fictitious radial acceleration, but not a Coriolis acceleration. Therefore, the movement of P_2 relative to P_1 is described by the equation:

$$\ddot{\mathbf{r}} = -G(m_1+m_2)\frac{\mathbf{r}}{r^3} = -GM\frac{\mathbf{r}}{r^3} \tag{12.17}$$

Equation 12.17 is the same as in the barycentric case, provided the reduced mass m is substituted by the sum of the two masses M. Geometrically, the difference between the relative and barycentric description is shown in Figure 12.2.

Therefore, we shall proceed with the relative description and with the astronomical notation, using the previous constant C (area), and the new constant h (energy in astronomical terms, in Latin the *vis viva* constant) given by:

$$h = \frac{1}{2}V^2 - \frac{GM}{r} \tag{12.18}$$

Notice that $2h$ is the square of the velocity at infinity.

The equation of the relative orbit can be derived by a second method. If (x, y) is a Cartesian system and (r, φ) a polar one, in the plane of the orbit and centered in P_1, the Newtonian differential equations are:

$$\ddot{x} = -GM\frac{x}{r^3}, \quad \ddot{y} = -GM\frac{y}{r^3}, \quad \ddot{z} = -GM\frac{z}{r^3}, \quad x^2 + y^2 = r^2$$

Recalling Binet's expression 12.10 for the radial acceleration, after simple passages the following equation can be derived:

$$-\frac{GM}{r^2} = -\frac{C}{r^2}\left[\frac{1}{r} + \frac{d^2(1/r)}{d\varphi^2}\right] \tag{12.19}$$

which is a linear nonhomogeneous differential equation with constant coefficients, in the unknown $1/r$ with respect to the variable φ. The solution is the conic curve of focus P_1 given by:

$$\frac{1}{r} = \frac{GM}{C^2}\left[1 + \sqrt{1 + 2h\left(\frac{C}{GM}\right)^2}\cos(\varphi - \varphi_0)\right] \tag{12.20}$$

where the constant term is a particular integral of the equation. As before, we indicate the constant with p (parameter of the conic), and eliminate the

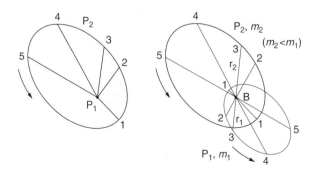

FIGURE 12.2
The relative and barycentric movements of two bodies, respectively, on the left and on the right panel (not to scale). The numbers refer to the same instants of time.

arbitrary constant φ_0 by counting the angles from the pericenter, thus using the true anomaly ν.

A second differential equation is obtained by again considering the radial acceleration and taking into account the area integral in Equation 12.8:

$$-GM\frac{\dot{r}}{r^2} = \ddot{r}\ddot{r} + r\dot{r}\dot{\varphi}^2 + r^2\dot{\varphi}\ddot{\varphi}$$

whose integral is given by the *vis viva* integral:

$$V^2 = \dot{r}^2 + r^2\dot{\varphi}^2 = 2\frac{GM}{r} + 2h$$

Summarizing these results:

$$p = \frac{C^2}{GM}, \quad e = \sqrt{1 + \frac{2hp}{GM}}, \quad r = \frac{p}{1 + e\cos\nu} = \frac{a(1 - e^2)}{1 + e\cos\nu} \quad (12.21)$$

$$V^2 = GM\left[\frac{2}{r} - \frac{1 - e^2}{p}\right] = GM\left[\frac{2}{r} - \frac{2h}{GM}\right], \quad 2h = \frac{GM}{p}(e^2 - 1) \quad (12.22)$$

Consider the instant of time when the distance of P_2 from P_1 is r and its velocity is V, and compare the measured value of V with the limit (parabolic) value V_P given by:

$$V_P = \sqrt{\frac{2GM}{r}} \quad (12.23)$$

V_P is the velocity in r by a body which moves toward P_1 starting from infinity with zero velocity, and therefore describing a parabolic orbit (because $e(h = 0) = 1$). The orbit of P_2 will be an ellipse, a parabola, or a hyperbola according to the ratio V/V_P being less than 1, equal to 1, or greater than 1.

Let us discuss again in more detail the elliptical case (the energy constant h is negative), identifying P_1 with the Sun. From the area integral, and from Equation 12.21 and Equation 12.22 we get:

$$\pi ab = \pi a^2\sqrt{1 - e^2} = \frac{1}{2}\int_0^P r^2\dot{\nu}\,dt = \frac{1}{2}CP,$$

$$C = r^2\dot{\nu} = \frac{2\pi}{P}\sqrt{1 - e^2}a^2 = n\sqrt{1 - e^2}a^2,$$

$$C^2 = pGM = aGM(1 - e^2), \quad a = \frac{GM}{\dfrac{2GM}{r} - V^2}, \quad 2h = -\frac{GM}{a}$$

Finally:

$$P = 2\pi GM\left(\frac{2GM}{r} - V^2\right)^{-3/2}, \quad 4\pi^2\frac{a^3}{P^2} = n^2a^3 \quad (12.24)$$

$$= GM = G(m_1 + m_2)$$

which again is Kepler's third law.

The quantity $n = 2\pi/P$, which has the dimension of an angular velocity, is called the mean motion; from n, another useful quantity is derived, namely the mean anomaly $M = n(t - t_0)$, which increases linearly with time. In several instances, it will be useful to substitute the time derivative, d/dt, with the derivative with respect to the mean motion, nd/dM.

Notice that the semimajor axis a depends on r and V, while the period P depends on a but not on e. Therefore, if two bodies P_2, P_3 pass through the same position r with identical V but different eccentricity, they will have the same a. After a time equal to the period P, they will be found again in the same place; therefore, if a planet disaggregates in several fragments, after the time P all fragments will be found together again in the fragmentation point, quite independent of their individual orbits. A similar focusing effect can be found in an association of stars orbiting around the Milky Way center, after a full galactic rotation (see Notes).

The square of the elliptical velocity is:

$$V^2 = GM\left(\frac{2}{r} - \frac{1}{a}\right) = n^2 a^2 \frac{1 + 2e\cos\nu + e^2}{1 - e^2}$$

from which we obtain:

$$V = na\sqrt{\frac{2a - r}{r}}, \quad V_{\text{aph}}^2 = \frac{GM}{a}\frac{1 - e}{1 + e}, \quad V_{\text{per}}^2 = \frac{GM}{a}\frac{1 + e}{1 - e},$$

$$V_{\text{aph}}/V_{\text{per}} = \frac{1 - e}{1 + e}$$

where V_{aph} and V_{per} are the velocities at the aphelion and perihelion, respectively. The velocity is equal to na on the semiminor axis.

We have already discussed (see Chapter 7) the convenience of decomposing the vector **V** along two nonorthogonal directions, namely perpendicular to the radial direction and perpendicular to the major axis. Those two components will maintain a constant modulus, respectively, $na/\sqrt{1 - e^2}$, $nae/\sqrt{1 - e^2}$, and the second also a constant direction.

For the angular velocity:

$$\dot\nu = na\frac{1 + e\cos\nu}{\sqrt{r(1 - e^2)}} = n\frac{a^2}{r^2}\sqrt{1 - e^2}, \quad \dot\nu_{\text{aph}}/\dot\nu_{\text{per}} = \left(\frac{1 - e}{1 + e}\right)^2$$

For the radial velocity:

$$\dot r = \frac{nae\sin\nu}{\sqrt{(1 - e^2)}}$$

To close this section, it will be useful to remember that in order to calculate the kinetic and total energy from the knowledge of the relative velocity V of P_2 with respect to P_1, we have to return to the inertial system. Namely, we must concentrate in P_2 the reduced mass of the system:

$$T = \frac{1}{2}mV^2, \quad E = T + U = -Gm_1 m_2 \frac{1 - e^2}{2p}$$

12.4 Planetary Masses from Kepler's Third Law

In the previous sections we have derived the three laws that Kepler empirically found analyzing the careful observations of Mars performed by Tycho Brahe. The first two laws were enunciated in 1607, the third in 1617. We have pointed out that the usual statement of the third law is slightly incorrect, because of the neglect of the mass of the planet with respect to that of the Sun, although for most applications the correction is small. Consider then two systems of two particles, e.g., the pairs Sun–Jupiter and Jupiter–Io. The third law gives:

$$n_J^2 a_J^3 = G(M_\odot + M_J) = GM_\odot(1 + M_J/M_\odot),$$

$$n_{Io}^2 a_{Io}^3 = G(M_J + M_{Io}) = GM_J(1 + M_{Io}/M_J)$$

With good approximation:

$$n_{Io}^2 a_{Io}^3 / n_J^2 a_J^3 = M_J(1 + M_{Io}/M_J)/M_\odot(1 + M_J/M_\odot) \approx M_J/M_\odot \quad (12.25)$$

Equation 12.25 provides a direct method to determine the mass of the planet in terms of the mass of the Sun, applicable however only if the planet has a small moon. Therefore, the masses of Mercury and Venus (which have no natural moons) could not be determined with this method, and they did not have the same precision of the masses of the other planets until very recently, when artificial spacecraft were put in orbit around them. Today, all planets (with the exception of the double system Pluto/Charon), have masses determined with great precision by man-made satellites.

Another case where this method cannot be applied with the same confidence, is that of the Earth–Moon system, because the mass of the Moon is not negligible with respect to that of the Earth. However, even in this case the correction is small, given the sensible equidistance of the Earth and of the Moon from the Sun. Therefore, the relative orbit continues to be approximately an ellipse, although the deviations are well measurable, as will be discussed in Chapter 14.

In the astronomical practice, all masses are expressed as fractions of the solar mass taken as unitary mass, the unitary distance is the AU and the unit of time is the mean solar day. The sum of the masses is written as:

$$M = M_\odot + m_2 = M_\odot\left(1 + \frac{m_2}{M_\odot}\right) \equiv 1 + \frac{1}{\mu}$$

Several almanacs give the quantity μ (obviously a very large number), instead of the mass of the planet or of another body of the solar system.

Using the astronomical system of units, a planet of negligible mass (and therefore unperturbed by external masses), in circular orbit around the Sun at the unitary distance, would have a revolution period P exactly equal to the sidereal year, $P_\oplus = 365.2563835...$; the constant:

$$k = \frac{P_\oplus}{\sqrt{2\pi}} = 0.0172021... \tag{12.26}$$

is named Gauss' constant. It formally coincides with the inverse of G, expressed though in astronomical, not in physical, units; in other words, the very high precision with which we know k does not transfer to the precision of G, because there is no direct connection of astronomical and laboratory units. The constant k also coincides with the mean diurnal motion:

$$k = 5348''.188...j^{-1}, \quad 1/k = 58^j.1324....$$

It is also the basis of the Gaussian year introduced in Chapter 10. Gauss' constant is so important for the system of astronomical units as to be called a defining constant. To appreciate the change in emphasis about the constants, until the IAU resolution of 1976 the velocity of light in vacuum was considered only a primary and not a defining constant.

12.5 Escape Velocity

We define now two useful limiting velocities, namely the escape velocity and the velocity in the circular orbit. Consider the total energy E of a particle P_2 of very small mass m_2 at the surface of a nonrotating spherical body P_1 of radius R and mass m_1:

$$E = \frac{1}{2}m_2V^2 - G\frac{m_1m_2}{r}$$

The limiting velocity V_e given by the expression:

$$V_e = \sqrt{\frac{2Gm_1}{R}} \tag{12.27}$$

is the escape velocity from body P_1. If, by some means we impart to P_2 a velocity V greater than V_e in whatever direction, P_2 will reach infinity with final velocity greater than zero.

Another useful limiting velocity is that on the circular orbit at distance $r > R$ from the center of P_1; from the equilibrium between centrifugal and gravitational forces we have:

$$\frac{m_2 V_c^2}{r^2} = G\frac{m_1 m_2}{r^2}, \quad V_c(r) = \sqrt{\frac{Gm_1}{r}}, \quad V_c(R) = \frac{1}{\sqrt{2}}V_e$$

Table 12.1 provides escape and circular velocities for the nine planets, neglecting their diurnal rotation (see Chapter 13 for data on inclinations, eccentricities, and sidereal periods). The third column gives the surface gravity in comparison with that at the Earth's surface (9.78 m/s^2). The first two velocities (fourth and fifth columns) pertain to the equator of each body; the other two velocities (sixth and seventh columns) to the circular orbit at the average distance of the body from the Sun.

These considerations on escape velocities from the planetary surfaces are useful not only for dynamical questions, but also for the understanding of their atmospheres. Let T (in Kelvin) be the temperature of such an atmosphere, supposed in thermal equilibrium; the distribution function of molecules of mass m_2 among the velocities is given by Maxwell's law:

$$dN(V) = \frac{4N}{\sqrt{\pi}}\left(\frac{m_2}{2kT}\right)^{3/2} V^2 e^{-m_2 V^2/2kT} dV \tag{12.28}$$

so that the mean square velocity of those molecules will be:

$$\frac{1}{2}m_2 V^2 = \frac{3}{2}kT, \quad V = \sqrt{\frac{3}{m_2}kT} \tag{12.29}$$

where $k = 1.38 \times 10^{-16}$ erg/K is the Boltzmann constant. For instance, the mass of the hydrogen atom H is $m_2 \approx 1.6 \times 10^{-24}$ g, so that:

$$V_H(T) \approx 1.6 \times 10^{-1}\sqrt{T} \text{ km/sec}$$

At the surface of the Earth, assuming $T \approx 290$ K, the thermal velocity of H is $V_H \approx 2.7$ km/sec $<< V_e$, and 1.4 times less for the molecule H$_2$, which is the most common form of hydrogen in our atmosphere. All other molecules being heavier than H$_2$, we conclude that the Earth is well capable of retaining a massive quasistationary atmosphere. However, Maxwell's distribution has a very long tail at high velocity, so that a fraction of the Earth's gases and, in particular, of atomic H, will continuously escape to the outer space. The observational evidence of such loss is the so-called geocorona, which is visible in the Ly-α spectral line at $\lambda = 1216$ Å.

Mercury and the Moon do not have such capability of retaining an atmosphere, because of much higher temperatures and smaller surface gravity. However, in both cases a tenuous atmosphere has been detected. Necessarily then these rarefied gases must be continuously lost by thermal escape, and replenished (in the absence of volcanic activity) by phenomena capable of extracting gases from the rocks, such as UV solar

TABLE 12.1

Typical Escape and Circular Velocities in the Solar System

Body	Distance (AU)	Mass (g)	Radius (km)	$g/g_\oplus$	V_e (km/s)	V_c (km/s)	$V_e(\odot)$ (km/s)	$V_c(\odot)$ (km/s)
Sun		1.99×10^{33}	6.96×10^5	27.9	618	437		
Mercury	0.387	3.3×10^{26}	2439	0.3	4.3	2.5	96	68
Venus	0.723	4.9×10^{27}	6051	0.9	10.4	7.3	49	35
Earth	1.000	6.0×10^{27}	6378	1.0	11.2	7.9	42	30
Moon	1.000	7.3×10^{25}	1738	0.2	2.4	1.6	42	30
Mars	1.524	6.4×10^{26}	3393	0.4	5.0	3.6	34	24
Jupiter	5.203	1.9×10^{30}	71492	2.3	59.6	42.5	18	13
Saturn	9.539	5.7×10^{29}	60268	0.9	35.5	25.0	14	10
Uranus	19.191	8.7×10^{28}	25559	0.8	21.1	15.5	10	7
Neptune	30.061	1.0×10^{29}	24764	1.1	23.6	16.0	7	5
Pluto	39.529	$\sim 1.3 \times 10^{25}$	~ 1150	~ 0.04	~ 1.1	~ 0.8	7	5

photons, solar particles impinging on the soil, or by meteoroid bombardment.

In the case of the Sun, the surface gravity is about 28 times that at the surface of the Earth, and the photospheric temperature is approximately 5800 K; in higher layers, namely in the cromosphere and in the corona, the temperature of the gases rises to tens, hundreds, and even millions of degrees, so that the thermal escape becomes conspicuous. However, observations prove that the loss of particles from the Sun (the so-called solar wind) is orders of magnitude larger than that accounted for by thermal losses: other more efficient mechanisms, whether magnetic or electric, must act to accelerate the ionized (electrically charged) particles escaping from the Sun.

12.6 Some Considerations on Artificial Satellites

Let us launch from the surface of a spherical nonrotating Earth of radius $a_\oplus$ a satellite of mass m_2 with initial velocity $V > V_e$. Its energy will be:

$$E = \frac{1}{2}m_2 V^2 - G\frac{m_2 M_\oplus}{a_\oplus} \qquad (m_2 \ll M_\oplus)$$

At an altitude H, the distance from the centre becomes $r = a_\oplus + H$, and the energy:

$$E = \frac{1}{2}m_2 V_r^2 - G\frac{m_2 M_\oplus}{r}$$

or else, equating the two values for the conservation of the energy:

$$E = \frac{1}{2}m_2 V^2 - G\frac{m_2 M_\oplus}{a_\oplus} = \frac{1}{2}m_2 V_r^2 - G\frac{m_2 M_\oplus}{r},$$

$$V_r^2 = V^2 - 2GM_\oplus\left(\frac{1}{a_\oplus} - \frac{1}{r}\right)$$

At infinity:

$$V_\infty^2 = V^2 - V_e^2 = (V - V_e)(V + V_e) \approx 2V_e\,\Delta V$$

In conclusion, if we launch with $\Delta V = +1$ km/sec, the satellite will reach infinity with a velocity of approximately 4.7 km/sec (ignoring the very small losses of energy due to the atmospheric drag). There are several practical consequences of this gain at infinity, for instance, one has to be careful not to reach the final destination with too high a velocity. We underline the convenience of using in space applications the parameter ΔV instead of the energy.

The circular velocity at the surface of the Earth is around 8 km/sec, which will also be the velocity of low altitude satellites (e.g., the International Space Station at 300 km). Their period is then of approximately 90 min. Suppose we place such a satellite in a polar orbit; it will go out of phase with the Sun by about 30 min at each orbit, and for several orbits it will see an almost constant illumination (day or night) of its nadir. Therefore, the low polar orbit is used for surveillance. At $H = 36,000$ km the orbital period becomes 24 h, so that a satellite placed on the equatorial plane at this altitude in a circular orbit (e.g., the Meteosat) will be practically stationary with respect to the ground observer. Actually, several satellites have simply a geosynchronous orbit (that was the case of the International Ultraviolet Explorer), slightly different from the rigorously defined geostationary one. At any rate, the two-body condition is a mathematical abstraction, several perturbing forces (such as the Earth–Moon and solar tides, the nonsphericity of the Earth's potential, the radiation pressure, etc.) will act to perturb the orbit, and appropriate corrections must be performed to keep the wanted position of the satellite, for instance by occasional firings of small thrusters.

Notes

- The present treatment is entirely traditional. For instance, no mention is made of the chaotic dynamics; an introduction to it can be found in: Baker, G.L., Gollub, J.P., 1996, *Chaotic Dynamics. An Introduction*, Cambridge University Press, Cambridge.

- The focusing properties discussed in the text were used by Encke, after the discovery of the first few asteroids in a small region of the solar system, to put forward the theory of the fragmentation of a progenitor planet for their origin.
- An interesting example of nonclosed orbit is provided by the Sun under the attraction of the other stars of the Milky Way. According to Oort, the overall force can be expressed by the sum of a point-like mass plus an extended disk of ellipsoidal shape:

$$F = \frac{q}{r^2} + sr \quad (s \ll q)$$

Actually, the movement of the Sun is not quite in a plane, because the force is not really a central one. In 250 million years the Sun spans a three-dimensional torus of inner radius 8 kpc, outer radius 9 kpc, thickness 0.3 kpc, with a velocity perpendicular to the galactic plane of approximately 7 km/sec (see Chapter 9).

Exercises

Let us add a small quantity δU to the potential energy $U(r) = -U_0/r$. The trajectory will not be closed, so that at each revolution a small quantity $\delta \nu$ will be added. Determine $\delta \nu$ in the two cases:

1) $\delta U = \beta/r^2$
2) $\delta U = \gamma/r^3$

(Adapted from Landau, L., Lifschitz, E., 1982, *Mechanics*, 4th ed., MIR, Moscow.)

When the mobile moves from r_{min} to r_{max} and again to r_{min}, the radius vector turns by an angle $\Delta\varphi$ given by:

$$\Delta\varphi = 2 \int_{r_{min}}^{r_{max}} \frac{\frac{K}{r^2}}{\sqrt{2m(E-U) - \frac{K^2}{r^2}}} dr$$

$$= -2\frac{\partial}{\partial K} \int_{r_{min}}^{r_{max}} \sqrt{2m(E-U) - \frac{K^2}{r^2}} \, dr \qquad (12.30)$$

Putting $U = U_0/r + \delta U$ in Equation 12.30, and developing according to the powers of δU, the term of order zero gives 2π, and the first term the wanted excess to 2π:

$$\Delta\varphi = -\frac{\partial}{\partial K}\int_{r_{\min}}^{r_{\max}}\frac{2m\delta U}{\sqrt{2m\left(E+\dfrac{U_0}{r}\right)-\dfrac{K^2}{r^2}}}\,dr = \frac{\partial}{\partial K}\left(\frac{2m}{K}\int_0^\pi r^2\delta U\,d\varphi\right)$$

In case (a), the integration gives:

$$\Delta\varphi = -\frac{2\pi}{K^2}\beta = -\frac{2\pi}{U_0 p}\beta$$

In case (b) the integration gives:

$$\Delta\varphi = -\frac{6\pi U_0 m}{K^2}\gamma = -\frac{6\pi}{U_0 p^2}\gamma$$

where p is the parameter of the unperturbed ellipse.

13

Orbital Elements and Ephemerides

In this chapter we discuss the problem of referring the theoretical orbit to the observations. First, we will examine how the apparent positions can be determined at every instant from the knowledge of the orbital elements; then, the inverse and much more difficult problem of how the orbital elements can be derived from the observed positions will be briefly examined. The treatment will be essentially limited to the case of the Solar System, with few considerations for binary stars.

13.1 Kepler's Equation

Let Q be the mobile particle on the ellipse of focus F and center O, and P_0 the perihelion (or the periastron for binary stars, or pericenter in general), where from the true anomalies ν are counted (see Figure 13.1).

It is convenient to draw through O an auxiliary circle of radius a and a new anomaly E, named eccentric anomaly (from the Latin *ex-centrum*), increasing as ν. Notice that the ellipse can be considered as the projection of the auxiliary circle on a plane passing through the line of the apses, and inclined by an angle ψ such as $e = \sin \psi$ (indeed, some authors directly write $\sin \psi$ instead of e). In the Cartesian heliocentric system (x, y) centered in F, with x parallel to the major axis, and y parallel to the minor axis, the coordinates of P are:

$$x = r \cos \nu = a(\cos E - e), \qquad y = r \sin \nu = a\sqrt{1 - e^2}\sin E$$

$$r = a(1 - e \cos E), \qquad \cos \nu = \frac{\cos E - e}{1 - e \cos E},$$

$$\sin \nu = \sqrt{1 - e^2}\,\frac{\sin E}{1 - e \sin E} \tag{13.1}$$

Introducing the semianomaly $\nu/2$, Equation 13.1 can be written as:

$$r\left(\cos^2 \frac{\nu}{2} - \sin^2 \frac{\nu}{2}\right) = a(\cos E - e), \quad r\left(\cos^2 \frac{\nu}{2} + \sin^2 \frac{\nu}{2}\right) = a(1 - e \cos E)$$

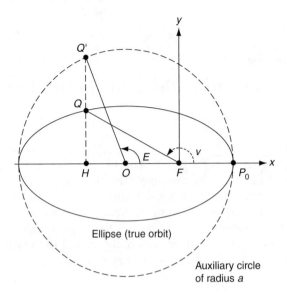

FIGURE 13.1
Ellipse and auxiliary circle, v = true anomaly, E = eccentric anomaly.

Then:

$$\tan\frac{v}{2} = \sqrt{\frac{1+e}{1-e}}\tan\frac{E}{2} \tag{13.2}$$

where $v/2$ and $E/2$ always have the same sign. Recalling the mean anomaly:

$$M = 2\pi(t - t_0)/P = n(t - t_0)$$

and the relationship between the angular velocity and the area constant, we obtain:

$$C = r^2\dot{v} = na^2\sqrt{1-e^2}, \quad C(t-t_0) = \int_{t_0}^{t} Cd\tau = \int_0^{2\pi} r^2 dv = Ma^2\sqrt{1-e^2} \tag{13.3}$$

If Q' is the projection of Q on the auxiliary circle, it is also:

$$Q'H = a\sin E, \quad QH = a\sqrt{1-e^2}\sin E, \quad \frac{Q'H}{QH} = \frac{a}{b} = \frac{1}{\sqrt{1-e^2}}$$

By virtue of the projectivity properties between the auxiliary circle and the ellipse, we obtain:

$$\text{area}\, FP_oQ' = \frac{1}{\sqrt{1-e^2}}\,\text{area}\, FP_oQ = \text{area}\, OP_oQ' - \text{area}\, OFQ'$$

$$\text{area } OP_oQ' = \frac{1}{2} aE^2, \quad \text{area } FP_oQ' = \frac{1}{2} \int_{t_0}^{t} C d\tau = \frac{1}{2} a^2 \sqrt{1-e^2}\, M,$$

$$\text{area } OFQ' = \frac{1}{2} a^2 e \sin E$$

so that:

$$E - e \sin E = M = n(t - t_0) \tag{13.4}$$

Equation 13.4, known as Kepler's equation, fixes the relationship between the eccentric anomaly and the time. The same result can be obtained in the following way:

$$\frac{dE}{dM} = \frac{dE}{ndt} = \frac{1}{1 - e \cos E} = \frac{a}{r} = 1 - e \cos \nu$$

By altering Equation 13.2, after some simple passages we arrive at:

$$d\nu = \frac{\sqrt{1-e^2}}{1 - e \cos E} dE \tag{13.5}$$

$$\frac{d\nu}{dE} = \sqrt{1-e^2}\, \frac{a}{r} = \frac{\sqrt{1-e^2}}{1 - e \cos E} = \frac{\sin \nu}{\sin E}, \quad \frac{dE}{dt} = n\frac{a}{r} = \frac{n}{1 - e \cos E}$$

Integrating the last differential equation, we obtain again Equation 13.4.

In either way, if the four orbital elements (a, n, e, t_0) are known, E is obtained at any instant t by solving Equation 13.4; from the knowledge of E, ν is known by virtue of Equation 13.2, and finally, r, x, and y can be calculated via Equation 13.1.

The solution of the transcendental Equation 13.4 is by no means a trivial one, unless e is small, as in the case of the Earth's orbit. More than a hundred methods have been proposed in the literature. In the simplest case, Newton's method can be used: take as the initial value $E_1 = M + e \sin M$ and calculate:

$$M_1 = M + e \sin E_1, \quad \Delta M_1 = M - M_1$$

Remembering that: $dE = (dE/dM)dM$, $dM = dE - e \cos E\, dE$, we must have approximately:

$$\Delta E_1 = \Delta M_1/(1 - e \cos E_1), \quad E_2 = E_1 + \Delta E_1$$

Calculate then:

$$M_2 = M + e \sin E_2, \quad \Delta M_2 = M - M_2, \quad \Delta E_2 = \Delta M_2/(1 - e \cos E_2),$$

$$E_3 = E_2 + \Delta E_2$$

and so on, until E becomes stable. It is also useful to remember that if e is sufficiently small:

$$\sqrt{1-e^2} \approx 1 - \frac{e^2}{2} - \frac{e^4}{8}$$

Several developments in series have been derived, following d'Alembert, as already discussed in Chapter 2 with reference to Equation 13.2; however, the convergence is not always guaranteed, as was shown by Laplace in the case $e > 0.66274$. We quote here the following expressions, written as Fourier series:

$$E = M + \left(e - \frac{1}{8}e^3\right) \sin M + \left(\frac{1}{2}e^2 - \frac{1}{6}e^4\right) \sin 2M + \frac{3}{8}e^3 \sin 3M + \cdots$$

$$v = M + \left(2e - \frac{1}{4}e^3\right) \sin M + \left(\frac{5}{4}e^2 - \frac{11}{24}e^4\right) \sin 2M + \frac{13}{12}e^3 \sin 3M + \cdots$$

$$\frac{a}{r} = 1 + \left(e - \frac{1}{8}e^3\right) \cos M + \left(e^2 - \frac{1}{3}e^4\right) \cos 2M + \frac{9}{8}e^3 \cos 3M + \cdots$$

$$\frac{r}{a} = 1 + \frac{e^2}{2} - \left(e - \frac{3}{8}\right)e^3 \cos M + \left(\frac{1}{2}e^2 - \frac{1}{3}e^4\right) \cos 2M - \frac{3}{8}e^3 \cos 3M + \cdots$$

The last equation provides the average value of r with respect to time:

$$<r> = a\left(1 + \frac{e^2}{2}\right)$$

while the average value with respect to v is b, and that with respect to E is a. For very small values of the eccentricity:

$$E \approx M + e \sin M, \quad v \approx M + 2e \sin M$$

The last equation was discussed in Chapter 4 for the geocentric orbit of the Sun as the equation of the center.

Regarding the Cartesian coordinates (x, y), sometimes referred to as reduced coordinates, the following expressions are easily found:

$$x = \frac{r}{a} \cos v = \cos E - e$$

$$= -\frac{3}{2}e + \left(1 - \frac{3}{8}e^2\right) \cos M + \left(\frac{1}{2}e - \frac{1}{3}e^2\right) \cos 2M + \frac{3}{8}e^2 \cos 3M + \cdots$$

$$y = \frac{r}{a} \sin v = \sqrt{1 - e^2} \sin E$$

$$= \left(1 - \frac{5}{8}e^2\right) \sin M + \left(\frac{1}{2}e - \frac{5}{12}e^2\right) \sin 2M + \frac{3}{8}e^2 \sin 3M + \cdots$$

13.2 Ephemerides from the Orbital Elements

Consider the (quasi) inertial heliocentric reference system, $X_\odot Y_\odot Z_\odot$, with axis $X_\odot$ toward the vernal point γ and axis $Y_\odot$ in the ecliptic plane, and let (X, Y, Z) be the instantaneous coordinates of a planet (or comet or asteroid).

The intersection of the planet's orbital plane with the ecliptic is a line passing through the Sun called the line of the nodes. Let N be the ascending node (that node through which the Z coordinate of the planet passes from negative to positive values along the orbital movement), and Ω the angle on the ecliptic plane between γ and N; this angle is called longitude of the ascending node. The angle i between the orbital plane and the ecliptic is said to be the inclination of the orbit; it varies between $(0°, +90°)$ for a direct movement, between $(90°, 180°)$ for a retrograde one. The pair (Ω, i) gives therefore the position of the orbital plane with respect to the ecliptic. Consider now, in the orbital plane, the angle ω between the ascending node N and the perihelion $P_\odot$; such angle is named argument of the perihelion; it fixes the orientation of the orbit in its plane. There will be cases when Ω cannot be defined, because $i = 0°$, so that a different choice of arguments must be made (e.g., sometimes it is useful to consider the angle $\Pi = \Omega + \omega$, even though, in general, this so-called longitude of the perihelion is the sum of two angles in two different planes).

We shall limit the discussion to the generic orbit for which the seven orbital elements $(a, e, P, n, t_0, \Omega, i, \omega)$ are known. In fact, by virtue of Kepler's third law, six elements are sufficient, either P or n, and only five for a parabolic orbit. In order to fix the position of the planet in the heliocentric system at a given time, several steps are necessary: first, derive the pair (r, v) or the reduced coordinates (x, y) in its orbit, and next, make a rotation of ω in the plane of the orbit. Then, project the orbit on the ecliptic according to i and with a further rotation of Ω, project the position of P on the heliocentric axes, finally to derive (X, Y, Z). Namely:

$$\begin{cases} X = r\left[\cos(\omega + v)\cos\Omega - \sin(\omega + v)\sin\Omega\sin i\right] \\ Y = r\left[\cos(\omega + v)\sin\Omega + \sin(\omega + v)\cos\Omega\sin i\right] \\ Z = r\sin(\omega + v)\sin i \end{cases} \tag{13.6}$$

Let $(X_\oplus, Y_\oplus, Z_\oplus)$ be the heliocentric coordinates of the Earth at the same instant (with $Z_\oplus \approx 0$); the geocentric coordinates of P, $(x_\oplus, y_\oplus, z_\oplus)$, will be:

$$\begin{cases} x_\oplus = X - X_\oplus \\ y_\oplus = Y - Y_\oplus \\ z_\oplus = Z - Z_\oplus \end{cases} \tag{13.7}$$

From $(x_\oplus, y_\oplus, z_\oplus)$ the ecliptic coordinates (λ, β) can be derived, and finally the equatorial coordinates (α, δ) and the distance of the planet from the Earth. This is not the final step of the procedure to calculate the apparent coordinates; we still have to apply the corrections due to aberration, precession, and nutation, position of the observer (namely from geocentric to topocentric coordinates), atmospheric refraction. Let us briefly discuss the precessional effect: ignoring the planetary perturbations, the plane of the orbit in inertial space will be fixed and so will the inclination i and the argument of the perihelion ω; instead, the direction of γ is function of time, and so will be Ω. Therefore, in first approximation: $di = d\omega = 0$, $d\Omega = -52''.3/\text{yr}$.

For many newly discovered comets, the orbit is essentially a parabolic one, so that we can safely put $e = 1$, $h = 0$, and write:

$$r = \frac{p}{1 + \cos \nu} = \frac{p}{2\cos^2 \frac{\nu}{2}} = q\left[1 + \tan^2 \frac{\nu}{2}\right] \qquad (13.8)$$

The constant $q = p/2$ is the distance to the Sun at the perihelion. It is usual to write:

$$s = \tan\frac{\nu}{2}, \quad x = r\cos \nu = q(1 - s^2), \quad y = r\sin \nu = 2qs$$

It is easy to find that:

$$C = \sqrt{2GM_\odot q} = 0.0243\sqrt{q}, \quad V = \frac{2GM_\odot}{r} = \frac{C}{q\sqrt{1 + s^2}},$$

$$t - t_0 = \frac{1}{3}\sqrt{\frac{2}{GM_\odot}} q^{3/2}(3s + s^2)$$

Therefore the orbital elements are $(q, t_0, \Omega, i, \omega)$. The following relationships are of interest in this case:

$$\nu \sec^2 \frac{\nu}{2} = \sqrt{2GM_\odot}\, q^{-3/2}, \quad \tan\frac{\nu}{2} + \frac{1}{3}\tan^3\frac{\nu}{2} = \sqrt{\frac{GM_\odot}{2q^3}}(t - t_0)$$

The hyperbolic case is seldom encountered; we can simply state that the expressions of $r\cos \nu$, $r\sin \nu$ contain the hyperbolic functions ($\cosh F$, $\sinh F$) of an appropriate anomaly.

13.3 Planetary Configurations and Titius–Bode Law

The upper part of Figure 13.2 shows the different configurations (projected on the ecliptic plane) with respect to the Sun for an internal planet such as Mercury or Venus, and for an external one such as Mars or Jupiter, with the usual terminology of elongations, conjuction, quadratures, and oppositions. The lower part of the figure shows the apparent paths among the fixed stars, with the predominance of direct movements, occasional points of stationarity, reversal of the motion, and even loops: for the terrestrial observer, therefore, the simplicity of the Keplerian orbits is masked by the complexity of the apparent configurations. Undoubtedly, this complexity did not favor the acceptance of the heliocentric vision of the solar system.

The maximum elongations of Mercury and of Venus are 23 and 46 degrees, respectively (note the fact that some authors call elongation the angle between Sun and planet, while strictly speaking the name refers to

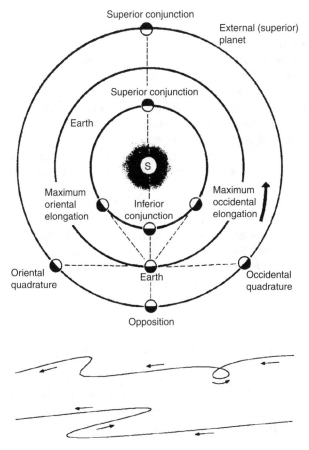

FIGURE 13.2
Planetary configurations and apparent paths.

the distance along the ecliptic, namely the difference in ecliptic longitudes). Their orbital inclinations are not zero, and therefore the two planets can be seen only occasionally in front of or behind the solar disk. For Mercury, this can happen every 7 or 13 years (the last inferior conjunction was on May 7, 2003). For Venus the phenomenon is much rarer, but when it happens there will be two passages 8 years apart (the last was in June 2004 and the next in 2012). Furthermore, according to the particular configuration, the disk of an inferior planet can appear partly illuminated, with phases similar to the ones observed on the Moon; the dividing line between the illuminated region and that in shadow is called a terminator. Phases of the outer planets can be noticed essentially only for Mars; for Jupiter and Saturn a modest limb darkening can be seen with careful observations. See the Notes for a fuller discussion.

Other interesting properties of the orbits of the planets are their small eccentricities and modest inclination to the ecliptic, the largest values being those of Mercury and of Pluto (see Table 13.1), and the direct sense of their motion. None of these properties derives from fundamental physical reasons, they must be associated with the mechanism of formation and subsequent dynamical evolution of the Solar System. Indeed, comets and asteroids can have very large eccentricities, and even revolve in retrograde sense (e.g., Halley's), and so behave many moons of the giant planets, for instance Triton, the large moon of Neptune, is in retrograde orbit, suggesting a capture by Neptune after formation in a different place.

Even the diurnal rotation of the planets is not determined by the basic laws; a great variety of situations is encountered, from the very slow retrograde rotation of Venus to the rotation around an axis lying almost in the orbital plane for Uranus.

Regarding periods, a distinction must be made between those observed by the terrestrial observer and the true ones with respect to a fixed direction. The first is the so-called synodic period, the second is the sidereal one. We can define the first by the interval of time between two similar configurations of Sun, Earth, planet, the second one by the mean motion n divided by 2π. If $n_\oplus$ is the mean sidereal motion of the Earth, we have:

$$P_{syn} = \pm \frac{2\pi}{n - n_\oplus} \tag{13.9}$$

where the plus sign applies to an internal planet and the negative sign to an external one (see again Table 13.1).

The distances of the planets from the Sun seem to show a remarkable regularity, which had already been noticed by two German astronomers, Titius and Bode, in the eighteenth century; this regularity was expressed by Wurm in 1787 (Uranus was discovered in 1781) with the mathematical formula:

$$a_n = 0.4 + 0.3 \times 0.2^n \tag{13.10}$$

where $n = -\infty$ for Mercury, $= 0$ for Venus, $= 1$ for Earth, $= 2$ for Mars, $= 4$ for Jupiter, $= 5$ for Saturn, and $= 6$ for Uranus. The lack of a planet for $n = 3$

TABLE 13.1

Orbital Elements of the Nine Planets

Planet	a (AU)	e	i (deg)	n (''/j)	P_{sid} (j)	P_{syn} (j)
Mercury	0.387	0.2056	7.004	14732.424	87.969	115.88
Venus	0.723	0.0068	3.394	5767.740	224.698	583.92
Earth	1.000	0.0167	0.000	3548.095	365.266	
Mars	1.524	0.0934	1.850	1886.619	686.943	779.94
Jupiter	5.203	0.0483	1.308	299.156	4332.177	398.88
Saturn	9.539	0.0560	2.488	119.993	10800.594	378.09
Uranus	19.191	0.0461	0.774	41.869	30953.629	369.66
Neptune	30.061	0.0097	1.774	21.301	60841.437	367.49
Pluto	39.529	0.2482	17.148	14.194	91304.976	366.73

motivated great efforts to locate one, until finally Giuseppe Piazzi in Palermo discovered Ceres on January 1, 1801. Ceres is the largest of the main belt asteroids, and this discovery prompted Carl Gauss to develop the theory of orbital motions and the method of the least squares for properly treating the observational errors. The discovery of Ceres was followed soon after by that of Pallas and Vesta.

Alternative formulations of Equation 13.10 were proposed after the discoveries of Neptune and of Pluto, e.g., $a_n = 1.53^n$ by Armellini ($n = -2$ Mercury, $= -1$ for Venus, $= 0$ for Earth, $= 1$ for Mars, $= 2$ and $= 3$ for two main groups of asteroids, $= 4$ Jupiter, $= 5$ Saturn, $= 6$ Uranus, $= 7$ lacking, $= 8$ Neptune, and $= 9$ Pluto).

Of course, there is no reason for the Solar System to stop at Pluto, and indeed, several large trans-Neptunian bodies are continuously being discovered in the so-called Kuiper belt. The largest of these bodies have diameters greater than 1500 km, even larger than that of Charon. The total mass of the trans-Neptunian objects exceeds by at least ten times the total mass of the asteroids in the main belt.

13.4 Orbital Elements from the Observations

The inverse problem of deriving the six orbital elements of a Solar System object from the observations, is much more complex than the direct one, so that only some basic notions (derived from the methods developed by Laplace and Gauss) can be dealt with here.

We have already said that the knowledge of six initial constants is required. In usual astronomical observations, that means three pairs of angular coordinates (α_i, δ_i) at three instants t_1, t_2, t_3 (radar data may instead provide full position and velocity vectors at a particular instant). At the beginning of the observations, the distance ρ of the new comet or asteroid is unknown, so we assume that the three angular pairs are geocentric ones. With reference to Figure 13.3, let $\oplus \xi \eta \zeta$ be the Cartesian geocentric ecliptic system, (λ, β) the ecliptic longitude and latitude, and $\odot XYZ$, ($\lambda_\odot, \beta_\odot$) the corresponding heliocentric values.

At each instant t_i we have:

$$\xi_i = \rho_i \cos \lambda_i \cos \beta_i, \quad \eta_i = \rho_i \sin \lambda_i \cos \beta_i, \quad \varsigma_i = \rho_i \sin \beta_i \quad (13.11)$$

For a generic instant t inside the interval (t_1, t_3), we can develop each variable as:

$$\xi_i = \xi(t) + \dot{\xi}_t(t_i - t) + \frac{1}{2}\ddot{\xi}_t(t_i - t)^2 + \cdots$$

It is convenient to start from the intermediate time $t = (t_1 + t_2 + t_3)/3$, because the third-order terms vanish. With the further assumption that the velocity and acceleration of the body remain constant during that interval of

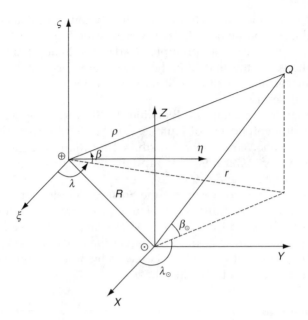

FIGURE 13.3
Heliocentric and geocentric coordinates of a new comet or asteroid, Q.

time, the development can thus be terminated at the second-order terms:

$$\xi_i = \xi(t) + \dot{\xi}_t(t_i - t) + \frac{1}{2}\ddot{\xi}_t(t_i - t)^2 \qquad (i = 1, 2, 3)$$

and similar for the other two coordinates. Therefore, from three observations we can derive the three (supposedly) geocentric components of position, velocity, and acceleration of Q at the intermediate time t, a part the still unknown distance ρ. The heliocentric position of the Earth at the same instant is obviously well-known, so that the heliocentric ecliptic coordinates of Q are:

$$X = X_\oplus + \xi, \quad Y = Y_\oplus + \eta, \quad Z = Z_\oplus + \zeta \approx \zeta$$

From the last relationship we obtain: $Z/r = \sin \beta_\odot$.
The equations of motion of Q are:

$$\ddot{X} + GM_\odot \frac{X}{r^3} = 0, \quad \ddot{Y} + GM_\odot \frac{Y}{r^3} = 0, \quad \ddot{Z} + GM_\odot \frac{Z}{r^3} = 0$$

where r is the distance of Q from the Sun. Similarly, for the Earth:

$$\ddot{X}_\oplus + GM_\odot \frac{X_\oplus}{a_\oplus^3} = 0, \quad \ddot{Y}_\oplus + GM_\odot \frac{Y_\oplus}{a_\oplus^3} = 0$$

Let us introduce three auxiliary variables (σ, μ, v):

$$\sigma = \rho \sin \beta, \quad \mu = \cos \lambda \cot \beta, \quad v = \sin \lambda \cot \beta \qquad (13.12)$$

connected to (ξ, η, ζ) by:

$$\xi = \mu\sigma = \mu\rho\sin\beta, \quad \eta = v\sigma, \quad \zeta = \sigma \tag{13.13}$$

The two variables (μ, v) are directly known from the observations, and so are their first and second derivatives at the intermediate time. The equations of motion become:

$$
\begin{cases}
\ddot{\mu}\sigma + \mu\ddot{\sigma} + 2\dot{\mu}\dot{\sigma} + \ddot{X}_\oplus + GM_\odot \dfrac{X_\oplus + \mu\sigma}{r^3} = 0 \\[2mm]
\ddot{v}\sigma + v\ddot{\sigma} + 2\dot{v}\dot{\sigma} + \ddot{Y}_\oplus + GM_\odot \dfrac{Y_\oplus + v\sigma}{r^3} = 0 \\[2mm]
\ddot{\sigma}_\oplus + GM_\odot \dfrac{\sigma}{r^3} = 0
\end{cases}
\tag{13.14}
$$

Multiply the first of Equation 13.14 by $\dot{v}$, the second by $\dot{\mu}$, and subtract the two; taking into account the third, we obtain:

$$\rho\sin\beta[\ddot{\mu}\dot{v} - \dot{\mu}\ddot{v}] + GM_\odot[X_\oplus\dot{v} - Y_\oplus\dot{\mu}]\left(\frac{1}{r^3} - \frac{1}{a_\oplus^3}\right) = 0 \tag{13.15}$$

The angle θ between Sun, Earth, and Q is known from the observations, and it is connected to r, ρ by:

$$r^2 = \rho^2 - 2\rho a_\oplus \cos\theta + a_\oplus^2 \tag{13.16}$$

Recalling the first of Equation 13.12, and the fact that the values of the auxiliary constants (μ, v) and of their first and second derivatives are also known by Equation 13.13, we can (at least in principle) derive (r, ρ) from Equation 13.15 and Equation 13.16. The solution is certainly not easy; proceeding in the elementary manner we would end up with an equation of the eighth degree!

We cannot discuss here the several methods developed for the determination of the possible solutions, but quite often the orbit remains poorly known, and a fourth observation is needed to resolve the ambiguities. At any rate, from ρ we have immediately σ, then $\dot{\sigma}$ from:

$$
\begin{cases}
\ddot{\mu}\sigma + 2\dot{\mu}\dot{\sigma} + GM_\odot X_\oplus\left(\dfrac{1}{r^3} - \dfrac{1}{a_\oplus^3}\right) = 0 \\[2mm]
\ddot{v}\sigma + 2\dot{v}\dot{\sigma} + GM_\odot Y_\oplus\left(\dfrac{1}{r^3} - \dfrac{1}{a_\oplus^3}\right) = 0
\end{cases}
\tag{13.17}
$$

The second derivative of σ is known from Equation 13.13.

This procedure, admittedly still approximate, nevertheless has provided a first estimate of the position, velocity, acceleration, and distance of Q at the intermediate time t, both in the heliocentric and geocentric systems. By comparing the heliocentric velocity at distance r with the parabolic value, we can recognize the type of conic and the constant h. If the orbit is an ellipse, we have immediately P (or n) and a. To derive e, we take advantage of Kepler's second law:

$$\dot{\varphi}^2 = \dot{\lambda}_\odot^2 + \dot{\beta}_\odot^2, \quad C = r^2 \dot{\varphi}$$

and then derive E, ν, t_0. In the subsequent iterations of the procedure, the light time corresponding to the geocentric distance and the topocentric position of the observer must be introduced.

As already noted, it is usual to assume at the start of the calculation that the orbit of a new comet is a parabola.

In the Notes available software packages are suggested, which can be employed to calculate the orbital elements.

13.5 Application to Visual Binary Stars

In the case of binary stars, the ratio of the two masses is close to 1, rarely being as small as 0.05. Furthermore, the orbit is projected on the plane tangent to the line of sight, and the tridimensional reconstruction of the system is not always possible. We limit the present considerations to well-resolved systems of two self-gravitating stars, observed for the entire orbital period (or at least for a good fraction of it). In reality, many systems have a multiplicity greater than 2 (for instance α Cen is triple), and most have been observed only for small arcs. There are also cases when only one of the two bodies is readily seen; classic in this respect is the case of Sirius, the brightest star in the sky. Bessel, in 1844, observed that the proper motion of the star is not along a straight line, but instead is undulated; there ought to be a much fainter but massive star, in order to maintain the center of mass along the required straight path (see Figure 13.4). This fainter star, Sirius B, was discovered by the telescope maker Alvan Clark three decades later. Finally, in 1914, Adams showed spectroscopically that the color, and thus the temperature, of Sirius B were very similar to those of Sirius A. According to the Stefan–Boltzmann law (see Chapter 16), the very low luminosity of Sirius B had therefore to be attributed to its much smaller radius. Indeed, the radius of Sirius B is even smaller than the radius of the Earth. Sirius B is therefore the prototype of the white dwarfs, namely of stars with radii much smaller than that of normal stars having the same temperatures and masses. According to the theories of the stellar structure, white dwarfs represent the end point of the evolution of those normal stars.

Visual observations are typically made comparing the fainter star B (secondary) to the brighter star A (primary, which in general is also the more massive one). A polar system (ρ, θ) centered on A is used to measure the relative position of B, with ρ in arcsec and θ in degrees from North through East. The observations usually take a long span of time, so that precession and proper motion must be removed from the data before analyzing the pairs (ρ, θ) for the orbital analysis. These corrections are of the form

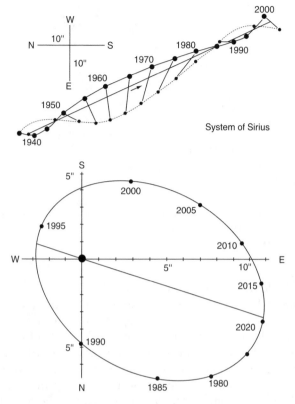

FIGURE 13.4
Upper panel: the undulated proper motion of Sirius A. Bottom panel: the relative orbit of B with respect to A.

(see Chapter 5 and Chapter 8, t in years):

$$\theta(t) = \theta(t_0) + (\Delta_1 + \Delta_2)(t - t_0), \quad \Delta_1 = -0°.0056 \sin \alpha \, \text{s} \, \delta \quad \text{per year}$$

$$\Delta_2 = -0°.00417 \, \mu_\alpha^s \sin \delta \quad \text{per year}$$

A further correction is due to the motion of the barycenter of the pair, which moves with respect to the observer during an orbital revolution of the secondary. This sort of planetary aberration modifies the apparent time intervals.

The methods of dealing with the observations are not discussed here, as we assume that the apparent orbit has been determined with great accuracy (Kepler's second law plays a very important role in this determination). For his didactic value, we first expound a graphical method, essentially due to Zwier, to derive the orbital elements. We have already underlined that the relative apparent orbit is an ellipse, which is the projection of the true one on the plane tangent to the celestial sphere (see Figure 13.5). In other words,

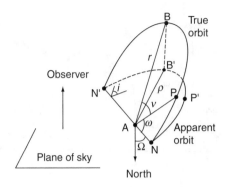

FIGURE 13.5
The apparent orbit is the projection of the true one on the plane perpendicular to the line of sight. A = primary star, B = true companion, B′ = projected companion, P = periastron, P′ = projected periastron. The line NN′ is the line of nodes. Notice that A is not in the focus of the apparent ellipse.

the observed elements are connected to the true one by a projective transformation, which conserves the center C of the ellipse, Kepler's second law, the period P, and the instant of passage through the periastron, but not the focus of the ellipse, nor the orthogonality of the axes or the area constant.

In Figure 13.6, let C be the center of both apparent and true ellipses, and A be the primary (namely the focus of the true orbit); the direction CA is therefore that of the true major axis, and point A_1 on that direction is the position of the true periastron. The ratio $CA/CA_1 = ae/a$ immediately gives the true eccentricity e. Let us now take the conjugate direction of CA_1 in the projectivity: draw any chord parallel to CA_1, identify its middle point, and join it with C. This is the direction of the true minor axis.

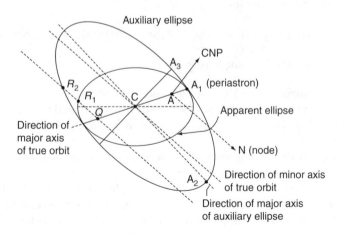

FIGURE 13.6
Apparent and auxiliary ellipses. CNP = celestial North pole.

Then draw a line parallel to the true minor axis through a generic point Q belonging to the line CA; this parallel intersects the apparent ellipse in R_1. Determine a segment QR_2 of such length that: $QR_2/QR_1 = 1/\sqrt{1 - e^2}$.

By repeating this construction for several points Q along CA, a series of points R_2 is obtained, which define an auxiliary ellipse, tangent to the true one in the periastron and apoastron. This auxiliary ellipse is also the projection on the sky of the auxiliary circle we have used to define the eccentric anomaly E. Therefore, the semimajor axis of the auxiliary ellipse is also the true semimajor axis:

$$CA_2 = a$$

The normal to CA_2 defines on the auxiliary ellipse the semiminor axis and the point A_3: the ratio CA_3/CA_1 gives therefore $\cos i$, and finally the modulus of i (but not its sign). The direction CA_2 is also the direction around which we have performed the projection, namely the line of nodes of Figure 13.5. Now, let CNP be the direction of the celestial North pole through A: the parallel to CA_2 through A points to the node N, so that the angle in A between the directions of CNP and N gives Ω. However, we do not know which node is the ascending one, so that $0° < \Omega < 180°$. The ambiguity can be resolved in the favorable case when the radial velocity V_r of B is known (at the descending node, B is approaching the observer and $V_r(B) < 0$).

In order to derive the angle ω between the ascending node and the periastron, consider both Figure 13.5 and Figure 13.6. P' being the projection of P from the pole of the apparent orbit, the spherical triangle $P'NP$ is rectangle in P', and the angle $P'NP = i$. On the other hand, angle NP' in Figure 13.5 is also angle A_1AN in Figure 13.6, so that:

$$\tan NP' = \tan NP \cos i = \tan \omega \cos i \tag{13.18}$$

We have thus derived e and the geometric elements (i, Ω, ω), which are called Campbell's elements.

Regarding the time arguments: in Figure 13.6 it is:

$$\text{angle } (CNP - A - A_1) = \theta(t_0), \quad \text{angle } (NA_1) = \theta(t_0) - \Omega,$$

$$\text{angle } (NAP) = \omega$$

and for any position of the true secondary B and projected secondary B':

$$\text{angle } (CNP - B' - A_1) = \theta(t), \quad \text{angle } (NB') = \theta(t) - \Omega,$$

$$\text{angle } (NB'P) = \omega + \nu$$

Finally:

$$\tan NB' = \tan NP \cos i, \quad \tan(\theta - \Omega) = \tan(\omega + \nu) \cos i \tag{13.19}$$

$$\rho \sin(\theta - \Omega) = r \sin(\nu + \omega) \cos i, \quad \rho = \frac{a(1 - e^2)}{1 + e \cos \nu} \frac{\cos(\nu + \omega)}{\cos(\theta - \Omega)} \tag{13.20}$$

From the knowledge of $\theta(t)$ we derive v and then E, and finally, by solving Kepler's equation also the mean motion $M = 2\pi(t - t_0)/P$. From two values of M, we immediately know t_0:

$$t_0 = \frac{M_1 t_1 - M_2 t_2}{M_1 - M_2}$$

and thus P.

Several analytical approaches have been developed to replace the above geometrical procedure. We quote some elements of the method named after Thiele and Innes. It has been seen that the primary star A is not in the focus of the projected ellipse, and that the fundamental direction in the observations is the axis through A toward CNP. In other words, the apparent ellipse referred to a system of axes centered in A, with axis x directed to the CNP and axis y toward east, can be expressed as:

$$\alpha x^2 + 2\beta xy + \gamma y^2 + 2\delta x + 2\varepsilon y = 1, \quad x = \rho \cos \theta, \quad y = \rho \sin \theta$$

where the five constants (α, β, γ, δ, ε) are known from the observations. On the plane of the true orbit, (X, Y) are the reduced coordinates:

$$X = \frac{r}{a}\cos v = \cos E - e, \quad Y = \frac{r}{a}\sin v = \sqrt{1 - e^2}\sin E \quad (13.21)$$

whose origin is A, while the direction of X is that of the periastron. Therefore, the apparent coordinates (x, y) are obtained by a rotation followed by a projection, so that they are linear combinations of the former:

$$x = \rho \cos \theta = AX + FY, \quad y = \rho \sin \theta = BX + GY$$

The four quantities A, F, B, G are the Thiele-Innes constants, given by:

$$A = a(\cos \omega \cos \Omega - \sin \omega \sin \Omega \cos i),$$

$$B = a(\cos \omega \sin \Omega + \sin \omega \cos \Omega \cos i)$$

$$F = a(-\sin \omega \cos \Omega - \cos \omega \sin \Omega \cos i),$$

$$G = a(-\sin \omega \sin \Omega + \cos \omega \cos \Omega \cos i)$$

The coordinates of the periastron and of the point having $v = 90°$ on the true orbit are, respectively:

$$\begin{cases} X_0 = 1 - e \\ Y_0 = 0 \end{cases}, \quad \begin{cases} X_1 = 0 \\ Y_1 = 1 - e^2 \end{cases}$$

while on the apparent one:

$$\begin{cases} x_0 = A(1 - e) = a(1 - e)(\cos \omega \cos \Omega - \sin \omega \sin \Omega \cos i) \\ y_0 = B(1 - e) = a(1 - e)(\cos \omega \sin \Omega + \sin \omega \cos \Omega \cos i) \end{cases}$$

$$\begin{cases} x_1 = F(1 - e^2) = a(1 - e^2)(-\sin \omega \cos \Omega - \cos \omega \sin \Omega \cos i) \\ y_1 = G(1 - e^2) = a(1 - e^2)(-\sin \omega \sin \Omega + \cos \omega \cos \Omega \cos i) \end{cases}$$

so that, after several manipulations:

$$\tan(\Omega + \omega) = \frac{B - F}{A + G} \quad (0° < \Omega < 180°),$$

$$\tan(\Omega - \omega) = \frac{B + F}{A - G} \quad (0° < \omega < 360°)$$

$$\tan^2 \frac{i}{2} = \frac{A - G}{A + G} \frac{\cos(\Omega + \omega)}{\cos(\Omega - \omega)} = \frac{B + F}{B - F} \frac{\sin(\Omega + \omega)}{\sin(\Omega - \omega)}$$

Let C be the area integral on the true orbit, and c that on the projected one:

$$c = C \cos i = x\dot{y} - \dot{x}y = (X\dot{Y} - \dot{X}Y)(AG - BF), \quad a^2 = \frac{AG - BF}{\cos i}$$

It is customary to furnish both Campbell's and Thiele-Innes' elements, and also the following:

$$C' = a \sin \omega \sin i, \quad H = a \cos \omega \sin i, \quad z = C'X + HY = r \sin(v + \omega) \sin i$$

where z is along the line of sight. Its derivative is the radial velocity (in arcsec per year) of the secondary with respect to the primary:

$$\dot{z} = C'\dot{X} + H\dot{Y} = n\left[C' \frac{dX}{dM} + H \frac{dY}{dM} \right]$$

If the parallax π of the binary is known, then it is possible to derive the radial velocity in km/sec:

$$V_r = 4.74 \left(\frac{2\pi}{P} \right) \frac{1}{\pi} \frac{a \sin i}{\sqrt{1 - e^2}} [e \cos \omega + \cos(v + \omega)] \tag{13.22}$$

This expression can also be read in the other way, in order to derive the dynamical parallax of the binary (see Chapter 8).

The knowledge of the two individual masses, and not only of their sum, must come from additional information, e.g., from the motion of the barycenter (such as for Sirius) or from the radial velocities of both stars.

The study of binary stars has made great progress by virtue of modern optical techniques, such as speckle interferometry and multiple telescope (array) interferometry. The angular resolution that can be achieved is approximately λ/D radians, where λ is the wavelength and D, respectively, the diameter of the mirror or the total length of the array; the resolution varies from a few hundredths to a few thousandths seconds of arc (see Notes). However, the great majority of binary (or multiple) stars remain unresolved. Their multiplicity can be recognized by photometric or spectral variations.

Notes

- Several sites provide ephemerides of planets, comets, and asteroids for each location on Earth, with a variety of possible choices. For instance, visit the JPL site: http://www.jpl.nasa.gov and implement the program Horizon (http://ssd.jpl.nasa.gov/horizons_doc.html)

- The transits of Mercury and Venus in front of the solar disk, the numerology associated with the dates, and images of the 2003 Mercury's transit can be found in the following sites:

  a) http://www.eso.org/outreach/eduoff/

  b) http://sunearth.gsfc.nasa.gov/eclipse/eclipse.html

  c) http://dipastro.pd.astro.it/planets/barbieri/

- Although Titius–Bode law tends to be considered not more than numerology, an interesting book about it was written by Nieto (1972). A variant of Armellini's formula (namely $a_n = r_0 K^n$, a geometric progression) has been discussed by F. Graner, and B. Dubrulle, in two papers: Titius–Bode law in the solar system, *Astronomy and Astrophysics*, **282**, p. 262 and *ibidem* p. 269, 1994.

- Several software packages allow the determination of the orbit from the observations. For instance OrbFit from Prof. A. Milani (University of Pisa, Italy). OrbFit is the computational engine of the online asteroid information services:

  http://newton.dm.unipi.it/neodys/ and: http://hamilton.dm.unipi.it/astdys/

- Another package (Find Orb) can be found at http://www.projectpluto.com/find_orb.htm

- Speckle interferometry can be found for instance at: http://www.mpifr-bonn.mpg.de/div/ir-interferometry/

- Array interferometry at: http://www.chara.gsu.edu/CHARA/ and http://www.eso.org/projects/vlti/

14

Elements of Perturbation Theories

This chapter provides a few elementary notions on the following arguments: perturbations of the planetary movements including the relativistic precession of the perihelion of Mercury; the case of a planet plus a small moon; the inequalities of the lunar movement; the Jupiter–Saturn interaction, the restricted circular 3-body problems, the 2-bodies of which one with finite dimensions, the librations. The present treatment will be limited to bodies of our Solar System. Important extensions of the theory have been made for the case of multiple stars and stellar systems, for which the reader is directed to the specialized literature.

From a historical perspective, after Newton's *Principia*, Clairaut examined the three-body interaction, d'Alembert paved the way for the study of the dynamics of extended bodies, Euler considered the polar motion of the Earth discussed in Chapter 6, the interaction Jupiter–Saturn, and the secular variation of the orbital elements. Euler's ideas were pursued by Lagrange and Laplace, among others. Great mathematicians, such as Poisson, Hamilton, Jacoby, and Poincaré followed in their footsteps in the nineteenth century, arriving at Albert Einstein (around 1915) with general relativity.

Today, precise ephemerides of the planets and of their moons are calculated with very refined methods, which take into account the perturbations briefly discussed in the following, and others not included here. We recall the fundamental ephemerides of the Sun, Moon, and planets constructed at the Jet Propulsion Laboratory (JPL, see Notes). The mathematical model used at JPL includes: point mass interactions among the Sun, planets, and moons; general relativity; Newtonian perturbations of selected asteroids; the Moon and Sun's action on the figure of the Earth; the Earth and Sun's action on the figure of the Moon; Earth tides on the Moon; and physical librations of the Moon. Chapter 5 of the *Explanatory Supplement* provides many references to other ephemerides.

14.1 Perturbations of the Planetary Movements

When N point-like bodies are under their mutual gravitational attraction, the difficulty of describing their movements greatly increases, even when excluding external forces. Consider three bodies, having masses m_0, m_1, and m_2, respectively. Let $\mathbf{r}_i$ be the vectors from the barycenter B and $\mathbf{d}_{ij}$ the mutual distances $(i, j = 0,1,2, i \neq j)$. We have three vector differential equations:

$$
\begin{cases}
m_0\ddot{\mathbf{r}}_0 = G\left(m_0 m_1 \dfrac{\mathbf{d}_{01}}{d_{01}^3} + m_0 m_2 \dfrac{\mathbf{d}_{02}}{d_{02}^3} \right) \\[2ex]
m_1\ddot{\mathbf{r}}_1 = G\left(m_1 m_0 \dfrac{\mathbf{d}_{10}}{d_{10}^3} + m_1 m_2 \dfrac{\mathbf{d}_{12}}{d_{12}^3} \right) \\[2ex]
m_2\ddot{\mathbf{r}}_2 = G\left(m_2 m_0 \dfrac{\mathbf{d}_{20}}{d_{20}^3} + m_2 m_1 \dfrac{\mathbf{d}_{21}}{d_{21}^3} \right)
\end{cases}
\tag{14.1}
$$

By projecting these equations on the three coordinate axes, we obtain 3×3 scalar differential equations of the second order, whose solution requires the knowledge of 18 initial constants (in general, for N bodies we would derive $3N$ scalar equations). The existence of three integrals of motion, namely the rectilinear motion of the barycenter, the conservation of the energy, and the conservation of the total angular momentum, can be easily demonstrated. The latter integral insures the existence of a so-called invariable plane (or Laplace plane) of the system passing through the barycenter. In the case of the Solar System, the invariable plane is inclined by $1.5°$ to the ecliptic, intermediate between the orbital planes of Jupiter and Saturn. Those integrals are, however, only equivalent to $6 + 1 + 3 = 10$ initial constants, while 18 are needed. Therefore, in the case of three bodies, the problem cannot be solved in general in an analytical way, although a fully analytical solution is possible in special cases (e.g., the restricted circular three-body problem treated in a later section).

In the relative description of the movements, we identify the Sun with the first body of mass $m_0 (= M_\odot)$, and draw through it a nonrotating, slightly accelerated, heliocentric reference system (X, Y, Z) parallel to the barycentric (inertial) one. Let us call Δ_1, Δ_2 the distances of two planets from the Sun, and d_{12} their mutual distance. The equations for the relative movement of the first planet are:

$$
\ddot{X}_1 = -G(M_\odot + m_1)\frac{X_1}{\Delta_1^3} + Gm_2\left[\frac{X_2 - X_1}{d_{12}^3} - \frac{X_2}{\Delta_2^3} \right]
\tag{14.2}
$$

and similar for Y_1, Z_1 and for the second planet. It is easily demonstrated that:

$$\ddot{X}_1 = -\frac{\partial V_1}{\partial X_1}$$

$$= -\frac{\partial}{\partial X_1}\left\{ G(M_\odot + m_1)\frac{1}{\Delta_1} + Gm_2\left[\frac{1}{d_{12}} - \frac{X_1 X_2 + Y_1 Y_2 + Z_1 Z_2}{\Delta_2^3} \right]\right\}$$

$$= -\frac{\partial}{\partial X_1}\left\{ G(M_\odot + m_1)\frac{1}{\Delta_1} + R_{12}\right\} \tag{14.3}$$

The first term of V_1 represents the gravitational interaction between the Sun and the first planet. The second term R_{12} can be called the perturbing function of the second planet on the first: its first term is the gravitational interaction between the two planets, while the second term is inversely proportional to the cube of the distance of the perturbing planet from the Sun:

$$R_{12} = Gm_2\left[\frac{1}{d_{12}} - \frac{X_1 X_2 + Y_1 Y_2 + Z_1 Z_2}{\Delta_2^3} \right] \tag{14.4}$$

Several different cases can then occur, according to the relative distances of the three bodies, as we will see in the following paragraphs.

Suppose that at a certain instant t_0 the perturbing function vanishes. The first planet would then follow a two-body trajectory whose initial conditions were also determined by the perturbation due to the other planet. Such hypothetical trajectory is called (with a slight formal imprecision) osculating orbit. True and osculating orbit are actually only tangent to each other, having at that instant, identical positions and velocity vectors. After a short time Δt, the positions on the two orbits will be identical, but not so the force or the velocity. Therefore, at $t_0 + \Delta t$ we can calculate a new osculating orbit, in other words a new set of osculating Keplerian orbital elements, slightly different from the previous set. The variations of the orbital elements can be expressed by the following Lagrange's planetary equations:

$$\dot{a} = \frac{2}{na}\frac{\partial R}{\partial M}, \quad \dot{M} = n - \frac{2}{na}\frac{\partial R}{\partial a} - \frac{1}{na^2}\frac{1-e^2}{e}\frac{\partial R}{\partial e},$$

$$\dot{e} = -\frac{1}{na^2}\frac{\sqrt{1-e^2}}{e}\frac{\partial R}{\partial \omega} + \frac{1}{na^2}\frac{1-e^2}{e}\frac{\partial R}{\partial M}$$

$$\dot{i} = -\frac{1}{na^2}\frac{1}{\sqrt{1-e^2}\sin i}\frac{\partial R}{\partial \Omega} + \frac{1}{na^2}\frac{\cos i}{\sqrt{1-e^2}\sin i}\frac{\partial R}{\partial i}, \tag{14.5}$$

$$\dot{\Omega} = \frac{1}{na^2}\frac{1}{\sqrt{1-e^2}\sin i}\frac{\partial R}{\partial i},$$

$$\dot{\omega} = \frac{1}{na^2}\frac{\sqrt{1-e^2}}{e}\frac{\partial R}{\partial e} - \frac{1}{na^2}\frac{\cos i}{\sqrt{1-e^2}\sin i}\frac{\partial R}{\partial i}$$

M being the mean anomaly. Notice that the variable n on the perturbed orbit is only the designation of $(GM_\odot)^{1/2}a^{3/2}$, because a is continuously varying. The six nonlinear equations above must be solved numerically, but solution is not always possible; there are, indeed, cases where the denominator becomes a very small number, or is even zero (e.g., a circular orbit, or one exactly on the ecliptic). Therefore, while it is very intuitive to regard the true orbit as a Keplerian one, with continuously changing osculating elements, from the mathematical point of view there are more convenient methods and variables; we can only quote here the variables associated with the names of Delaunay and of Poincaré. For more clarity, the osculating elements change from time to time, and so they do not represent the average orbit described by the body.

A particularly simple but instructive case, is when the second body has negligible mass (e.g., a small comet) and the third body is a large planet such as Jupiter. The planet will then follow an essentially unperturbed Keplerian orbit, while the comet acceleration will be given by Equation 14.3, where (X_1, Y_1, Z_1) are the heliocentric coordinates of the comet, (X_2, Y_2, Z_2) are the heliocentric coordinates of Jupiter, d_{12} is the distance of the comet to Jupiter, Δ_1 is the heliocentric distance of the comet, and Δ_2 is the heliocentric distance of Jupiter.

The perturbing force can be conveniently broken down into different directions, for instance the radial one, the tangential one, or the one perpendicular to the orbital plane. Let us discuss only the radial perturbation, and write:

$$U = GM\left[\frac{1}{r} + g\frac{1}{r^2} \right] \tag{14.6}$$

with g being very small. The perturbed orbit will have the expression (see Chapter 12):

$$\frac{1}{r} = GM\frac{1}{C^2(1-q)^2} + A\cos(1-q)(\varphi - \varphi_0), \qquad (1-q)^2 = 1 - \frac{gM}{C^2}$$

namely an ellipse rotating in its plane by $2\pi q$ at each revolution.

The difficulty of taking radial perturbations into account has led several times to suspecting that Newton's law is not fully representative of the reality. Clairaut in 1750 was probably the first to question the validity of the $1/r^2$ law, owing to his inability to reproduce the motion of the line of apses of the Moon. In the following century, it was the turn of LeVerrier, because of the unexplained advance of Mercury's perihelion. Later, the discrepancy of the ephemerides of Uranus (60 years after its discovery by Herschel in 1781, the error was already larger than 2′) was tentatively attributed to a failure of Newton's law. More recently, doubts were raised in order to account for observations at very small accelerations, for instance the unexplained accelerations of the spacecraft Pioneer beyond 20 AU, or the rotation curves of galaxies; a modified Newtonian dynamics (MOND) has even been proposed (see Notes). However, until now the difficulty that led to the most fruitful results was the excess of the advance of Mercury's perihelion. The observed advance of the perihelion is approximately 573″/century, of which 43″/century cannot be explained by planetary perturbations. LeVerrier suspected the existence of another planet internal to the orbit of Mercury (often called Vulcan; vulcanoids is a term used to indicate still unknown small bodies possibly orbiting inside Mercury's orbit), while others suggested a nonspherical distribution of mass. Finally, the effect was explained by general relativity, which actually represented one of the decisive proofs for the correctness of the theory. According to general relativity, the force can be written as:

$$F(r) = -GM\frac{1}{r^2}\left[1 + \frac{3C^2}{c^2r^2}\right] \tag{14.7}$$

where C is the area constant and c is the velocity of light in vacuum, namely a physical entity completely extraneous to classical celestial mechanics. Instead of the instantaneous action-at-distance of the Newtonian mechanics, general relativity foresees a transmission of forces with the velocity of light (and the radiation of gravitational waves by suitably accelerated bodies, or by a body collapsing in a strongly asymmetric configuration). From Equation 14.7, after several elaborated calculations, one can derive the following variation of ω:

$$\dot{\omega} = 6\pi\frac{GM_\odot}{Pa(1 - e^2)c^2} \text{ (rad/s)}$$

or else:

$$\Delta\omega = 6\pi\frac{n^2a^2}{(1 - e^2)c^2} \text{ (rad/revolution)}$$

As we have already remarked in Chapter 7 and Chapter 13, the astronomical measurements provide the so-called "Keplerian mass" $GM_\odot$ through the equivalence $GM_\odot = n^2a^3$, which can also be expressed by means of Gauss' defining constant k. The value for Mercury's excess is 43″.0/century (see Notes). It would be 8″.6/century for Venus, 3″.8/century for the Earth, and 1″.4/century for Mars.

A perturbation tangent to the orbit, and so parallel to **V**, would alter the velocity modulus but not its direction; this is the case, for instance, of a body traveling in a resisting medium (for example, an artificial satellite in the Earth's atmosphere). The consequence would be a gradual diminution of the semimajor axis and a corresponding (and at first sight counterintuitive) increase of the mean motion. This situation of movement in a resisting medium was suspected in the past for comets, when several of them appeared to increase their mean motion; but on the contrary, other comets displayed a decrease. Finally, an internal source for these unexplained accelerations was found, namely sudden directional losses of mass, namely of jets (a sort of a rocket effect). More properly, we can refer to nongravitational forces. Owing to the rotation of the body, to the inclination of the rotation axis to the orbital plane and to the thermal inertia, these jets can occur in any direction. In general, a tangential perturbation will alter the elements (a, e, n, ω).

A perturbation perpendicular to $\mathbf{V}$ cannot alter the modulus, but only the direction of the velocity vector: e and ω will be affected, but not a or n. Should this perturbation have a component normal to the orbital plane, then (i, Ω) will also change.

14.2 Planet Plus Small Moon

Consider, for instance Jupiter, of mass M_J and one of its distant (in order to remain in the point-like approximation) and small moons, of mass $m \ll M_J$. The distance planet-Moon d is much smaller than the distance to the Sun, so that both Jupiter and moon can be considered at equal distances from the Sun, $\Delta \approx \Delta_J$. We can therefore translate the origin of the reference system to the center of Jupiter, and consider the perturbations caused by the Sun, assumed in a fixed Keplerian orbit around Jupiter, to the orbit of the small moon around Jupiter.

If (x, y, z) be the planetocentric coordinates of the moon, the perturbing function then becomes:

$$R = GM_\odot \left[\frac{1}{\Delta} - \frac{x_\odot x + y_\odot y + z_\odot z}{\Delta_J^3} \right] \tag{14.8}$$

We also have:

$$\Delta^2 = (x - x_\odot)^2 + (y - y_\odot)^2 + (z - z_\odot)^2 = \Delta_J^2 - 2(xx_\odot + yy_\odot + zz_\odot) + d^2$$

Call θ the angle Sun–Jupiter–Moon:

$$\cos \theta = \frac{xx_\odot + yy_\odot + zz_\odot}{\Delta_J d}$$

and therefore, from the binomial theorem:

$$\frac{1}{\Delta} = \frac{1}{\Delta_J} \left[1 - 2\frac{d}{\Delta_J} \cos \theta + \frac{d^2}{\Delta_J^2} \right]^{-1/2}$$

$$\approx \frac{1}{\Delta_J} \left[1 + \frac{d}{\Delta_J} \cos \theta + \frac{d^2}{\Delta_J^2} \left(-\frac{1}{2} + \frac{3}{2}\cos^2 \theta \right) + \cdots \right] \tag{14.9}$$

Notice, however, that in the equations of motion, the partial derivatives of R contain only the terms depending on the coordinates of the perturbed body, so that we can conclude:

$$R = GM_\odot \frac{1}{\Delta_J} \left[\frac{d^2}{\Delta_J^2} \left(-\frac{1}{2} + \frac{3}{2}\cos\theta \right) + \frac{d^3}{\Delta_J^3} \left(-\frac{3}{2}\cos\theta + \frac{5}{2}\cos^2\theta \right) + \cdots \right]$$

(14.10)

(the complete expression can be given in terms of Legendre polynomials P_n of argument $\cos\theta$, see Chapter 6). The leading term:

$$R_1 = GM_\odot \frac{d^2}{\Delta_J^3} \left(-\frac{1}{2} + \frac{3}{2}\cos\theta \right)$$

(14.11)

is sufficiently small to leave almost Keplerian the planetocentric orbit of the moon.

14.3 Case Earth–Moon

We can extend the considerations of the previous paragraphs to the Earth–Moon system. Although the approximation will certainly be insufficient, the examination of the case of the Moon perturbed by a hypothetical Sun describing a fixed Keplerian orbit around the Earth, is of such great interest to be called the principal problem of the lunar orbit.

With reference to Figure 14.1, and considering the spherical triangle Sun–ascending node N–Moon, after several tedious but not difficult passages, we arrive at the approximate expression:

$$R_1 = n_\odot^2 a^2 \left(\frac{a_\odot}{\Delta_\odot} \right)^3 \left(\frac{d}{a} \right)^2 \left\{ \begin{array}{l} -\frac{1}{2} + \frac{3}{8}\left[\cos 2(\omega + v) + \cos 2(\omega_\odot + v_\odot - \Omega)\sin^2 i \right] \\[2mm] +\frac{3}{4}\left[1 + \cos 2(\omega - \omega_\odot + v - v_\odot + \Omega) \right]\cos^4 \frac{i}{2} \\[2mm] +\frac{3}{4}\left[1 + \cos 2(\omega + \omega_\odot + v + v_\odot - \Omega) \right]\sin^4 \frac{i}{2} \end{array} \right\}$$

(14.12)

We can still simplify Equation 14.12 assuming a circular orbit for the Sun, namely $a_\odot/\Delta_\odot = 1$, and $v_\odot = M_\odot$ (mean anomaly, do not confuse it with the mass of the Sun). Furthermore, because i_D is fairly small ($i_\mathrm{D} \approx 5°.2$), we put $\sin i = i_\mathrm{D}$ (in radians) and ignore all terms in i_D^4; because e_D is also small ($e_\mathrm{D} \approx 0.055$), we ignore all terms in e_D^2 and higher. Finally, we arrive at an expression containing the mean motion, the argument of the perigee, and the mean anomaly of the Sun (all quantities known with great accuracy), and the six osculating elements of the Moon. The planetary equations of Lagrange would thus allow derivation of the variation of those elements with time.

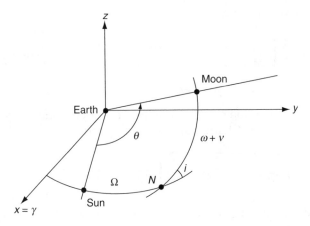

FIGURE 14.1
Geocentric elliptic orbit of the Moon in the geocentric ecliptic reference frame. The plane xy is the ecliptic. N = ascending node, Ω = angle γN, $\omega + \nu$ = angle N − Moon, $\omega_\odot + \nu_\odot$ = angle γ − Sun.

For instance, examine the nonperiodic part of R_1, namely:

$$R = n_\odot^2 a_{\mathbb{D}}^2 \left[\frac{1}{4} + \frac{3}{8}(e_{\mathbb{D}}^2 - i_{\mathbb{D}}^2) \right] \tag{14.13}$$

From Lagrange's equations, we can derive the secular motion of the perigee and of the nodes, obtaining periods of 9.21 and 17.86 yr, respectively. The observations provide, instead, the values of 8.85 and 18.6 yr, a discrepancy which gives an indication of the amount of approximations made in the previous discussion. The true orbit is appreciably different from an arc of ellipse even after a few days; the method of osculating elements simply provides a convenient intuitive visualization, but it is not appropriate for the precise calculations of lunar ephemerides. The theory of the movement of the Moon is, indeed, one of the most challenging problems of celestial mechanics, and several great mathematicians (Plana, Hansen, Delaunay, Hill, Brown, von Zeipel, etc.) have been confronted by it.

In our simple visualization, we can think of a basic elliptical orbit with $e_{\mathbb{D}} = 0.0549$ undergoing a double rotation, one in its plane around its focus (retrogradation of the perigee with a period of $8^y.85$) and one of the orbital plane around the normal to the ecliptic (retrogradation of the line of nodes with a period of $18^y.6$). Since the eccentricity $e_{\mathbb{D}}$ is so much larger than that of the Sun, the corresponding equation of the center is also extremely large: $EC_{\mathbb{D}} \approx 2e_{\mathbb{D}} \approx 6°17$. A whole spectrum of perturbing terms with different periods and amplitudes (terms called inequalities, and mostly of solar origin) deforms this basic orbit. For instance, the passage of the Sun through the lines of the apses occurs every 206 days. As a consequence, $e_{\mathbb{D}}$ varies between 0.045 and 0.065 within the same period. Also, the inclination i does not remain constant, it varies by approximately $\pm 9'$ around the mean value of $5°8'43''$, with the period of 173^j corresponding to the passage of the Sun through the nodes (it is slightly shorter than 6 months because of the retrogradation of the nodes). When the Sun passes through the nodes, the inclination displays its larger value.

We can visualize the total effect in longitude by superimposing on the equation of the center $EC_{\leftmoon}$ a number of inequalities of different periods and amplitudes:

- The largest is so great, $\pm 1°.274$, that it had already been detected by Hipparchus, and named evection. Its expression is: $1°.274 \sin[(M - 2M_\odot) + 2(\varpi - \varpi_\odot) + 2\Omega)]$. The evection displaces the Moon in longitude by 2.5 times its diameter, every 31.812 day.

- Tycho Brahe detected the second largest, namely the variation, of amplitude $\pm 39'.5$ and period $14^j.78$ (half the lunar month, namely half the synodic period). The amount of the variation is zero both at the so-called syzigies (new Moon and full Moon) and at the quadratures, it is maximum at the octants. Intuitively, the variation will be a function of the angle Sun–Earth–Moon, having as argument the angle $2[(M - M_\odot) + (\varpi - \varpi_\odot) + \Omega]$. Isaac Newton provided its first theoretical explanation; it is truly admirable that Newton was able to derive a fairly good theory of the lunar orbit by means of purely geometric reasoning.

- The parallactic inequality, with an amplitude of only $\pm 125''$, is a function of the varying distance to the Sun. It has twice the period of the variation, therefore its argument is $[(M - M_\odot) + (\varpi - \varpi_\odot) + \Omega]$. The name derives from the fact that by measuring the varying angular dimension of the Moon, the mean distance to the Sun can also be derived, which constitutes another method of deriving the solar parallax. The parallax of the Moon is therefore composed of a constant mean value: $\pi_{\leftmoon} = 3422''.70$ ($d_{\leftmoon} = 384,000$ km) (the corresponding average angular diameter is $\alpha_{\leftmoon} = 932''$, that is 1738 km for a circular disk), plus a number of periodic terms. At the new moon, the parallax is approximately $1''$ (110 km) smaller than the average, and correspondingly larger at the full moon. We shall consider again the apparent diameter in Chapter 15, when discussing eclipses and occultations. Note that the monthly variation of the geocentric diameter depends on the motions of the two bodies around the common barycenter, in other words by the ratio of the two masses. This ratio also appears in the equations of motion of the Earth–Moon system around the Sun. Therefore, the agreement between theory and observations constitutes another proof of the equality between gravitational and inertial masses.

- The annual equation has amplitude $\pm 12'$ and a period of 1 year (argument $M_\odot$). This term is due to the periodic variation of the distance of the Earth–Moon system to the Sun caused by the eccentricity of the Earth orbit. It was discovered by Tycho Brahe, who noticed a periodic variation in the instants of the eclipses. From January to July eclipses occur with a noticeable delay with respect to the expected time, the maximum deviation of 20 min occurring in April. The phenomenon reverses from July to January, with eclipses advancing by the same amount. In order to describe in more depth the occurrence of eclipses during the recorded millennia, one must consider two additional points. First, the secular variation of the terrestrial eccentricity $e_\oplus$, which causes a secular (and not only periodic) increase of the distance to the Sun, and therefore also a secular decrease of the mean motion of the Moon. Second, is the slowing down of the Earth's rotation, as already noted in Chapter 10. In 1693 Halley noticed that the mean motion of the Moon at that present time had to be smaller than in past centuries, otherwise the instants of the eclipses measured by Ptolemy, and later by Arab astronomers, could not be reproduced. Lalande estimated the amount of this acceleration to be $10''/\text{century}^2$ (the modern value is more like $12''.4/\text{century}^2$). In 1787 Laplace made a theoretical discussion of the lunar acceleration as being entirely due to the secular variation of $e_\oplus$, coming very close to Lalande's value. However, Laplace's conclusions were seriously questioned by Delaunay and Adams in the following century, because they could account for no more than $6''/\text{century}^2$ as being due to that reason. Energy dissipation by the tides, which lengthen the day, must play an equally important role. Furthermore, attention must also be paid to the fact that the other planets of the Solar System exert both a direct perturbation of the lunar orbit, and also an indirect one through the changes of the Earth orbit: so to speak, the Earth amplifies the planetary perturbations on the lunar orbit.

Similar to the line of the apses, the line of the nodes presents several inequalities. The largest has the same period (173^j, namely the passage of the Sun through the nodes) of the variation of inclination. The double inequality of the nodes and of the inclination is reflected by a great variation in latitude, with amplitude $\pm 10'24''$ (± 550 km), and period $32^j.38$, very close to that of evection but not to be confused with it.

The number of periodic terms is exceedingly high; in longitude, there are some 150 terms larger than $1''$. Modern observations have added a third dimension. While in the past the distance to the Moon was known with far lesser accuracy than the angular coordinates, today, radar and laser lunar ranging give the radial coordinate with a precision of approximately 1 cm. Incidentally, this is also the dimension of the Schwarzschild radius for the Earth, so that general relativity considerations must appear in present theories. The number of periodic terms in longitude, latitude, and distance (or diurnal parallax) has risen from several hundreds used by mathematicians such as Plana and Delaunay in the 18th and 19th centuries, to several thousands in contemporary computer programs.

14.4 The Lunar Month and the Librations

The very complex situation discussed in the previous section influences both the definition of the lunar month and the visibility of the lunar surface from a terrestrial site. We shall discuss in Chapter 15 the eclipses. Regarding the lunar month, we have the following definitions:

- *Synodic month*: the apparent geocentric period, namely the interval of time between two identical phases of the Moon, in particular between two new moons (Sun and Moon in conjunction). This is also the lunar months in civil calendars. The mean value is $P_{syn} = 29^j 12^h 44^m = 29^j.53059$. A more refined value is given by Chapront-Touzè and Chapront (see Notes):

$$P_{syn} = 29^j.5305888531 + 0.00000021621.T$$

where $T = (JD - 2451545.0)/36525$. Any particular cycle can vary around this average value by up to 7 h.

- *Sidereal month*: by the knowledge of the solar mean motion, we derive the lunar mean sidereal motion:

$$n_{syn} = n_{\mathbb{D}} - n_{\odot} = 1296000/29.5305881 = 43886.6979''/j$$

$$n_{\mathbb{D}} = 43886.6979 + 3548.1928 = 47434.8907''/j$$

The sidereal period $P_{\mathbb{D}}$ is therefore the mean interval of time needed for the lunar longitude to increase by $360°$ with respect to the fixed equinox, $P_{\mathbb{D}} = 1296000/n_{\mathbb{D}} = 27^j 07^h 43^m 11^s = 27^j.3216609$. Consequently, the Moon moves $13°10'25''$ eastwards each day with respect to a fixed star. This is the fundamental period in stellar occultations.

- *Tropical month*: the interval of time needed for the mean lunar longitude to increase by $360°$ with respect to the mean equinox. Recalling the value of the general precession, namely $0.1376''/j$, we derive $P_{trop} = 27^j 07^h 43^m 05^s = 27^j.3215816$, which is also the interval of time for the right ascension of the Moon to increase by 24^h.

- *Anomalistic month*: the interval of time between two passages for the mobile perigee. Its value is $P_{anom} = 27^j 13^h 18^m 33^s = 27^j.5545502$. Therefore, the mean anomaly of the Moon increases by $360°$ in a time slightly greater than the period of sidereal revolution.

- *Draconitic month*: the time between two passages of the Moon through the ascending node. It regulates the declination and also the eclipses. Owing to the secular retrogradation of the nodes, its value is the shortest: $P_{drac} = 27^j05^h05^m35^s = 27^j.2122178$. Owing to the irregularities of the retrogradation (the largest having the same period of 173^j affecting the inclination), the duration of the draconitic month has an appreciable variation.

We close this section with a discussion regarding the visibility of the Moon's surface by a terrestrial observer. The rotation of the Moon is approximately synchronous with the orbital period, so that from the Earth we always see one face but not exactly so, as telescopic observations have proved since the time of Galileo Galilei. Very accurate measurements were made by Cassini, whose name is associated with three basic laws, enunciated in 1693, regarding the lunar figure, as explained later in this section. Every lunar month, the surface appears to nod, in both North–South and East–West directions. Namely, the lunar figure exhibits the phenomenon of lunar librations.

Consider three vectors through the center O of the Moon: the rotation axis OR (inclined by $1°32'.1$ to the ecliptic), the normal to the orbit OP (inclined on average by $5°8'.7$ to the ecliptic, but undergoing a precession of $9'.0$), and the normal to the ecliptic OE, as in Figure 14.2. According to the first of Cassini's laws, the rotation axis is fixed in the Moon, and the period of rotation is equal to the sidereal period (the Moon is in synchronous rotation with the Earth). According to the second law, the angle EOR (the inclination of the spin axis to the ecliptic) is constant, about $1°35'$. According to the third law, while the node of the orbit regresses, the inclination EOP remains constant at about $5°9'$, and the three vectors always lie in the same plane (in other words, OR

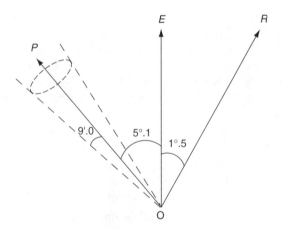

FIGURE 14.2
The three vectors connected with the lunar figure. OP is perpendicular to the orbit, E is perpendicular to the ecliptic, R is the direction of rotation.

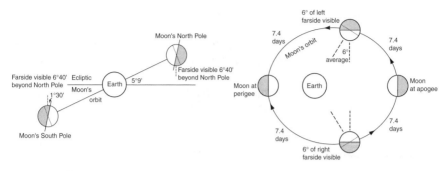

FIGURE 14.3
The libration in latitude (left) and in longitude (right).

precesses around OE in $18^y.6$). Therefore, every 27.2 days (2 days shorter than the lunation) the Moon's north and south poles are alternatively tilted towards the Earth by $6°40'.8$. This is the libration in latitude (see Figure 14.3, left panel).

Owing to the strong variation of the lunar orbital velocity, while the rotational velocity is essentially constant, the part of the lunar surface visible from the Earth undergoes slight changes during the lunation (see Figure 14.3, right panel). The greatest variations occur halfway between the perigee and apogee. The mean value of this libration in longitude is $6°9'$, but orbital perturbations can raise this value to approximately $8°$. The mean period is 27.6 days, somewhat greater than that of the libration in latitude. The combination of the two librations is analogous to the combination of two periodic motions at right angles with slightly different periods, namely a Lissajous figure; the maximum displacement of the center of the lunar disk can amount to about $10°.5$.

Furthermore, depending on diurnal rotation the terrestrial observer moves his vantage points from moonrise to moonset. At both these times almost $1°$ above the upper limb can be seen, but the exact amount depends on the lunar altitude (this is the so-called diurnal libration).

Finally, because of the unsymmetrical distribution of internal masses, the whole body of the Moon undergoes very small oscillations (physical librations), never exceeding $0°.04$.

The overall result is that approximately 59% of the lunar surface can be seen from the Earth, although in an irregular manner, while 41% remains invisible to the terrestrial observer.

The first images of the lunar far side were delivered by the Soviet spacecraft *Luna 3* in 1959. The first complete map came from the American *Lunar Orbiters*, and more recently from *Clementine* and *Lunar Prospector*. Together with the surface details, we now know, with great precision, both the altimetry and the internal distribution of masses. The barycenter is displaced by some 2 km towards the Earth with respect to the geometrical center. Just below the surface, on the Earth side there are strong concentrations of masses, the so-called lunar mascons. Probably the Moon has a small iron core.

The Sun, being an extended source, illuminates more than one hemisphere at each time. Light extends for arcs approximately 180°.5 long; in particular, near the Moon's south pole there is a peak which is always in sunlight, and has been given the fairly pompous name of peak of eternal light by the popular writer Camille Flammarion.

14.5   The Case Planet Plus Planet

Consider Jupiter and Saturn, of masses m_1 and m_2; in the present approximate treatment, the two planets have the same orbital plane. Their mutual distance d can be larger than their distances to the Sun Δ_1, Δ_2. If φ is the angle between the two planets as seen from the Sun, the perturbing function of Saturn on Jupiter can be written as:

$$R = Gm_2 \left[\frac{1}{d} - \frac{\Delta_1 \cos \varphi}{\Delta_2^2} \right]$$

$$= Gm_2\Delta_2 \left[\frac{1}{\sqrt{1 + \left(\frac{\Delta_1}{\Delta_2}\right)^2 - 2\frac{\Delta_1}{\Delta_2}\cos \varphi}} - \frac{\Delta_1}{\Delta_2^3} \cos \varphi \right] \qquad (14.14)$$

where $\Delta_1 < \Delta_2$. Therefore, we consider the perturbation of an external planet on an inner one; the same treatment could be applied to Jupiter upon Mars or upon an asteroid. Appropriate methods have been developed for the symmetric case, e.g., perturbation of Jupiter on Saturn. R is a periodic function of the true anomalies of the two planets. Indeed, after p_1 laps of Jupiter and p_2 laps of Saturn, the position will be identical to the starting one. Equation 14.14 is formally identical to that of the Moon, but there are important differences. First, in no cases can the ratio Δ_1/Δ_2 be considered arbitrarily small, indeed, it can reach 1 for the perturbation of Jupiter upon one of its Trojans. Second, the secular motions of the nodes and of the perihelion of the planets are fairly slow, with periods exceeding 18,000 years. Therefore, we can look for an approximate solution valid only for a few centuries. Then, we can assume that the perturbing planet moves on a fixed Keplerian ellipse, and we can develop each orbital element of the perturbed body as a series of secular, long period, and short period inequalities.

Using analytical methods, it can be shown that the semimajor axes and the mean motions of the great planets do not have appreciable secular inequalities. The same conclusion can be reached for the eccentricities and the inclinations, so that the present configuration of the Solar System must be stable for a very long time (this result is associated with the names of Delaunay and Tisserand). The stability certainly holds for long times, but it cannot be demonstrated for arbitrarily long periods, as was already shown by Poincaré. Modern digital methods have allowed the orbital evolution for a hundred million years to be followed (see e.g., Milani et al., Notes).

Some periodic terms can become of very high amplitude (resonances) when the mean motions n_1, n_2 of the two planets are commensurable. A large amplitude resonance occurs when two integer numbers i, j (negative or positive) can be found for which:

$$in_1 + jn_2 = n_r \approx 0 \qquad (14.15)$$

In the case of Saturn–Jupiter:

$$n_J = 299''.1283\,j^{-1}, \quad n_S = 120''.4547\,j^{-1}, \quad 5n_J - 2n_S = n_r \approx 4''.0169\,j^{-1}$$

which corresponds to a period of 883 years. The largest effect is in longitude, and amounts to approximately $20'$ for Jupiter and $50'$ for Saturn. The amplitude is so large that Kepler was able to detect it; he interpreted the effect as a secular acceleration of Jupiter and deceleration of Saturn. Laplace found the correct explanation in 1783.

An immediate application of the theory of perturbations was the discovery of Neptune in 1846. The situation was, so to speak, the reverse of the one previously discussed: from the observed perturbations of a given body (Uranus), the mass, position, and orbital elements of a still unseen perturbing body were inferred.

The explanation for the large inequalities of Uranus was investigated almost contemporarily by Adams in England and by LeVerrier in France. LeVerrier's calculations were completed in August 1846, and by September 23, Galle in Berlin discovered Neptune at only $52'$ from the predicted position. It was a great triumph for celestial mechanics. However, the true orbital elements of Neptune are appreciably different from those calculated by both LeVerrier and Adams, as seen in Table 14.1.

Note in particular the small mass of Neptune. The discovery was therefore largely accidental, and due to the contemporary presence of some favorable circumstances: Neptune and Uranus are in heliocentric conjunction (when the mutual perturbations are the largest) every 171 years;

TABLE 14.1

Orbital Elements of Neptune

Element	Adams	LeVerrier	True
a	37.2	36.1	30.1
e	0.12	0.11	0.009
ω	299	284	46
m	0.00015	0.00011	0.00005

the 1822 occurrence was only 24 years before the discovery. Therefore, any reasonable mean motion and small inclination would have produced a position correct to approximately 1°. The question of the perturbations of Uranus are therefore not entirely explained by the real Neptune; the next planet, Pluto, discovered by Clyde Tombaugh in 1938, has a mass far too small to induce such large perturbations (see also Notes). From time to time, the existence of a tenth planet with sufficiently high mass has been revived, but none has been found. Alternatively, we are led to suspect strong and systematic errors in the timescales used in the 17th and 18th centuries.

14.6 The Restricted Circular Three-Body Problem

This very special case was resolved in analytical form by Lagrange (1772). Consider two bodies, of mass m_1 and m_2, respectively, $(m_1 > m_2)$, on a relative orbit of circular form, and therefore at a constant relative distance. It is customary to write:

$$G = 1, \quad m_1 + m_2 = 1$$

so that the first (heavier) body has indicative mass $\mu_1 = 1 - \mu_2$ and the second $\mu_2 < 1/2$, and to use as unit of distance the fixed distance between the two bodies. It also follows that the common mean motion n is unity. It is also useful to introduce the parameter α defined by:

$$\alpha = \left(\frac{\mu_2}{3\mu_1} \right)^{1/3} \tag{14.16}$$

Let (x, y) be a barycentric reference system in the orbital plane, with the x-axis always coincident with the line joining the two bodies (in other words, a synodic reference system). The invariable coordinates of the two masses in it will be $(x_1, y_1) = (-\mu_2, 0)$ and $(x_2, y_2) = (\mu_1, 0)$. Therefore, this system rotates with respect to the inertial barycentric system, with an angular velocity equal to the mean motion of the two bodies n (which is $= 1$ in the present unit system, but that we keep explicitly for clarity).

Let us add to the system a third body of negligible mass m' (so that it is gravitationally attracted by the other two, but does not exert any attraction on them), orbiting in the same plane of the relative orbit, and therefore with zero velocity component perpendicular to that plane. Its distances from the two particles will be, respectively:

$$r_1^2 = (x + \mu_2)^2 + y^2, \quad r_2^2 = (x - \mu_1)^2 + y^2$$

It can be demonstrated that the accelerations of the test particle are given by:

$$\ddot{x} - 2n\dot{y} = \frac{\partial U}{\partial x}, \quad \ddot{y} + 2n\dot{x} = \frac{\partial U}{\partial y} \tag{14.17}$$

where U is a pseudopotential:

$$U(x, y) = \frac{n^2}{2}(x^2 + y^2) + \frac{\mu_1}{r_1} + \frac{\mu_2}{r_2} \tag{14.18}$$

As expected, the introduction of a noninertial system has produced the presence of the Coriolis terms ($-2n\dot{y}$ and in $-2n\dot{x}$ in Equation 14.17) and of the centrifugal acceleration (the terms in n^2x and n^2y in the equations of movement derived from Equation 14.18).

After some manipulation of the previous equations and their integration, the modulus of the total velocity V of the test particle in the rotating frame can be written as:

$$V = \dot{x}^2 + \dot{y}^2 = 2U - C \tag{14.19}$$

Equation 14.19 is the so-called Jacobi integral; the constant $C = 2U - V$ is the Jacobi constant (notice that the Jacobi integral is not an energy integral). From this, it follows that the movement of the test particle in the rotating frame can occur only in the regions of the (x, y) plane for which $2U > C$: inside the line for which $2U = C$ the particle has a stable position.

Without entering into the very interesting but very complex discussion of the topology of the curves $2U = C$ as a function of the value of C (usually referred to as Roche's curves), we state the following results:

- Five points of stability (zero velocity in the rotating frame) can be identified; these points are called Lagrangian points, and are designated L_1, L_2, L_3, L_4, L_5.
- $L_1, L_2,$ and L_3 are collinear on the line joining the two bodies (they all have $y = 0$). L_2 is outside the two bodies and opposite to that of the smaller mass, L_1 is between the two bodies and nearer to that of the smaller mass, L_3 is outside the two bodies and opposite to that of the larger mass (be advised that some authors interchange the designation of L_1 and L_2). The precise derivation of their x-coordinates is not a trivial exercise; approximately, their distances from the two masses are:

$$r_2(L_2) \approx \alpha + \frac{1}{3}\alpha^2, \quad r_2(L_1) \approx \alpha - \frac{1}{3}\alpha^2,$$

$$r_1(L_3) \approx 1 - \frac{7}{12}\left(\frac{\mu_2}{\mu_1} - \frac{\mu_2^2}{\mu_1^2}\right)$$

Therefore, L_1 and L_2 are almost symmetric with respect to the smaller body.

- L_4 and L_5 are at the vertices of the equilateral triangle for the two bodies, L_4 on the preceding (leading) point and L_5 on the following (trailing) one. Their coordinates are:

$$x = \frac{1}{2} - \mu_2, \qquad y = \pm \frac{\sqrt{3}}{2}$$

Several noticeable examples of Lagrangian points can be given:

- In the Sun–Jupiter system, the regions around L_4 and L_5 are occupied by many small bodies, collectively called Jupiter's Trojans; some 1500 of them are known, of which 900 are in the preceding cloud L_4, and 600 in the following L_5 cloud.
- In the Sun–Earth system (see Figure 14.4), L_1 is occupied by the Solar Heliospheric Observatory (SoHO), while L_2 is occupied by WMAP (the Wilkinson Microwave Anisotropy Probe), a satellite observing the cosmological background radiation, and could in the future be occupied by the successor of the Hubble Space Telescope.

On closer examination, those mathematical "points" of stability become regions of small extent, inside which the test particle moves with respect to the ideal point (the particle librates inside this region); the stability can also

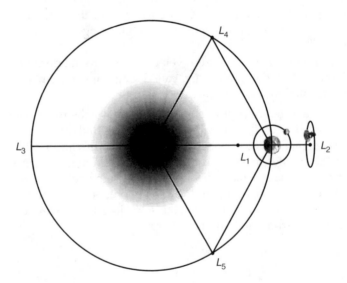

FIGURE 14.4

The Lagrange points of the Earth–Moon system. Since 1991, the cosmological satellite WMAP librates around L_2. To the same region will be directed the successor to the Hubble Space Telescope.

be very weak and readily perturbed, as for instance in the Lagrangian points of the Earth–Moon system. Also the stability if the L_1 and L_2 points of the Sun–Earth system is not very long, orbital corrections every 22 days or so are needed to keep a satellite parked there.

The same considerations can be applied to the case of binary stars, with the important difference that the two stellar masses are essentially equal. Examination of the Roche curves allows identification of the regions through which the two stars can exchange mass.

We will briefly examine three consequences of the restricted three-body problem, namely Tisserand's criterion for the stability of the orbital elements, the concept of the Hill sphere, and the technique of spacecraft gravitational assist:

1. *Tisserand criterion*: the Jacobi integral can be used to follow the changes of the orbital elements of a lighter particle after an encounter with a heavier body, for instance a comet after a near pass with Jupiter. Call (a, e, i) the initial elements and (a', e', i') those after the encounter; it can be shown that:

$$\frac{1}{2a} + \sqrt{a(1 - e^2)} \cos i = \frac{1}{2a'} + \sqrt{a'(1 - e'2)} \cos i'$$

   where the unit of length is the Sun–Jupiter distance, the unit of mass is the Sun's mass, and the timescale is such that Jupiter's angular velocity is unity. Tisserand's criterion is not a rigorous one, nor does it provide a definite proof that the comet observed after the encounter is the same observed before; nevertheless, if the orbital elements are widely different it can be concluded that the two sets do not belong to the same comet.

2. *Hill sphere*: the Hill sphere can be loosely defined as the volume around a planet where its gravitational influence overcomes that of the Sun. If Δ is the distance of the planet from the Sun, then the radius of its sphere R_H can be expressed as:

$$R_H = \alpha \Delta \tag{14.20}$$

   where the parameter α is given in Equation 14.16. Actually, several definitions of R_H are found in the literature, with different values of the constants and of the exponent. However, all have a linear dependence on the distance to the Sun, which leads to an increasing distance between the planets (in qualitative agreement with Titius–Bode law), and only a weak dependence on the mass of the planet. For the Earth, $R_H \approx 0.01$ AU, for Jupiter it is some 30 times larger. Should a comet pass inside Jupiter's Hill sphere, two different outcomes are possible: either the comet is captured in the inner Solar System, or it is gravitationally scattered outwards. Indeed, we know of the existence of a Jupiter family of comets, but also of an outside cloud (Oort's cloud) beyond 50,000 AU. During the initial formation phases of the Solar System, both mechanisms of accretion and of outward scattering of the original mass elements (known as planetesimals) by the large planets must have occurred in order to shape its present structure.

3. *Gravitational assist*: the previous considerations about Tisserand criterion and Hill sphere can be applied to the so-called gravitational slingshot effect, or gravitational assist. This is a navigational technique, by which the initial orbit of a spacecraft is purposely strongly perturbed by means of a passage inside the Hill sphere of a planet. Following this close encounter, the spacecraft is directed to reach another planet, in a position where the rocket energy would not be sufficient to go directly. Tisserand criterion will give an indication of the orbital elements after the passage. Many missions to planets and comets (e.g., Giotto, Galileo, Cassini, Rosetta) took advantage of this technique. A noticeable example is the trajectory of Ulysses, a spacecraft orbiting between Jupiter and the Sun along an orbit essentially perpendicular to the ecliptic plane, namely on an orbit where previously only comets of the Jupiter family were known (see Figure 14.5).

As has already been noted, strictly speaking the energy of the third body is not conserved in the general three-body problem,

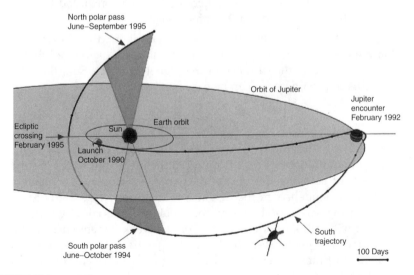

FIGURE 14.5
The initial orbit of Ulysses.

and therefore the same violation occurs in the gravitational assist: the increase of the semimajor axis of the third body must be compensated by a decrease of the semimajor axis of the second one. However, in practice, the mass of the spacecraft is so small that no effect can be detected on the planet orbit. The same reasoning cannot be applied to follow the original formation phases of the Solar System, because the planetesimals could have had masses comparable to those of the protoplanets.

14.7 A Nonspherical Body Plus a Small Nearby Satellite

In the case of an artificial satellite in low-Earth's orbit, the real shape of the Earth cannot be ignored. The following considerations will apply to other similar configurations, e.g., the two small moons Phobos and Deimos close to Mars, or an inner moon of Jupiter. The gravitational potential of the Earth (in the present simplified assumption an elliptical body of equatorial radius $a_\oplus$ and total mass $M_\oplus$ having rotational symmetry) on an external point-like body at distance r and latitude φ with respect to the Earth equator can be expressed as (see also Chapter 6):

$$U(r) = G\frac{M_\oplus}{r}$$

$$\times \left[1 + J_2 \frac{1}{r^2}\left(\frac{1}{2} - \frac{3}{2}\sin^2\varphi\right) + J_4\frac{1}{r^4}\left(\frac{3}{8} - \frac{15}{4}\sin^2\varphi + \frac{35}{8}\sin^4\varphi\right) + \cdots \right]$$

$$= G\frac{M_\oplus}{r}\left[1 - \sum_{n=1}^{\infty} \frac{1}{r^{2n}}J_{2n}P_{2n}(\sin\varphi) \right] \tag{14.21}$$

where the P_{2n} are Legendre's polynomials of degree $2n$.

In our qualitative discussion, we can truncate the development after the second term, writing the perturbing function as:

$$R = G\frac{M}{2r^3}J_2[1 - 3\sin^2\varphi] \tag{14.22}$$

where: $J_2 = -(C - A)/M_\oplus = -\frac{3}{2}Ja_\oplus^2$, $\quad J = \frac{3}{2}\frac{C - A}{M_\oplus a_\oplus^2}$.

The perturbing force will contain radial and nonradial terms. In other words it will be a noncentral one, with the accompanying consequences of nonclosed orbit of the satellite, of orbital plane precession, and advance (or retrogradation) of the pericenter on the orbital plane. According to Lagrange's planetary equations, we must expect the presence of secular, long-term and short-term periods in the orbital elements. The previously cited theorems of Delaunay and Tisserand insure that the semimajor axis, the inclination, and the eccentricity do not have secular terms.

The largest long-term periods have amplitudes of the order J_2 and periods half of that of the pericenter; higher-order terms in J_2 exist, with periods $1/4$, $1/6$ etc., of the fundamental one.

The short-term periods will deform the basic orbit, that which turns and undergoes long-term period perturbations. In particular, the line of the nodes has a retrograde motion with a period given by:

$$P_{\text{retr}} = \frac{4}{3}\,\pi\frac{1}{J_2}\frac{(1-e_0^2)^2 a_0^2}{n_0\cos i_0} \tag{14.23}$$

where (a_0, e_0, i_0) are the initial orbital elements of the satellite. The motion is more rapid when the inclination tends to zero, and it does not exist for polar orbits; it is slower for large eccentricities and for large semimajor axes.

Let us consider, for example, the motion of the perigee according to Lagrange's Equation 14.5. Even in the present simplified treatment, at the first order in J_2 we would find a fairly elaborate expression for $\omega(t)$:

$$\omega(t) = \omega_0 + \frac{3n_0 J_2}{4a_0^2}(-1+5\cos^2 i_0)t + f\left(\frac{J_2}{a_0^2}, \omega_0, n_0(t-t_0), e_0\right)$$

where f is an appropriate function of its arguments. In the coefficient of the secular term (the second one), there is a particular inclination for which $-1+5\cos^2 I_1 = 0$ ($I_1 = 63°26'$) dividing two cases: for $i_0 > I_1$ the movement of the pericenter is retrograde, for $i_0 < I_1$ it is direct. Should we push the treatment to the second order in J_2^2, we would find long-term periods with amplitudes of the order of J_2: this is a major problem in celestial mechanics, the long-term periods of the first order are found only in the second approximation.

Regarding the mean motion, it can be shown in the same manner that there is a second inclination I_2 for which $-1+3\cos^2 I_2 = 0$ ($I_2 = 35°15'$) again dividing two cases: if $i_0 > I_2$ the mean motion of the satellite is slower than the Keplerian value $2\pi/n_0$, if $i_0 < I_1$ the mean motion is faster.

In the case of an equatorial satellite (or small moon) in circular orbit, a simple formula expresses the influence of the nonsphericity of the planet of mass m on Kepler's third law; if f is the flattening as in Chapter 2, then:

$$\frac{P^2}{a^3} = \frac{4\pi^2}{Gm}(1-\kappa), \quad \kappa = \left(\frac{R}{a}\right)^2\left(f - \frac{\sigma}{2}\right)$$

where R/a is the ratio between the radius of the planet and the semimajor axis of the orbit of the small moon, and σ is the ratio between the

TABLE 14.2

Data on the Rotation of the Sun and Planets

Body	f	P_{rot}	P_{rot} (internal)	Obliquity (°)
Sun	0.0	25.4 j[a]		7.25
Mercury	0.0	58.65 j		0
Venus	0.0	(R)243.01 j		177.4
Earth	0.0034	23.9345 h		23.45
Moon	0.0020	27.32 j		6.67
Mars	0.0069	24.623 h		25.19
Jupiter	0.0649	9.841 h	9.925 h (system III)	3.12
Saturn	0.0980	10.233 h	10.656 h (system III)	26.73
Uranus	0.0229	(R) 17.9 h	(R) 17.240 h	97.86
Neptune	0.0170	19.0 h	16.11 h	29.56
Pluto	0.0?	(R) 6.387 j		119.6

(R) Means retrograde rotation; the obliquity is the tilt of the equator with respect to the orbital plane (for the sun with respect to the ecliptic).

[a] This is the adopted period at 16° latitude. The actual rotation rate varies with the latitude Λ as: $(14.37 - 2.33 \sin 2\Lambda - 1.56 \sin 4\Lambda)$ deg/day.

centrifugal force and the gravity at the equator of the planet ($\sigma = \omega^2 R^3 / Gm$). Consider for instance the case of Mars and Phobos:

$$f = 1/192, \quad \sigma = 1/218, \quad a(\text{Phobos})/R(\text{Mars}) \approx 2.8, \quad \kappa \approx 1/2400$$

a small deviation indeed. This result can be derived from Clairaut's theorem, which expresses the gravity at the distance a from a rotating body of ellipsoidal shape (see Chapter 6).

In addition to the low precision of the qualitative discussion dealt with in the present section, the case of a real artificial satellite is physically more complex, for the presence of nongravitational forces such as atmospheric drag, solar radiation pressure (a perturbation proportional to the ratio of the surface to the mass of the satellite), the Earth's magnetic field, and so on.

In order to complete the information on the Solar System, Table 14.2 provides some data on the flattening and rotation characteristics of the Sun, the planets, and the Moon. Note that the giant gaseous planets have different rotation periods according to how they are measured, namely, the period measurable by the visible surface details is not the same given by the internal rotation, which is manifested through the magnetic field and associated radio-frequency emission (system III).

By definition, the North pole of any planet is that above the ecliptic plane; therefore the rotation is considered retrograde for Venus, Uranus, and Pluto.

14.8   Other Interesting Cases

The Solar System offers many interesting problems, of which a few examples are given in the following.

Mercury: There is a very stable spin-orbit resonance 2:3 between rotation (58.65 days) and revolution (87.97 days). In this case the orbit is very elongated and so is the figure of the planet (whose Greek name is Hermes, so that the adjective Hermean is often used to refer to its characteristics). The major axis of the figure points to the Sun at each passage through the perihelion, but on successive passages opposite faces are presented to the Sun (see Figure 14.6). At aphelion, the major axis of the planet is at right angle to the Sun, but the planet is about 50% more distant than at perihelion. Owing to this motion, we see the same side of Mercury each time the planet comes closest to Earth. The study of the librations of the figure cannot be carried out from Earth, space missions such as NASA Messenger and ESA BepiColombo are needed.

Venus–Earth: Venus has a very slow retrograde rotation of 243 days, while revolving in 225 days. At each inferior conjunction it presents the same face to the Earth.

Synchronous rotation: As in the case of Moon–Earth, most natural satellites are in synchronous rotation, always keeping the same face pointed towards the planet (the so-called spin-locking effect). Most of the observational evidence came from the Voyagers, the twin space mission to the outer Solar System. This corotation situation is probably not the initial one, satellites were trapped in the synchronous rotation because of their elongated shapes, in a period of time much shorter than the age of the Solar System. Cassini's laws could thus be generalized to any spin-locked satellite undergoing orbital precession at a uniform rate. However, for the general case particular values of the inclinations are required, the so-called Cassini's states

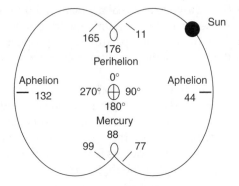

FIGURE 14.6
The apparent path of the Sun in a Hermean reference system. The ellipsoidal figure of the planet has been exaggerated. The numbers are terrestrial days; the whole figure is described in a Hermean day.

(see Colombo, Notes). Among all the moons, Hyperion is certainly a very peculiar one. This moon of Saturn is the largest irregularly shaped satellite ever observed, suggesting a major collision in the past with another body which blew part of Hyperion away. Its orbit is fairly eccentric ($e = 0.1$), and its radial distance from Saturn is large, approximately 25 Saturn' radii, so as to make tidal trapping into corotation unlikely. To complicate the matter, the revolution of Hyperion is in 4:3 resonance with Titan. The rotational period varies from one orbit to the next, a situation described as chaotic. Even the attitude of the moon is unstable, the spin axis tumbles in space. Tumbling has also been observed on a few asteroids, and probably results from collisions (see Notes).

Notes

- For the theory of the JPL Ephemerides, see in:
  Standish, E.M., 1982, *Astronomy and Astrophysics*, **114**, p. 297 and **233**, p. 252, 1990, and in: http://ssd.jpl.nasa.gov/iau-comm4/README(2002).
- The modified Newtonian dynamics (MOND) has been proposed by Milgrom, M., 1983, *Astrophysics Journal*, **270**, p. 365. See also in Sanders, R.H., McGaugh, S.S., 2002, *Annual Review Astronomy Astrophysics*, **40**, p. 263, and in Scarpa, R., Marconi, G., Gilmozzi, R., 2003, Using globular clusters to test gravity in the weak acceleration regime, *Astronomy and Astrophysics*, **405**, pp. L15–L18.

  For the anomalous acceleration of Pioneer 10 and 11 see: Anderson, J.D., Laing, P.A., Lau, E.L., Liu, A.S., Nieto, M.M., Turyshev, S.G., 2002, Study of the anomalous acceleration of Pioneer 10 and 11, *Physics Review*, D65.
- Regarding Mercury's perihelion advance and general relativity, see in Nobili, A.M., Will, C.M., 1986, The real value of Mercury's perihelion advance, *Nature*, **320**, pp. 39–41.

  However, the relativistic effects on Mercury's orbit are not limited to the one described here. According to the Lense–Thirring effect, a precession of the plane of the geodesic orbit of a test particle around a rotating mass might also be detected. It arises from the coupling of the rotation of the central mass with the orbital angular momentum of the test particle. This precession is described as resulting from the dragging of inertial frames. The theory is reviewed, for instance by Mashhoon, B. *Gravitoelectromagnetism*, a brief review, General Relativity and Quantum Cosmology, abstract gr-qc/0311030 (http://arXiv.org/abs/gr-qc/0311030), 2003. The Lense–Thirring effect is also present in Earth's artificial satellites. For its measurement, see for instance: Ciufolini, I., 2000, The 1995–1999 measurements of the Lense–Thirring effect using laser-ranged satellites, *Classical Quantum Gravity*, **17** (21 June), pp. 2369–2380.

- For an historical review of the lunar characteristics see:
  Kopal, Z., 1971, *Physics and Astronomy of the Moon*, Academic Press, New York.

- For the dynamical justification of the laws found by Cassini concerning the rotation of the Moon see Danby, J.M.A., 1988, *Fundamentals of Celestial Mechanics*, Willmann-Bell. The Cassini laws and the Cassini states were examined, for instance by Colombo, G., 1966, Cassini's second and third law, *Astronomical Journal*, **71**, pp. 891–896.

- Chapront-Touzé, M., Chapront, J., 1983, Developed a lunar theory (Ephémérides Lunaires Parisiennes) in two papers: the lunar ephemeris ELP 2000, *Astronomy Astrophysics*, **124**, pp. 50–62, and Chapront-Touzé, M., Chapront, J., 1988, ELP 2000-85. A semi-analytical lunar ephemeris adequate for historical times, *Astronomy Astrophysics*, **190**, pp. 342–352.

- Recent data on space missions to the Moon are given on the following websites:
  http://sse.jpl.nasa.gov/missions/moon_missns/moon-lp.html
  For Clementine: http://www.nrl.navy.mil/clementine/
  For the Lunar Prospector: http://lunar.arc.nasa.gov/
  For the European mission SMART-1 see:
  http://sci.esa.int/science-e/www/area/index.cfm?fareaid = 10#

- For a discussion regarding the long-term stability of the Solar System and numerical integration of the orbits see: Nobili, A.M., Milani, A., Carpino, M., 1989, Fundamental frequencies and small divisors in the orbits of the outer planets, *Astronomy Astrophysics*, **210**, pp. 313–336.

- In 1979, Pluto was discovered to be a double planet, with a fairly large moon named Charon; many authors today consider this double body not as a real planet, but simply as the largest in a fairly numerous and massive population of trans-Neptunian objects orbiting in the so-called Kuiper belt.

- Most data on the outer solar system still come from the Voyagers, see: http://voyager.jpl.nasa.gov/index.html

- For the SoHO orbit: http://sohowww.estec.esa.nl/, http://sohowww.nascom.nasa.gov/
  For the Ulysses orbit: http://helio.esa.int/ulysses/
  http://ulysses-ops.jpl.esa.int/ulysses/
  For the WMAP orbit: http://map.gsfc.nasa.gov/index.html

- The NASA Mercury Messenger and ESA BepiColombo websites are: http://messenger.jhuapl.edu/

http://solarsystem.nasa.gov/missions/merc_missns/merc-msgr.html
http://esa.int/science/bepicolombo/

- Regarding Hyperion motion, see: Black, G.J., Nicholson, P.D., Thomas, P.C., 1995, Hyperion: rotational dynamics, *Icarus*, **117**, pp. 149–171.
- Regarding tumbling asteroids: Paolicchi, P., Burns, J.A., Weidenschilling, S.J., 2002, *Side Effects of Collisions: Spin Rate Changes, Tumbling Rotation States, and Binary Asteroids*, Asteroids III, W.F. Jr. Bottke, A. Cellino, P. Paolicchi, R.P. Binzel, eds., University of Arizona Press, Tucson.

Exercise

An artificial satellite is usually launched with an easterly component in order to obtain a speed advantage from Earth's rotation from west to east. Using Equation 14.23 show that for a circular orbit the node line regresses westward by:

$$\psi = \frac{6\pi}{5}\left(\frac{C-A}{C}\right)\left(\frac{R}{a}\right)^2 \cos i = 2\pi J_2\left(\frac{R}{a}\right)^2 \cos i$$

$$\approx 0.01\left(\frac{R}{a}\right)^2 \cos i \ (\text{radians/revolution})$$

where ψ is measured with respect to a fixed direction in the equatorial plane.

15

Eclipses and Occultations

In this chapter, we will briefly examine the solar and lunar eclipses, and the occultations of the stars by the Moon. The treatment will be limited essentially only to the geometric aspect, even if eclipses also have important physical implications. For instance, when the Moon enters the terrestrial shadow, the flux of solar photons on its surface is suddenly turned off. In the same way, the thermal balance of a satellite will be dramatically altered by an eclipse. We have already commented on the usefulness of ancient eclipses to determine the rate of change of the Earth diurnal rotation. The solar eclipses also deserve the merit of having permitted the discovery of the very hot solar corona by visual observations. For instance, Halley in 1715 noticed both the corona and red prominences, but he tended to attribute the luminous gases to the heated lunar surface. In 1836 Bailey discovered, during an annular eclipse, the bright rays coming from the valley separating the mountains on the lunar limb (the so-called Bailey's beads), and in 1842 he noticed again the corona and the prominences. Finally, during the eclipse of August 18, 1860, Father A. Secchi, S.J., and W. de la Rue for the first time could photograph it from two separate places, and thus it became clear that the gases were indeed around the Sun.

In the same manner, during a lunar eclipse a very tenuous lunar atmosphere can be observed, especially in the light of the yellow Na doublet (see Notes).

The generic term occultation indicates the situation of an opaque body passing across the line of sight to a planet or to a star, therefore the term transit (say of Mercury or of Venus in front of the solar disk, see also Chapter 13) means actually a partial occultation; also a solar eclipse is a form of occultation. Occultations can provide very important information on the occulted body, such as its position. Lunar occultations were, indeed, the most precise way to determine the positions of radio or x-ray sources in the early days of these new astronomies (two examples: the FK4 position of 3C 273B by Hazard et al. (1971, see Notes to Chapter 6) and the European satellite EXOSAT, launched with the specific purpose of x-ray occultations). Furthermore, occultations have allowed determination of the structure of the occulted source, such as the duplicity, the apparent size, or the limb darkening of several stars, and also of the occulting body, for instance the rings of Uranus were first detected observing the occultation of a star, and subsequently confirmed by Voyager's images (see Notes).

15.1 Moon's Phases

The Moon is an opaque body illuminated by the Sun, with a day and a night side. When the angle between the Moon and the Sun is at its minimum value as seen by the geocentric observer, in other words when the night side faces the Earth, it is the astronomical new moon. Following the new moon, we see a gradual increase in the illuminated portion of the disk (waxing moon). The crescent moon is seen at sunset on the western horizon. After approximately 7.4 days, the ecliptic longitude of the Moon is 90° greater than that of the Sun; this is the astronomical first quarter. The Moon is in the meridian of each place at approximately 6 p.m. local time. The first quarter marks the end of the crescent phase and the start of the gibbous phase. After another 7.4 days, the longitude of the Moon is 180° larger than that of the Sun; the lunar disk is fully illuminated, it rises when the Sun sets, it reaches the meridian around local midnight, and finally sets when the Sun rises. Notice that full moon is not equal to phase angle (namely the angle Earth–Moon–Sun) $\alpha = 0$, otherwise an eclipse would occur; the phase angles of full moons are usually larger than $1°.5$, because the lunar latitude is different to zero. The next two periods of 7.4 days lead, respectively, to the last quarter and again to the new moon. This half lunation is also called the waning moon. We have already noted in Chapter 14 that the solar perturbations influence the duration of the lunations; for instance in January 1974 there was one of the longest, $29^{j}19^{h}55^{m}$, while in June 2035 there will be one of the shortest, $29^{j}06^{h}39^{m}$ (see Meeus, 1991).

The disk of the Moon can be taken as circular, so that the illuminated surface will appear as the intersection between the circular limb and an elliptical terminator, which is the border of the circular illuminated area as projected on the surface. Let us call k the fraction of the illuminated area, and α the phase angle. Then:

$$k = \frac{1}{2}(1 + \cos \alpha) \tag{15.1}$$

Notice that k is also the fraction of the illuminated diameter perpendicular to the terminator. According to the formulae of Chapter 3, the phase angle α can be derived from the geocentric equatorial or ecliptic coordinates of the Sun and of the Moon, by the intermediate of the angular distance ϑ between the two bodies (which we can call the elongation of the Moon from the Sun along the great circle joining them):

$$\cos \vartheta = \sin \delta_{\odot} \sin \delta_{\leftmoon} + \cos \delta_{\odot} \cos \delta_{\leftmoon} \cos(\alpha_{\odot} - \alpha_{\leftmoon})$$
$$= \cos \beta_{\leftmoon} \cos(\lambda_{\odot} - \lambda_{\leftmoon}) \tag{15.2}$$

$$\tan \alpha = \frac{\sin \vartheta}{\Delta - \cos \vartheta} \quad (15.3)$$

Δ being the distance Earth–Moon normalized to the instantaneous Earth–Sun distance.

The cusps (popularly, horns) of the terminator always point away from the Sun; therefore if C is the middle point of the illuminated disk, its position angle p (from North through East, and paying attention to the true quadrant) is given by:

$$p = \frac{\cos \delta_\odot \sin(\alpha_\odot - \alpha_{\leftmoon})}{\sin \delta_\odot \cos \delta_{\leftmoon} - \cos \delta_\odot \sin \delta_{\leftmoon} \cos(\alpha_\odot - \alpha_{\leftmoon})} \quad (15.4)$$

The cusps are then at $p \pm 90°$. For any particular observer, the zenith distance of C will be given by $p - q$, where q is the parallactic angle (see Equation 3.18 and Equation 3.19). The best chance to see the very first crescent phase after the new moon, is when the ecliptic is as perpendicular as possible to the horizon, namely in March for a Northern observer; for similar reasons, the winter full moon will be higher in the sky for the same observer.

15.2 Conditions for the Occurrence of an Eclipse

Solar eclipses occur when the Moon passes in front of the Sun (Moon and Sun in conjunction), therefore necessarily around new moon phases. In the same manner, lunar eclipses happen when the shadow of the Earth falls on the Moon (Moon and Sun in opposition), namely around full moon. However, because of the inclination i of the lunar orbit on the ecliptic, the conditions of new moon or of full moon are not sufficient for the occurrence of the eclipse. In addition, the Sun (or the antisolar point) and the Moon must be close to the same orbital node, as shown in Figure 15.1. Therefore, the Moon must be as close as possible to the ecliptic (hence the name *ecliptic*, the locus around which eclipses can occur). In other words, the ecliptic latitude of the Moon, the distance of the Sun from one of the nodes, and the angular distance between the two bodies (or between the Moon and the antisolar point), must all be smaller than given maximum values, whose exact amounts vary from eclipse to eclipse essentially due to the varying conditions of the lunar orbit.

From Equation 15.2, at new moon (or full moon) the angular distance ϑ between the Moon and the Sun (or the antisolar point) will reach a minimum value σ given approximately by:

$$\sigma \approx |\beta_{\leftmoon}| \quad (15.5)$$

In order to verify the possibility of an eclipse, this minimum angular separation must be confronted with the angular diameters of the Sun, of the Moon, and of the Earth as seen from the Moon. Actually though, the Sun is

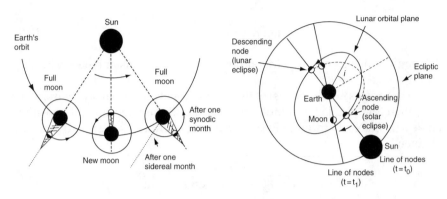

FIGURE 15.1
Possibility of solar and lunar eclipses. On the right panel, the regression of the lines of nodes is indicated.

an extended source, so that more precise considerations are needed regarding the shadows projected by the Moon and by the Earth: the shadows are composed of two cones, one of total obscuration, and one of partial obscuration (in Latin, *umbra* and *penumbra*, respectively). In the following section, we demonstrate that a solar eclipse cannot take place if the latitude of the Moon is larger than 1°35′, or else if the distance of the Sun from the node is greater than approximately 18°, while a lunar eclipse cannot occur if the distance of the Sun from the node exceeds approximately 10°. These angles (and the angular diameters of the two bodies) are small enough that plane trigonometry is sufficient for the present approximate treatment.

15.3 Solar Eclipses

To gain more insight into the circumstances of a solar eclipse, let us derive some preliminary indications from a very simple *ad hoc* model in which the observer is on the equator, the Sun and the Moon are both on the celestial equator and follow circular geocentric orbits, and the eclipse happens at noon. The observer would then see the disk of the Moon moving eastwards and approaching the west limb of the Sun, with an angular velocity whose value can be easily computed in the following way:

- The Sun moves eastward with respect to the fixed stars with an angular velocity $n_\odot = 0°.9856/\text{j}^{-1}$; the Moon moves in the same direction with its sidereal velocity of $n_{\mathbb{D}} = 13°.1763/\text{j}^{-1}$, so that the Moon overtakes the Sun with a relative angular velocity of $0′.5079$ per minute. In order to cross the entire solar disk of $32′$, the eastern limb of the Moon will then take approximately 62 min; in total, the eclipse will last about 2 h, of which only a few minutes are really of totality.

- The shadow projected by the eastern border of the Moon will run eastward above the Earth's surface with a linear velocity of approximately 0.93 km/sec, but the equatorial observer is moving in the same direction with a velocity of 0.46 km/sec, so that the apparent velocity is 0.47 km/sec, again toward east.
- If the Sun was a point source at infinity, the shadow of the Moon (here, a disk with a diameter of 3480 km) would be a disk with a cross-section decidedly smaller than the Earth itself, so that a particular solar eclipse would be seen only by a fraction of terrestrial observers, namely by those located inside this circle. In reality, the Sun is a source at finite distance with an angular extension $\theta_\odot$ of approximately 32′, so that the shadow can be divided into two conical regions of different obscuration. The vertex of the umbra is located at approximately 400,000 km from the lunar center, a value very close to the Earth–Moon distance; owing to the strong ellipticity of the lunar orbit at certain times the vertex will be just inside the surface of the Earth and at other times it will be just outside it. In its turn, the Earth can appear in these cones in a variety of relative positions (see Figure 15.2, all angles are exaggerated). Therefore, a total solar eclipse will be a rare event; more common will be the occurrence of a partial eclipse.

To carry out precise calculations of the circumstances and times of a particular eclipse, several additional factors must be taken into account:

- The elliptical orbits, and consequently the different apparent diameters of the Moon and the Sun. The apparent solar semidiameter $\theta_\odot/2$ varies between 15′44″ and 16′18″, the apparent lunar semidiameter $\theta_{\text{☽}}/2$ between 14′41″ and 16′44″.
- The obliquity ε of the ecliptic on the equator.
- The position of the observer on the Earth surface. The linear velocity of the shadow on the tangent plane through the observer

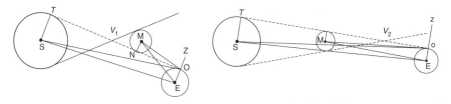

FIGURE 15.2

Different circumstances of a solar eclipse as seen by the observer O having his zenith in Z. On the left panel, the Moon occults the limb of the Sun in T (on the horizon of O). On the right panel, the Moon is fully inside the solar disk. V_1 is the vertex of the penumbral cone, V_2 the vertex of the umbral one. The angle ESO is the solar parallax $\pi_\odot$, the angle EMO is the lunar parallax $\pi_{\text{☽}}$, the angle SOT is the angular semidiameter of the Sun $\theta_\odot/2$, the angle MON is the angular semidiameter of the Moon $\theta_{\text{☽}}/2$.

increases with the latitude ϕ, from 2009 km/h at the equator, to 2514 km/h at $\phi = 45°$, to 2844 km/h at $\phi = 60°$, so that the eclipses are longer at the equator. Furthermore, at noon the velocity will be smaller because the Sun is higher in the sky. From these numbers, it appears that the velocities of the shadow on the ground are comparable with those of high-speed airplanes, so that the shadow can be followed for a considerable time by flying telescopes.

The projection of the shadow over the Earth surface is therefore a locus of complex shape (see for instance the *Astronomical Almanac*, Section A), according to several possible configurations. If the instantaneous direction joining the centers of the Moon and of the Sun intercepts the Earth, the eclipse is said a central one. However, even a central eclipse does not necessarily imply totality, because the disk of the Moon can be slightly smaller than that of the Sun (Moon at apogee, Earth outside the vertex of the umbral cone); a bright rim is then observed (annular eclipse). Nor is a central eclipse absolutely mandatory for a total eclipse; there are (rare) circumstances when observers around the poles of the Earth can see a total noncentral eclipse. The width of the strip of total eclipse usually varies between 40 and 100 km, but it can shrink to zero or rise to 700 km. We have already seen that the duration of totality is maximum at the equator around noon, when the Earth is at aphelion and the Moon at perigee. The maximum possible duration is 7^m31^s, but this rare event will not happen before the year 2186. In July 2009, a solar eclipse will occur with duration of more than 6^m.

The other following points are worth mentioning:

- The lunar barycenter (used in the ephemerides) does not coincide with the center of the disk; there are minute differences between the times calculated from dynamical considerations and the actual times.

- The calculations of an eclipse are usually performed several years in advance; dynamical calculations use TD as the time argument, but observations refer to UT1. The difference $\Delta T = \text{TD} - \text{UT1}$ is therefore only tentative (see Chapter 10).

- A source of larger time differences is the atmospheric refraction (see Chapter 11), which is usually not included in the published circumstances of an eclipse.

15.4 Lunar Eclipses

Many of the previous considerations could be repeated in the present context. The Earth produces a cone of total shadow extending for approximately 0.01 AU in the antisolar direction (therefore, to the

Lagrangian point L2), and a region of penumbra, as shown in Figure 15.3 (all angles are exaggerated).

Analogous to Figure 15.2, the angle OME is the horizontal parallax π_{D} of the Moon, the angle SET is the apparent angular radius $\theta_\odot/2$ of the Sun, the angle ETO is the solar horizontal parallax $\pi_\odot$, and the segment ES is the geocentric distance to the Sun. All these quantities are known with great precision. Calling f_1 the angle LEM on the left panel (namely the angular radius of the penumbra as seen from the geocenter), and f_2 the angle V_2EM (the corresponding radius for the umbra), we obtain:

$$f_1 = \pi_{\mathrm{D}} + \pi_\odot + \theta_\odot/2 \qquad f_2 = \pi_{\mathrm{D}} + \pi_\odot - \theta_\odot/2 \qquad (15.6)$$

two relations that were known to Hipparchus.

To be more precise, the Earth has to be considered as a flattened disk (see Chapter 2), and the disk of the umbra as slightly more flattened than the disk of the Earth. To compensate for this effect, the equatorial parallax π_{D} is substituted with that at 45° latitude, which is $0.998\pi_{\mathrm{D}}$. Furthermore, the extinction caused by the terrestrial atmosphere enlarges the width of the angles f_1 and f_2 by approximately 2%.

The angular width of the Earth's shadow at the distance of the Moon varies from 75' to 90', so that a lunar eclipse can be seen by all observers having the Moon above the horizon, and the duration of totality can exceed 2 h, two aspects indeed greatly different with respect to the solar case. The particular geometric circumstances produce total or partial penumbral, and total or partial umbral eclipses. For the Moon to be totally occulted by the umbra, the Sun cannot be more than 4°.6 from the node. During a penumbral eclipse, the dimming of the lunar disk is usually so slight as to go unnoticed by visual observations.

Our atmosphere causes another remarkable phenomenon, namely the faint illumination of the lunar disk seen during a total eclipse (the color of this faint light varies appreciably from eclipse to eclipse, but it is usually a deep red). The reason is the diffusion of solar radiation by the lower atmospheric layers, where there is a variable content of clouds, haze, pollutants, ozone, meteoric dust, etc. Scattering of the solar light by clouds of the Earth's atmosphere is obviously present also outside the eclipses, and it

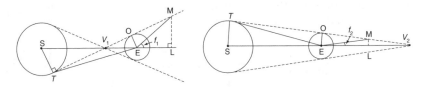

FIGURE 15.3
Penumbra (left panel) and umbra (right panel) cast by the Earth. S = center of the Sun, E = center of the Earth, M = point on the penumbral or umbral cone at the distance Earth–Moon (not necessarily the center of the Moon), LM = semidiameter of the cone at the distance of the Moon. Points V_1 and V_2 are both on the ecliptic.

is seen as a faint gray illumination beyond the terminator; because of its color, it is usually called ashen light (in Latin, *cinerea lux*).

15.5 Besselian Elements and Magnitude of the Eclipse

The precise calculation of the circumstances of the solar and lunar eclipses can be made with a method devised by Bessel. It can be adapted to calculate the circumstances of a stellar occultation by the Moon or by a planet.

Firstly, let us discuss a solar eclipse. Consider the line between the center of the Sun and the center of the Moon, namely the axis of the shadow, and its intersection with a perpendicular plane passing through the Earth's center. This is Bessel's fundamental plane; on it, the umbra and the penumbra are projected as two concentric circles which are at a given distance and position angle from the geocenter. A so-called fundamental Cartesian geocentric system (x, y, z) is then instituted, with the z-axis parallel to the shadow axis and positive toward the Moon, the x-axis parallel to the equator and positive toward east, and the y-axis positive toward north. At every instant, the geocentric equatorial coordinates of the Moon $\mathbf{R}_{\leftmoon}$ and of the Sun $\mathbf{R}_{\odot}$ are known with great precision. For construction, the shadow axis (and thus also the z-axis) is parallel to the vector $\mathbf{G} = \mathbf{R}_{\odot} - \mathbf{R}_{\leftmoon}$, and it intercepts the celestial sphere in a point Z having equatorial coordinates (α_Z, δ_Z). Denoting as usual with the letter π the horizontal parallaxes, and normalizing to the Sun–Earth distance, we also have:

$$r_{\leftmoon} = \frac{R_{\leftmoon}}{R_{\odot}} = \frac{\sin \pi_{\odot}}{R_{\odot} \sin \pi_{\leftmoon}}, \quad \mathbf{g} = \frac{\mathbf{G}}{R_{\odot}} = \mathbf{r}_{\odot} - \mathbf{r}_{\leftmoon}, \quad g = \frac{G}{R_{\odot}} \tag{15.7}$$

$$\mathbf{r}_{\odot} = \begin{pmatrix} \cos \alpha_{\odot} \cos \delta_{\odot} \\ \sin \alpha_{\odot} \cos \delta_{\odot} \\ \sin \delta_{\odot} \end{pmatrix}, \quad \mathbf{r}_{\leftmoon} = \frac{\sin \pi_{\odot}}{R_{\odot} \sin \pi_{\leftmoon}} \begin{pmatrix} \cos \alpha_{\leftmoon} \cos \delta_{\leftmoon} \\ \sin \alpha_{\leftmoon} \cos \delta_{\leftmoon} \\ \sin \delta_{\leftmoon} \end{pmatrix},$$

$$\mathbf{g} = \begin{pmatrix} \cos \alpha_Z \cos \delta_Z \\ \sin \alpha_Z \cos \delta_Z \\ \sin \delta_Z \end{pmatrix} \tag{15.8}$$

The vector distance of the Moon in the fundamental system, and in units of the terrestrial radius, is then given by:

$$\mathbf{r}_{\leftmoon \, \text{Fund}} = \frac{1}{\sin \pi_{\leftmoon}} \begin{pmatrix} \cos \delta_{\leftmoon} \sin(\alpha_{\leftmoon} - \alpha_Z) \\ \sin \delta_{\leftmoon} \cos \delta_Z - \cos \delta_{\leftmoon} \cos(\alpha_{\leftmoon} - \alpha_Z) \sin \delta_Z \\ \sin \delta_{\leftmoon} \sin \delta_Z + \cos \delta_{\leftmoon} \cos(\alpha_{\leftmoon} - \alpha_Z) \cos \delta_Z \end{pmatrix} \tag{15.9}$$

The x,y components of this vector are the intersection of the axis of the shadow with the fundamental plane. For practical uses, the right ascension of Z is substituted by its Greenwich hour angle, indicated for this application with the letter μ: $\mu_Z = ST_{\text{Greenw}} - \alpha_Z$.

The calculation of the radii of the penumbra and of the umbra on the fundamental plane, l_1 and l_2, respectively, in units of the Earth radius, can be carried out without difficulty from the basic geometry of the system (notice that $l_1 = l_2 + 0.546$). Finally, the almanacs provide tables (or interpolating formulae) of the quantities $x, y, \sin \alpha_Z, \cos \delta_Z, \mu, l_1, l_2$ and the time derivatives of α_Z, δ_Z, indicated with α'_Z, δ'_Z (time in this context usually means Universal Time), which are the Besselian elements of the eclipse. At this stage, we insert into the procedure the position of the observer in the fundamental reference system, which can be derived by means of a suitable rotation of its ellipsoidal coordinates (ρ, Λ, ϕ'), given in Chapter 2, and the velocity with respect to the shadow. This relative velocity can be computed by knowing the Earth diurnal rotation and the lunar motion. The fairly intricate procedure (see the *Explanatory Supplement*) is carried out through a set of auxiliary Bessel elements. Regular and auxiliary Bessel elements are the basis for deriving the general circumstances of a given solar eclipse in a given place on the Earth surface.

For lunar eclipses, the procedure can be considerably simplified, because the observer is on the body casting the shadow, and the circumstances are identical for all terrestrial observers having the Moon above the horizon. Therefore, the z-axis is not used, and the origin of the fundamental coordinates is the center of the umbra. The Besselian elements are not even published by the *Astronomical Ephemerides*. See an example in the Exercises.

For a lunar occultation of a star, the fundamental plane is geocentric and perpendicular to the line joining the center of the Moon and the star; the star being essentially a point source at infinity, the lunar shadow is a cylinder having cross-section equal to the diameter of the Moon. The method can be adapted to the occultation of a star by a planet, by scaling down the size of the shadow.

Now, let us discuss the so-called magnitude of an eclipse (the term magnitude here has a different meaning from that employed for stellar magnitudes, to be introduced in Chapter 16). In the case of a solar eclipse, the magnitude is the fraction of the solar diameter covered by the Moon at the instant of greatest phase, as seen by a given observer (the term obscuration means more rigorously the fraction of covered area, but the two terms are loosely used as synonyms). Notice that for total eclipses the magnitude can be larger than 1.

The same can be said for lunar eclipses. Let us introduce the quantity:

$$g_i = \frac{f_i - (\sigma - \theta_{\mathrm{D}}/2)}{\theta_{\mathrm{D}}} \qquad (i = 1, 2)$$

where $f_{1,2}$ are the quantities defined in 15.6, $g_1 = 0$ at the beginning of the partial phase, $0 < g_1 < 1$ during partiality, $g_2 = 1$ at the beginning of totality, and $g_2 > 1$ during totality. Therefore, g_1 corresponds to the eclipsed fraction of the lunar diameter during partiality, and g_2 to the penetration inside the umbra during totality. The maximum of the eclipse occurs at the instant of minimum distance σ, in the middle of the eclipse; g is then called magnitude of the lunar eclipse, and its value is published in almanacs.

15.6 Number and Repetitions of Eclipses

To determine the yearly number and repetition cycles of eclipses, we have to recall the values of the several lunar months discussed in Chapter 14. Eclipses are possible around the beginning of every draconitic month, which is the time interval of 27.21 d between two passages through the node.

Suppose a lunar eclipse (phase of full moon) occurred. The Sun moves away from the node with a velocity of $30°.67$ per synodic month of 27.21 d, so that after six synodic months (177.18 d) the Sun has moved in longitude by about $174°.64$. On the other hand, the node regresses at a speed of $-0°.53/d$, namely of $-9°.38$ after six synodic months, and the opposite node is located at longitude $180° - 9°.38 = 170°.62$. Therefore the full moon is at $174°.64 - 170°.62 = 4°.02$, and another lunar eclipse can occur; if all circumstances are favorable, a third lunar eclipse can happen before the end of that year. Notice though that there is no warranty that the first eclipse happens; there are indeed years with no umbral eclipses.

Now, consider solar eclipses (phase of new moon). The motion of $30°.67$ per synodic month of the Sun away from the node is less than twice the minimum ecliptic distance, so that at least one solar eclipse is inevitable at each node.

Therefore, the minimum number of eclipses in one year is two, and in that case, both will be solar eclipses. At the other extreme, if all orbital circumstances are favorable, two solar eclipses can take place at each node, and actually if the first is in early January, a fifth can occur, because there can be 13 lunations in that year (this rare circumstance will not happen before the year 2160). Therefore, 2, 3, 4, or even 5 solar and 3 lunar eclipses can occur in 13.5 lunations. This interval of time is larger than the Julian year, so that in conclusion the maximum yearly number is 7, either 5 solar and 2 lunar, or 4 solar and 3 lunar.

Regarding the repetitions, the Chaldean and Babylonian astronomers had detected a cycle of 18 yr and 11 d, after which the eclipses would repeat themselves in the same sequence and at the same node under almost identical conditions. This basic cycle is called Saros (Greek for repetition). To understand its value of 18 y 11 d, consider that the ratio between the durations of the draconitic and the synodic months is numerically very close to the ratio of the integer numbers 242 and 223: if a given date is full moon, after 223 synodic months, namely after $6585^j.32$, the Moon again will be full. Those 6585.32 days correspond to 18 yr and 11.3 d if there were 4 bissextile years, or to 18 yr and 10.3 d if the number of bissextile years was 5. This time interval is practically equal to 242 draconitic months: although there is a difference of $0^j.46$, the Sun will not have moved out of the node by more than $29'$.

TABLE 15.1

Three Solar Eclipses at Intervals of One Saros

			Central Point of Path	
Date	Type	Duration	Longitude	Latitude
June 11, 1983	Total	5^m11^s	$-126°91$	$-4°70$
June 21, 2001	Total	4^m56^s	$-8°.73$	$-10°.97$
July 02, 2019	Total	4^m32^s	$+104°.94$	$-17°.48$

Furthermore, the cycle also matches 239 anomalistic months, which guarantees almost the same position of the Moon in its orbit.

Table 15.1 shows an example of three total solar eclipses at intervals of one Saros.

Regularities are evident, the spacing is of 18 years + 11 days, because the eclipses occur at the same ascending node, and each is a total one, with similar duration. Each eclipse moves about 120° west of its predecessor,

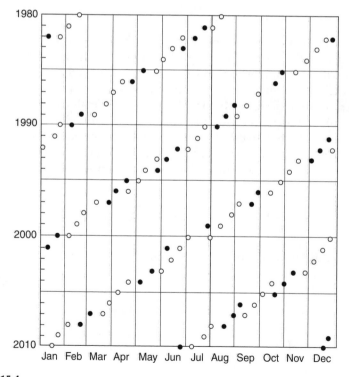

FIGURE 15.4

The eclipses from 1980 to 2010 arranged along 1.5 Saros period. Filled circles: lunar eclipses; open circles: solar eclipses. (Adapted from Roth, *Compendium of Practical Astronomy*, Vol. 3, Springer-Verlag, 1994.)

because the fraction $0^j.32$ corresponds to a diurnal rotation of 120°, so that the zones of totality move toward west by the same quantity at each Saros, and after three periods (54 years) returning approximately to the same longitude. Furthermore, the totality moves south (it would move north for an eclipse at the descending node). Figure 15.4 shows all eclipses from 1980 to 2010, namely over 1.5 Saros cycles. By examining these patterns, we can see that eclipses occur every year at an earlier date due to the retrogradation of the nodes. The same sequence repeats after one Saros.

It is possible to find other periodicities and commensurabilities between the lunar months and the year. For instance, it was noticed that 235 lunations closely correspond to 19 tropical years, a period called the Metonic cycle. Hipparchus is credited with having discovered cycles of 20.2, 345, and 441.3 yr, but many more can be found by moving to high integer numbers. These regularities are not simply numerology, because they can be used in dynamical theories of the Sun–Earth–Moon system (see the Notes).

15.7 Stellar Occultations

When a star is occulted by the Moon, we have the means to determine its position and structure with great accuracy, at least in the direction along the path of immersion and reemersion. The method could actually be used to determine the position of the Moon, or the longitude of the observer, or to improve the lunar orbit theory, or finally, to ameliorate the knowledge of the rotation of the Earth. Consider indeed the spherical triangle between the celestial North pole P, the center of the Moon M and the contact point X between the star of coordinates (α, δ) and the lunar limb (namely, the immersion or emersion point). The following conditions apply:

$$\text{arc } PN = 90 - \delta_{\leftmoon}, \quad \text{arc } XN = 90 - \delta, \quad \text{arc } MX = \alpha_{\leftmoon} - \alpha$$

Let the position angle of M with respect to X be p. From Equation 15.4 we have:

$$\frac{1}{2}\theta_{\leftmoon}\sin p = (\alpha - \alpha_{\leftmoon})\cos\delta, \quad \frac{1}{2}\theta_{\leftmoon}\cos p = \delta - \delta_{\leftmoon} \tag{15.10}$$

$$\frac{1}{2}\theta_{\leftmoon} = (\alpha - \alpha_{\leftmoon})\cos\delta\sin p + (\delta - \delta_{\leftmoon})\cos p,$$

$$\tan p = \frac{(\alpha - \alpha_{\leftmoon})\cos\delta}{\delta - \delta_{\leftmoon}} \tag{15.11}$$

If the observed values of angular diameter and position angle do not coincide with those calculated by the ephemerides (and corrected for the slight difference between the barycenter and center of the apparent disk), then the theory of the Moon, or the position of the observer, or the stellar coordinates are in error.

Attention is concentrated here on determining the structure of the source, recalling the well-known phenomenon of Fresnel's light diffraction by an opaque straight screen. The Fresnel diffraction fringes produced by occultations have a typical length x and angle ψ given by:

$$x = \sqrt{\frac{\lambda d}{2}}, \quad \psi = \sqrt{\frac{\lambda}{2d}}$$

where λ is the wavelength of the light, and d is the distance Earth-screen. The screen is not necessarily the Moon, it could be an asteroid in the main belt, or even a trans-Neptunian object in the Kuiper belt.

Table 15.2 gives several values of x, ψ, and of the typical crossing time of a Fresnel fringe, calculated first for the Moon ($d = 3.8 \times 10^5$ km, and a typical relative transverse velocity $v = 0.9$ km/s), then for bodies in the asteroidal main belt ($d \sim 3 \times 10^8$ km, $v \sim 15$ km/s), and finally for bodies in the Kuiper belt ($d \sim 6 \times 10^9$ km, $v \sim 25$ km/s). The values have been calculated at four wavelengths (0.5, 1, 5, and 10 nm).

In the present context, we concentrate on the lunar occultations. A single star is the source at infinity, and the limb of the Moon acts in first approximation as a straight screen producing a system of clear and dark bands on the Earth surface. During immersion, the relative motion of the Moon and of the observer carries these bands across the telescope aperture, so that in a few milliseconds the observer will notice regular fluctuations of intensity before the geometrical condition of dark is reached. The effect

TABLE 15.2

Fresnel Fringe Length, Angular Size, and Crossing Times

	0.5 nm	1.0 nm	5.0 nm	10.0 nm
Moon				
x (m)	10	14	32	45
ψ (arcsec)	$0''.005$	$0''.007$	$0''.016$	$0''.022$
τ (sec)	0.012	0.017	0.038	0.053
Main belt asteroids				
x (m)	273	383	857	1210
ψ (arcsec)	$0''.00019$	$0''.00027$	$0''.00059$	$0''.00084$
τ (sec)	0.018	0.0256	0.058	0.081
KBO				
x (m)	1220	1710	3833	5411
ψ (arcsec)	$0''.000043$	$0''.000060$	$0''.00013$	$0''.00018$
τ (sec)	0.1776	0.216	0.39	0.512

begins to occur when the star is about $0''.01$ from contact; 20 ms elapse between the first and the second diffraction minimum. The contrast between the first bright and the first dark band reaches 50%. The phenomenon repeats in reverse order at emersion. Should the source be a double star, or an extended disk, the diffraction pattern will be decidedly different, because the fringe modulation decreases to zero. Therefore, from the observations one can determine the separation of the binary or the diameter of the star, or even its limb darkening (stars are not necessarily uniformly illuminated disks; our Sun is decidedly darker at the limb than at the center, and so are all normal stars). An example of lunar occultation is provided by Figure 15.5, which shows the light curves at emersion of the binary star 71 Tau, both in blue and red light. We quote this example to show the sensitivity of the technique to the actual shape of the lunar limb; a boulder only 20 m high above the lunar surface could have caused the distortion of the light curve noticeable some 250 ms after the reappearance of the primary.

If a circular planet with an atmosphere occults a star, a so-called central flash of the star is observed, caused by the light refraction at the planet's limb.

It is worth noting that the lunar occultation technique does not require large telescopes to achieve high-spatial resolution (as are necessary for speckle interferometry, or adaptive optics, or for interferometric techniques), because the diffraction takes place not in the telescope but at the lunar limb. Furthermore, the achievable resolution does not critically depend on the local atmospheric seeing. The obvious disadvantages are that the occultation circumstances are not under the control of the observer, and the Moon covers no more than 10% of the sky.

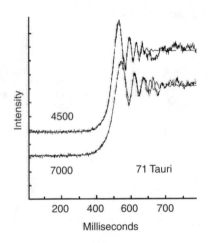

FIGURE 15.5
The lunar occultation of 71 Tau at emersion. The upper curve was taken in blue light ($\lambda = 4500$ A), the lower curve in red light ($\lambda = 7000$ A). (Adapted from Peterson, 1981, *Astronomy Journal*, **86**, p. 1090.)

Notes

- A computational masterpiece for eclipses was the *Canon der Finsternisse (Catalog of Eclipses)* produced by von Oppolzer (1887). The Catalog contained detailed information about 8000 solar and 5200 lunar eclipses between 1208 B.C. and 2161 A.D.

- An extremely useful web site is Espenak's: http://sunearth.gsfc.nasa.gov/eclipse/eclipse.html, which gives also the Six Millennium Catalog of Solar Eclipses from year − 1999 to year + 4000.

- For lunar eclipses, the circumstances of recent and upcoming lunar eclipses for any location worldwide can be obtained from: http://aa.usno.navy.mil/data/docs/LunarEclipse.html

- For lunar occultations in the visible, see for instance:
  Evans, D., 1971, Photoelectric measurements of lunar occultations, *Astronomy Journal*, **76**, p. 1107.
  Schmidtke, P.C., KPNO lunar occultation summary. II, *Astronomy Journal*, **97**, p. 909–916, March 1989 a paper which discusses the results from 65 lunar occultation observations.
  Richichi, A., 2002, *Astronomy and Astrophysics*, **382**, p. 178.

- The graph about 71 Tau in Figure 15.5 has been taken from:
  Peterson, D.M., 1981, Lunar occultations of the hyades II, *Astronomy Journal*, **86**, p. 1090.

- Other useful web sites are:
  http://tdc-www.harvard.edu/occultations/occultations.html, which gives information about stellar occultations, in general, and separately for each planet, and: http://www.lunar-occultations.com/entersite.htm, which is more specific for lunar occultations.

- For the Uranian rings, discovered for the first time in 1977 from the Airborne Kuiper Observatory when the planet occulted the star KM 12, see the paper by Elliot, J.L., French, R.J., Frogel, J.A., Elias, J.H., Mink, D.J., Liller, W., 1981, Orbits of nine Uranian rings, *Astronomy Journal*, **86**, p. 444.

- The occultation of the radio source of 3C 273B is described in:
  Hazard, C., Sutton, J., Argue, A.N., Kenworthy, C.M., Morrison, L.V., Murray, C.A., 1971, Accurate radio and optical positions of 3C 273B, *Nature*, **233**, p. 89.

- On February 20, 1998 the brightest x-ray source in the sky, namely Sco X-1 (which is a neutron star binary system) was eclipsed by the Moon. The event was captured by the German satellite ROSAT (http://wave.xray.mpe.mpg.de/rosat). See also:

http://heasarc.gsfc.nasa.gov/docs/objects/heapow/archive/
compact_objects/rosat_scox1_occult.html

- A discussion of the Saros and of the other periods and commensurabilities is given by Steves, B.A., 1998, The cycles of selene, *Vistas in Astronomy*, **41**, pp. 541–571.

- The dynamical implications of these periodicities for the lunar orbit theory can be found in:
  Valsecchi, G.B., 2001, *On the orbit of the Moon, in Earth, Moon, and Planets*, **85/86**, pp. 443–443.

- Occultations provide not only astrometric information (diameters, duplicity, etc., see for instance the already quoted paper Schmidtke et al., 1989). They can, indeed, be a powerful instrument for astrophysical considerations. For instance, Gies, D.R., 2004, Time-resolved H-alpha spectroscopy of the Be star PLEIONE during a lunar occultation, *Astronomy Journal*, **100**, pp. 1601–1609, obtained low-resolution spectra with a time sampling rate of 7 ms of the H-alpha emission line in the spectrum of the Be star Pleione. See also Richichi, A., 2004, Combining optical interferometry with lunar occultations, *Spectroscopically and Spatially Resolving the Components of the Close Binary Stars*, Proceedings of the Workshop held 20–24 October 2003 in Dubrovnik, Croatia, R.W. Hidlitch, H. Hensberge, K. Pavlovski, eds., ASP Conference Series Vol. 318, pp. 148–156.

- For Kuiper belt objects see:
  Cooray, A., 2003, Kuiper belt object sizes and distances from occultation observations, *The Astrophysical Journal*, **589**, pp. L97–100.

- Studies of the lunar atmosphere have been reported for instance by Verani, S., Barbieri, C., Benn, C., Cremonese, G., Mendillo, M., 2001, The 1999 Quadrantids and the lunar Na atmosphere, *MNRAS*, **327**, pp. 244–248.

Exercises

1. Calculate the arc on the spherical surface of the Earth corresponding to the linear diameter of the Moon if projected by a point-like source at infinity.

2. From a table of solar and lunar ephemerides, verify that the product between the angular diameters and the angular velocities of each body is approximately constant (Kepler's third law), and discuss why the constancy is better for the Sun than for the Moon. This exercise indicates how to use apparent diameters as distance indicators.

3. Here we sketch a method for calculating a lunar eclipse by the assumed known instantaneous values of the umbral semidiameter $\theta_{1/2}$, of the lunar and solar apparent semidiameters and distances (or parallaxes), without recurring to the Besselian elements. Call l the instantaneous distance between the center of the Moon M and the center of the umbra O; at the external and internal contacts we have, respectively:

$$l = \sigma_{ext} = \frac{1}{2}\theta_1 + \frac{1}{2}\theta, \qquad l = \sigma_{int} = \frac{1}{2}\theta_1 - \frac{1}{2}\theta_{\mathbb{D}}$$

while:

$$\frac{1}{2}\theta_1 = \pi_{\mathbb{D}} + \pi_{\odot} - \frac{1}{2}\theta_{\odot}$$

Then the two values of σ can be derived. Now, let (α_1, δ_1) be the coordinates of the center of the umbra, which can be derived from those of the Sun:

$$\alpha_1 = \alpha_{\odot} + 12^h, \qquad \delta_1 = -\delta_{\odot}$$

From the spherical triangle POM (where P is the celestial North pole), with a procedure analogous to that employed to derive Equation 15.10 and Equation 15.11, we also obtain the position angle of the center of the umbra with respect to the center of the Moon:

$$\sin l \sin p = \sin(\alpha_1 - \alpha_{\mathbb{D}}) \cos \delta_1,$$

$$\sin l \cos p = \sin \delta_1 \cos \delta_{\mathbb{D}} - \cos(\alpha_1 - \alpha_{\mathbb{D}}) \sin \delta_{\mathbb{D}} \cos \delta_1$$

All angles, except p, being small quantities, the following approximations will be sufficiently good:

$$x = l \sin p = (\alpha_1 - \alpha_{\mathbb{D}}) \cos \delta_1, \qquad y = l \cos p = \delta_1 - \delta_{\mathbb{D}} \quad (15.12)$$

The two auxiliary variables (x, y) can be calculated for a set of instants around opposition, and so their time derivatives (x', y') can be obtained (*note*: it is customary in the almanacs to indicate time derivatives with primed symbols). Now, insert in Equation 15.12 the two previously found values of σ. After some manipulation, we will obtain the position angle and the instant of the external and of the internal contact, respectively. It might well happen that no solution can be found for the position angle; if this happens for σ_{ext}, then no eclipse can occur; if it happens only for σ_{int} then the eclipse will be a partial one.

4. Calculate the magnitude of a central lunar eclipse $(\sigma = 0)$ when: $\pi_{\mathbb{D}} = 57', \theta_{\mathbb{D}}/2 = 16', l_1 = 63', l_2 = 41'$.

16

Elements of Astronomical Photometry

The relative apparent luminosities of the heavenly bodies can be measured even with the naked eye; Hipparchus and then Ptolemy had subdivided the visible stars in six classes of decreasing splendor (more properly magnitudes), indicated by the Greek letters α, β, γ, δ, ε, ζ and followed by the abbreviated Latin name of the constellation, e.g., α Cyg, or β Lyr. The ideal criterion is that passing from one class to the next, the eye senses the same difference of visual stimuli. However, the human eye can reliably judge this difference only over small angular distances and for objects of approximately identical color, so that the classification is not coherent over large arcs on the celestial sphere. This problem has not been completely solved even by modern detectors; great caution must be exerted in transferring a photometric sequence from one area of the sky to a distant one.

In the following sections, we shall essentially examine point-like, self-luminous objects (the stars), with brief considerations of extended luminous objects, such as nebulae or galaxies, and of illuminated objects, either point-like, such as asteroids or extended, such as planets. Furthermore, the treatment will be limited mostly to visible radiation (essentially from 3,000 to 10,000 Å, a spectral band we might call "extended visible", with short excursions into the near UV and near IR), with the caveat that in other spectral bands the celestial bodies can change their photometric characteristics appreciably. For instance, Jupiter in the infrared emits more radiation than it reflects from the Sun.

Photometric quantities will be defined according to the astronomical tradition and nomenclature, which is not always the same as that used by physicists or radiometrists. Furthermore, no account will be made of a possible refracting medium, so that we always assume that wavelength and frequency are related by:

$$\lambda = \frac{c}{\nu}, \quad d\lambda = -\frac{c}{\nu^2} d\nu = -\frac{\lambda^2}{c} d\nu \qquad (16.1)$$

As the unit of wavelength, the angstrom will be generally employed ($1\,\text{Å} = 10^{-10}\,\text{m} = 0.1\,\text{nm}$), while the frequency will be expressed in hertz (Hz).

16.1 Visual Magnitudes

For millennia, and until less than 200 years ago, the human eye (firstly naked, and after Galileo in 1609 at the focus of a telescope) was the only receptor of radiation. If S_1 and S_2 are the luminous stimuli produced by two stars, the eye produces a difference of stimuli $V_1 - V_2$, which is an (almost) logarithmic function of the ratio of the two stimuli:

$$V_1 - V_2 = k \log \frac{S_1}{S_2} \qquad (16.2)$$

This physiological law was found by Fechner and Weber in the nineteenth century. The true behavior of the eye is not described so simply by a logarithmic function, but Equation 16.2 has been assumed correct in all astronomical applications. In particular, the English astronomer Pogson found in 1856 that the law reproduced the astronomical system of magnitudes, provided the constant k was made equal to -2.5 and the logarithms were on the decimal base. Thus, by definition, the difference of magnitude between two stars whose intensities are S_1 and S_2 is given by:

$$m_1 - m_2 = -2.5 \log_{10} \frac{S_1}{S_2} \qquad (16.3)$$

Suppose then that a particular star, for instance Polaris, is used as common reference. For any other star we can write:

$$m = m_0 - 2.5 \log_{10} S \qquad (16.4)$$

an equation which must be treated with some caution in order not to forget that S must be a dimensionless quantity, and that the logarithm is not the natural one ($\log_{10} e = 0.43429\ldots, \log_e 10 = \ln 10 = 2.30259\ldots$). Therefore, $\Delta m = -1$ corresponds to $S_2/S_1 = 10^{0.4} = 2.51188\ldots$ (not to be confused with Pogson's constant which has no decimal), and $dm = 1.086 \, dS/S$. With slight approximation, a small difference of magnitude corresponds to the percent variation of the luminous intensity. Engineers use a quantity analogous to the magnitude, namely the decibel (dB), where the constant is 10 instead of 2.5 (4 dB = 1 mag).

A ratio of 10 in intensity produces a difference of magnitudes $\Delta m = -2.5$; a ratio of 100 a difference $\Delta m = -5.0$, and so on. The minus sign means that the fainter is the star, the larger is the magnitude. On the scale with Polaris at $m = 2.1$, Sirius, the brightest star in the sky, has approximately $m = -1.5$. Only three other stars have negative magnitudes, namely Canopus, Rigil Kent, and Arcturus). Vega has $m = +0.03$. The unaided eye, if well adapted and trained, can discern the 7th or even the 8th magnitude in very dark sites; at the focus of a large telescope it can reach the 20th (see Bowen, 1947, Notes). The faintest stars visible from the HST on long exposures can be as faint as the 28th. In other words, the observable stars span an impressive range of 29 magnitudes, which means a ratio of intensities $S_1/S_2 \approx 10^{12}$, 12 orders of magnitude due to several factors, namely the different intrinsic luminosities, the widely different distances, and the absorption by a diffuse material (dust) in the interstellar space. For objects at cosmological distances, the expansion of the Universe causes an additional dimming, because the light is not only red-shifted, but also spread over a larger spectral band (see Chapter 9).

Regarding Solar System objects, the Sun has $m = -26.7$, the full moon $m = -12.7$. At their maximum brightness, the planets reach the following approximate values: Mercury $m = -0.1$, Venus $m = -4.4$, Mars $m = -2.8$ (very rarely, it happened in 2003), and Jupiter $m = -2.6$. The largest asteroid Ceres is seen at $m = 6.8$, and Io, the moon of Jupiter, at $m = 5.5$.

However, the planets are extended sources, and so are comets, nebulae, and galaxies. Therefore, their apparent splendor is judged by integrating the light over a finite area, and one has to distinguish between the total magnitude and the magnitude per unit area (the surface brightness). We shall come back to this point in a later paragraph.

16.2 Extension of the Definition of Magnitude

The concept of magnitude derives from visual observations. However, we can generalize the definition in order to apply it to other types of detectors, e.g., photographic emulsion, or solid state detectors, and to other spectral regions. Let us suppose that our detector is sensitive to radiation in the spectral band (λ_1, λ_2), with $\lambda_1 < \lambda_2$, and $T(\lambda)$ be the function describing the sensitivity of our apparatus in that band. Further, if $f(\lambda)$ is the quantity of stellar radiation impinging on the detector per unit time per unit band pass, the total energy collected per unit time will be:

$$i(\lambda_c) = \int_{\lambda_1}^{\lambda_2} f(\lambda)T(\lambda)d\lambda \approx i_0 f(\lambda_c) \tag{16.5}$$

where λ_c is a wavelength inside the spectral band (λ_1, λ_2) usually close to the midpoint, and i_0 is a normalization constant which also takes care of the physical units of i. We can therefore define magnitude at λ_c the quantity:

$$m(\lambda_c) = m_0(\lambda_c) - 2.5 \log_{10} i(\lambda_c) \tag{16.6}$$

where the constant m_0 changes with λ_c. Notice that $m(\lambda_c)$ is only apparently a monochromatic quantity, it is actually derived from the integration of the luminous flux weighted by the instrumental sensitivity over a more or less wide spectral band; the designation eterochromatic is often employed to indicate this situation. If, ideally, a detector with 100% efficiency were available, we could measure the quantity:

$$m_{bol} = m_0 - 2.5 \log \int_0^\infty f(\lambda)d\lambda \tag{16.7}$$

which is called bolometric magnitude. In practice, this measure can be carried out only outside the Earth's atmosphere, measuring the heat coming from the Sun, the Moon, or bright stars. In general, we must be content with a bolometric correction C_V to apply to the eterochromatic visual magnitudes, assuming that the shape of $f(\lambda)$ is well-known from zero to infinity: $C_V = m_{bol} - m_{vis}$.

The practical realization of $T(\lambda)$ implies the consideration of several factors, such as:

- The reflectivity (or transmissivity) of the optics of the telescope, $R(\lambda)$ and the transmissivity of the filters in front of the detector, $K(\lambda)$
- The efficiency of the detector $Q(\lambda)$
- The transmissivity of the terrestrial atmosphere $A(\lambda)$

so that $T(\lambda) = R(\lambda) \cdot K(\lambda) \cdot Q(\lambda) \cdot A(\lambda)$. The first three functions are under the direct control of the astronomer, the fourth depends on the characteristics of the site and of the particular observational circumstances. We will now briefly examine these factors.

16.2.1   The Reflectivity of the Optics and Transmissivity of Filters

The basic optical elements of large telescopes are mirrors, whose surfaces are made reflective by means of a film of aluminum (Al) (approximately 900 Å thick), deposited in vacuum. The reflectivity of Al, $r(\lambda)$, is very good, approximately 90%, from the UV to the near IR, so that it can also be used in space telescopes, as in the HST, where the Al is additionally protected with a coating of MgF_2. However, the Al coating has a noticeable wide dip of approximately 10% around $\lambda = 8000$ Å; $r(\lambda)$ also depends on the age of the coating. In a given telescope there might be more than one mirror (two at the Cassegrain focus, three at the Nasmyth focus, etc.), so that the overall reflectivity becomes $R(\lambda) = r^n(\lambda)$, where n is the number of mirrors (see an example in Figure 16.1).

The use of Al coating on mirrors became widespread in the astronomical field after 1930. Before then, Ag coatings were employed; polished Ag also reflects very well from the green to the IR, and it lacks the dip at 8000 Å. However, its blue and violet response is decidedly worse than that of Al; therefore, caution must be exerted when comparing older blue magnitudes with modern ones.

On occasions, there are also refracting elements in the telescope, for example a corrector of optical aberrations made of several lenses whose transmissivities must be factored in the overall $R(\lambda)$. Note that the percentage of light transmitted by a refractive element depends on its chemical composition and thickness, but also on the reflection at the air-glass surface. For instance, the absorption by 1-cm thick Crown glass is 3% at

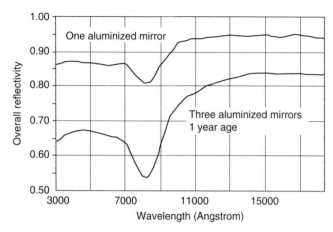

FIGURE 16.1
The overall reflectivity of a one-mirror (upper curve) and of a three-mirror telescope (such as the TNG) one year since aluminizing.

3600 Å, and rapidly decreases to less than 1% going to the red, while the reflection from the first surface is around 4%, unless a special antireflective coating is applied.

Regarding filters, there are several possible choices, from colored glasses to gelatin films to interferential devices. All colored glasses have a drawback; they tend to transmit the infrared even if their color is blue or yellow. This so-called red-leakage can cause considerable problems with solid state detectors such as CCDs. Therefore, each filter must be accurately characterized in the laboratory before being used at the telescope, in a spectral band as wide as possible.

As an example, we consider the widely used photometric system UBV introduced by Johnson and Morgan (1951, see Notes). The original realization employed the following glass filters: U: Corning 9863; B: Corning 5030 plus Schott GG13; V: Corning 3384. The detector was a blue-green sensitive photomultiplier 1P21, so that red-leakage was not important. The relative sensitivity stated in the original paper is given in Figure 16.2. The telescope had two aluminized mirrors, whose reflectivities were not included in the graph. Owing to its usefulness, the UBV system was largely adopted by other observers. However, by necessity many astronomers had to use different filters (e.g., the Schott filters, such as Schott UG2 for the U, Schott BG12 + plus Schott GG13 for the B, Schott GG11 or GG14 for the V), and different detectors (e.g., blue-green sensitive photographic emulsions such as the −O types produced by Eastman Kodak). Obviously, the telescopes and sites were also different. Therefore, each UBV realization has its own peculiarities, and care must be exerted to compare the magnitudes if great precision, for instance at the level of a few hundredths of magnitude, is required. Notice also that the colors of the stars play a role in Equation 16.5, so that stars of different colors will define a different λ_c. In other words,

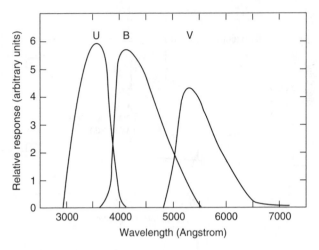

FIGURE 16.2
Response of the UBV photometer to an ideal lamp emitting equal energy at all wavelengths.

the relations between a given photometric system and another one, usually contain color terms that must be accurately calibrated.

Table 16.1 shows a set of indicative characteristics of the generalized UBV system. The third column of the table shows the largely used parameter FWHM (full width half maximum), namely the width in angstrom of the filter at its 50% transmissivity. The next three columns give three definitions of the indicative wavelength of the system; λ_{max} is where the efficiency of the system is maximum, while $<\lambda>$ is defined as:

$$<\lambda> = \frac{\int \lambda T(\lambda)d\lambda}{\int T(\lambda)d\lambda}$$

(some authors call this $<\lambda>$ isophotal). Both λ_{max} and $<\lambda>$ can be calibrated in the laboratory. Instead, λ_{eff} depends on the color (namely on

TABLE 16.1

Indicative Characteristics of the UBV System

Band	$\Delta\lambda$	FWHM	λ_{max}	$<\lambda>$	λ_{eff}
U	3,100–4,000	600	3,670	3,680	3,550 at $T = 2.5 \times 10^4$ K
					3,650 at $T = 1.0 \times 10^4$ K
					3,800 at $T = 4.0 \times 10^3$ K
B	3,750–5,350	1,000	4,295	4,450	4,330 at $T = 2.5 \times 10^4$ K
					4,400 at $T = 1.0 \times 10^4$ K
					4,500 at $T = 4.0 \times 10^3$ K
V	4,950–6,350	850	5,450	5,460	5,470 at $T = 2.5 \times 10^4$ K
					5,480 at $T = 1.0 \times 10^4$ K
					5,510 at $T = 4.0 \times 10^3$ K

the temperature) of the particular star:

$$\lambda_{\text{eff}} = \frac{\int \lambda T(\lambda) f(\lambda) d\lambda}{\int T(\lambda) f(\lambda) d\lambda}$$

as shown in Table 16.1 by changing the stellar surface temperature from 2.5×10^4 K to 4.0×10^3 K; λ_{eff} is therefore more representative of a real observation. The differences between the three wavelengths, and the shift to the red of λ_{eff} with decreasing stellar temperature, are so important because the UBV filters are very wide.

Let us introduce the other useful parameter $\Delta\lambda/<\lambda>$: in the present case $\Delta\lambda/<\lambda>$ is greater than 10%, so that we might call the UBV system a wide-band system. Intermediate band systems are those for which $\Delta\lambda/<\lambda>$ varies between 2 and 10%, narrow-band systems have $\Delta\lambda/<\lambda>$ less than 2%. The inverse parameter $<\lambda>/\Delta\lambda$ can be called the spectral resolution $r(\lambda)$ of the system. Johnson completed the wide-band photometric system in the (extended) visible with the addition of the R and I bands; later on, he added other IR bands called JKLMN. The current scheme also comprises the Y, H, and Q bands, as indicated in Table 16.2 (see also Figure 16.7).

These near-IR photometric bands can be obtained with different combinations of filters and detectors, so that each realization can differ from another one. Furthermore, the atmospheric transmission is very severe and selective, as discussed in the following sections. For further details on photometric systems and their different realizations, see the papers quoted in the Notes.

16.2.2 The Efficiency of the Detectors

The human eye is a very good detector, even though it is not a linear device and it lacks the capability to integrate the flux for long times and to store information. From physiology, we know that low-light level vision (scotopic vision) is insured by the rods, while the cones function in bright-light level vision (photopic vision). Only the cones are color sensitive (see Figure 16.3, left panel).

While we can clearly see the colors of bright stars (Sirius, Arcturus, Betelgeuse, etc.) with the naked eye, all faint stars look gray. The rods are

TABLE 16.2

Red and Near IR Wide Bands

Band	λ_c (nm)	$\Delta\lambda$ (nm)	Band	λ_c (μm)	$\Delta\lambda$ (μm)
R	700	220	K	2.2	0.48
I	900	240	L	3.4	0.70
Y	1,000	250	M	5.0	1.20
J	1,250	380	N	10.2	5.70
H	1,630	370	Q	20.1	7.80

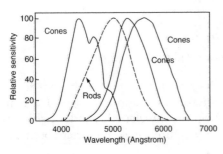 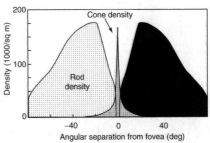

FIGURE 16.3

Left panel: sensitivity of the eye, arbitrarily normalized. Night vision (low light levels) is insured by the rods (dotted curve), whose sensitivity is approximately three times higher than that of the cones. Right panel: the cones are located at the center of the fovea, the rods at the periphery.

approximately three times more sensitive than cones, they can detect flashes of even a few photons, and finally determine the acuity of vision (the smallest angle the eye can resolve is typically from $40''$ to $60''$). However, their response time is decidedly slower than that of the cones. For maximum sensitivity, the averted vision technique can be tried, directing the star to approximately $18°$ from the optical axis of the eye (fovea), where the density of the rods reaches about 160,000 rod cells per square millimeter (see Figure 16.3, right panel). On the fovea, the density of the cones reaches about 140,000 cells per square millimeter. In reality, each observer has his own preferred angle (see also Notes).

Photographic emulsion was introduced in the astronomical field in the second half of the nineteenth century. It played a major role until about 1990, when solid-state detectors became available.

The great advantages of photographic emulsion were its capability to integrate luminous flux (exposures of many hours could be achieved), the large sensitive area (plates of up to 36×36 cm were used in Schmidt-type telescopes), the ease of handling, and the moderate cost. Although the response of the emulsion to light is strongly nonlinear and wavelength dependent (see Figure 16.4), and even varies with the ability in developing

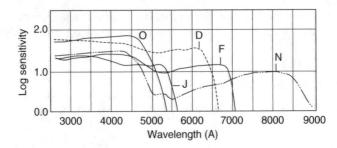

FIGURE 16.4

Sensitivity of Eastman–Kodak photographic emulsions. Normal emulsions (type O) respond only to blue–green light; the extension to the near IR (types N, I-N) can be obtained, although with lower sensitivity, with appropriate dyes.

and fixing it, an immense amount of quantitative measurements were performed thanks to the photographic plate.

In parallel with the photographic plate, the photocathode was employed for maximum photometric precision. The cathode quantum efficiency is at its best in the blue region, although there are products reaching the near IR (see Figure 16.5).

The strict linearity between the current and the light flux insures, with relative ease, photometric precisions better than a few thousandths of magnitudes. Therefore, the most precise astronomical photometry relied for a long time on photomultipliers. Furthermore, the capability to operate in photon counting mode was essential to study rapidly time-varying phenomena such as lunar occultations.

The main disadvantage of the cathode is its limitation to single object photometry. Finally, a linear, high-QE, extended-area detector become available at the end of the 1980s, namely the charge-coupled device, made of a rectangular matrix of pixels on a silicon substrate. An example of CCDs QE, referring to the device employed by the TNG, with pixel size of 13.5×13.5 μm for 4000×2000 pixels, is shown in Figure 16.6.

CCDs can also be made sensitive to UV light, and even to x-rays, so that they are the chosen detectors in many ground and space telescopes. They are also very effective in collecting cosmic rays and particles produced by radioactive decay inside the filters and lenses, a nuisance indeed for astronomical applications.

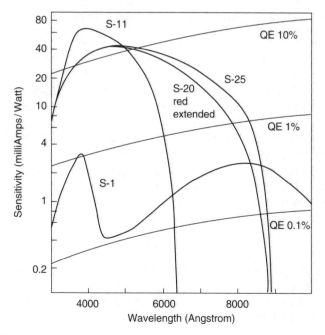

FIGURE 16.5
Sensitivity of selected photoelectric cathodes.

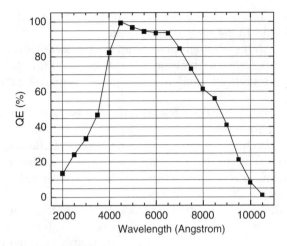

FIGURE 16.6
Quantum efficiency of the TNG CCD.

Other important classes of detectors are image intensifiers, multichannel plates, CMOS and SPAD devices, etc., but we refer for those to the specialized literature on low-light level astronomical detectors (see Notes).

16.3 Extinction by the Earth's Atmosphere

In the visible, the Earth's atmosphere absorbs and diffuses the radiation coming from the stars through a number of processes that give rise to both continuous and selective extinction. Molecular oxygen, in particular, is so effective in blocking radiation around 6800 and 7600 Å, that Fraunhofer could detect by eye two dark absorption bands in the far red of the solar spectrum, bands he called B and A, respectively (he examined the spectra from red to blue, current astronomical practice is from blue to red).

Let us consider the absorption due to a thin layer of atmosphere at height between h and $h + dh$, in the usual simple model of a plane-parallel atmosphere. The light beam from the star makes an angle z with the zenith, so that the traversed path is $dh/\cos z = \sec z \cdot dh$. If $I_\lambda(h)$ is the intensity at the top of the layer, at the exit it will be reduced by the quantity:

$$dI_\lambda = -I_\lambda(h) \cdot k_\lambda(h) \cdot \sec z \cdot dh \tag{16.8}$$

The variable k_λ (unit, cm^{-1}) represents the absorption per unit length of the atmosphere.

In total, if $I_\lambda(\infty)$ is the intensity outside the atmosphere, at the elevation h_0 of the observatory the intensity will be reduced to:

$$I_\lambda(h_0) = I_\lambda(\infty)e^{-\sec z \int_{h_0}^{\infty} k_\lambda(h) \cdot dh} = I_\lambda(\infty)e^{-\tau_\lambda(\infty) \cdot \sec z} \tag{16.9}$$

where we have introduced the unitless quantity τ_λ called optical depth:

$$d\tau_\lambda = k_\lambda(h)\cdot dh, \quad \tau_\lambda = \int_{h_0}^{\infty} k_\lambda(h)\cdot dh$$

By recalling that the magnitude is defined by the decimal logarithms, we then have:

$$m_{\text{ground}} = m_{\text{outside}} - 2.5\, D_\lambda(\infty)\cdot\sec z \qquad (16.10)$$

D_λ is called the optical density of the atmosphere, while the variable $X(z) = \sec z$ is called air-mass. The minimum value of the air-mass is 1 at the zenith, and 2 at $z = 60°$ (the limit of validity of the present approximate discussion).

Suppose we start observing the star at its upper transit, and then keep observing it while its hour angle (and therefore also its zenith distance) increases: we would notice a linear increase of its magnitude in agreement with Equation 16.10, namely a straight line with slope $2.5\, D_\lambda$ in a graph (m, $\sec z$). It is common practice to plot the m-axis pointing down. This straight line is known as the Bouguer line, from the name of the eighteenth century French astronomer who introduced it. The extrapolation of this line to $X = 0$ (a mathematical absurdity) gives the so-called loss of magnitude at the zenith, and therefore the magnitude outside the atmosphere.

According to the formulae derived in Chapter 2 and Chapter 3, we have:

$$\sec z = \frac{1}{\sin\varphi\sin\delta + \cos\varphi\cos\delta\cos HA} = X(z) \qquad (16.11)$$

where φ is the latitude of the site, δ and HA the coordinates of the star.

For a spherical atmosphere with exponentially decreasing density the following approximation can be used:

$$X(z) = \sec z\left(1 - \frac{H}{R}\sec^2 z\right) \approx \sec z\left[1 - 0.012\,\sec^2 z\right]$$

where H is a convenient scale (approximately 8 km) and R the radius of the Earth. Another useful formula was given by Garstang (1989):

$$X(z) = \frac{1}{\sqrt{1 - 0.96\sin^2 z}}$$

Table 16.3 shows the continuous extinction of the atmosphere above Mauna Kea, whose elevation above sea level is higher than that of most observatories, so that the transparency of the sky is at its best.

In the violet region, the transparency quickly drops to zero, essentially because of the ozone O_3 molecular absorption; at the other end of the spectrum the transparency is reasonably good until about 2.4 μm, when the H_2O and CO_2 molecules heavily absorb the light (see Figure 16.7).

TABLE 16.3

The Extinction at Mauna Kea

Wavelength (nm)	Extinction (Magnitude/Air Mass)	Wavelength (nm)	Extinction (Magnitude/Air Mass)
310	1.37	500	0.13
320	0.82	550	0.12
340	0.51	600	0.11
360	0.37	650	0.11
380	0.30	700	0.10
400	0.25	800	0.07
450	0.17	900	0.05

Therefore, astronomy can be performed from the ground in the spectral range from 3200 Å to 2.5 μm, then in three windows around 5, 10, and 20 μm, and again above the millimeter region up to kilometric waves (millimeter astronomy, radio astronomy). For the UV, X, Gamma, and IR astronomies, stratospheric airplanes and platforms, high-flying balloons and suborbital rockets, and finally, extraterrestrial space above 300 km are absolutely mandatory.

To complete these considerations about the influence of the atmosphere on photometry (and also on spectroscopy) of the celestial bodies, we must add that the atmosphere contributes radiation, both by spontaneous emission and by scattering of natural and artificial lights. If the observatory is close to populated areas, bright emission lines of mercury and sodium from street lamps are observed: Hg at $\lambda\lambda$ 4046.6, 4358.3, 5461.0, 5769.5, 5790.7; Na at 5683.5, 5890/96 (the yellow D-doublet), 6154.6; Ne at 6506, and so on. Natural lines come from the atomic oxygen in forbidden transitions (designated with [OI], see Chapter 17 for an explanation of this terminology) at $\lambda\lambda$ 5577.4, 6300 and 6367, and especially from the molecular radical OH which provides a wealth of spectral lines and bands filling the near-IR region above 6800 Å. The OH comes from the dissociation of the water vapor molecule under the action of the solar UV radiation. Therefore, the atmosphere is a diffuse source of radiation, whose intensity strongly depends on the observatory site; to set an indicative value in the visual band, a luminosity equivalent to one star of 20th mag per square arcsec can be assumed.

16.4 The Black Body

A very useful concept is that of the black body, of such importance in the development of all physics that we assume it well-known to the reader.

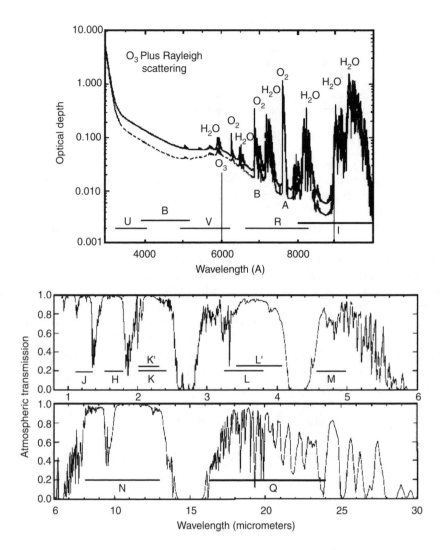

FIGURE 16.7

Upper panel: the optical depth of the atmosphere in the extended visible region. The upper curve is relative to an average sea-level observatory, the lower one (dotted) to Mauna Kea. Fraunhofer's A and B, and the extended Johnson system bands UBVRI are indicated. The heavy H_2O absorptions around 8200 and 9000 Å are the ρ and γ bands, respectively. Lower panel: the transmission of the atmosphere (linear scale) in the infrared, together with the main photometric bands. (Adapted from Allen, C.W., 2000, *Astrophysical Quantities* 4th ed.)

Only some basic facts, useful for astronomical applications will be recalled here. The black body is an ideal body of uniform temperature T (measured in Kelvin, K) that perfectly absorbs all incident radiation, and emits, isotropically, unpolarized thermal radiation according to

Planck's law:

$$\pi B_\lambda(T)d\lambda = \frac{2\pi hc^2}{\lambda^5}\frac{1}{e^{hc/k\lambda T}-1}d\lambda \ (\mathrm{erg\cdot cm^{-2}\cdot s^{-1}}) \tag{16.12}$$

where $h = 6.57 \times 10^{-27}$ erg·s^{-1} is Planck's constant and $k = 1.38 \times 10^{-16}$ erg·s is the Boltzmann constant. Expressing λ in cm:

$$2\pi hc^2 = c_1 = 3.742 \times 10^{-2} \ \mathrm{erg\cdot cm^2\cdot s^{-1}}, \quad hc/k = c_2 = 1.439 \ \mathrm{cm\cdot K}$$

where c_1 and c_2 are called the radiation constants. Therefore, the quantity $\pi B_\lambda(T)d\lambda$ represents the energy emitted in the interval $(\lambda, \lambda + d\lambda)$ by the unit surface in the unit time in the outward hemisphere.

Note the presence of π in the definition of $\pi B_\lambda(T)$, as is customary in astronomy but not in physics. The function $\pi B_\lambda(T)$ is called the emittance of the black body, while $B_\lambda(T)$, namely the emissivity into the unit solid angle, is called intensity or surface brightness (erg·cm^{-2}·s^{-1}·sr^{-1}). The reason for having π, and not 2π, resides in the projection effect of the surface at different angles, as will be demonstrated in a following section.

Two representations of Equation 16.12 are given in Figure 16.8.

All curves go to zero at zero wavelength, and they never cross: the unit area of a hotter black body emits at all wavelengths more than a cooler one, even in the infrared. The emissivity of the solar surface is approximately that of a black body at 5800 K, the emissivity of the Earth that of a black body at 290 K. Changing from wavelength to frequency:

$$\pi B_\nu(T)d\nu = \frac{2h\nu^3}{c^2}\frac{1}{e^{h\nu/kT}-1}d\nu \ (\mathrm{erg\cdot cm^{-2}\cdot s^{-1}}) \tag{16.13}$$

Equation 16.13 also gives the energy emittance, because $E = h\nu$.

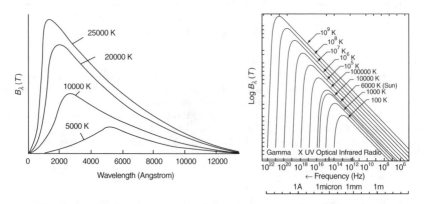

FIGURE 16.8
The black body curve for some temperatures of astrophysical interest, on linear axes (left) and logarithmic axes (right). Notice that at temperatures higher than 9000 K or lower than 4000 K, the peak of emissivity is outside the visible range.

Planck's law has two limiting cases:

1. If $(hc/k\lambda T) << 1$, then $\pi B_\lambda(T) = 2\pi kcT\lambda^{-4} = \frac{c_1}{c_2}T\lambda^{-4}$ (namely $\pi B_\nu(T) = 2\pi c^{-2}kT\nu^2 = 2\pi kT\lambda^{-2}$), the so-called Rayleigh–Jeans approximation valid in the infrared and radio domain.

2. If $(hc/k\lambda T) >> 1$, then $\pi B_\lambda(T) = 2\pi hc^2\lambda^{-5}e^{-hc/k\lambda T} = c_1\lambda^{-5}e^{-c_2/\lambda T}$ (namely $\pi B_\nu(T) = 2\pi hc^2\nu^3 e^{-h\nu/kT}$), the so-called Wien approximation, which is sufficiently apt to describe the spectral distribution of radiation from the surfaces of stars hotter than about 4000 K, in the visible and UV.

By integration of the monochromatic emittance over all wavelengths, or frequencies, we obtain the Stefan–Boltzmann law:

$$\pi \int_0^\infty B_\lambda(T)d\lambda = \pi \int_0^\infty B_\nu(T)d\nu = \sigma T^4 \qquad (16.14)$$

where $\sigma = 2\pi^5 \cdot k^4/15 \cdot h^3 \cdot c^2 = 5.6696 \times 10^{-5}$ erg·cm^{-2}·s^{-1}·K^{-4} is the Stefan–Boltzmann constant. Or else:

$$B = \frac{\sigma}{\pi}T^4 = 1.8047 \times 10^{-5}(\text{erg·cm}^{-2}\cdot\text{s}^{-1}\cdot\text{sr}^{-1}\cdot\text{K}^{-4})$$

From Equation 16.12, the Wien law (or displacement law) is easily seen: the wavelength of the maximum value of $B_\lambda(T)$ is inversely proportional to the temperature:

$$\lambda_{max}\cdot T = 0.28979 \text{ cm·K} \qquad (16.15)$$

so that the peak of emissivity of a star behaving like a black body of $T = 1 \times 10^4$ K (for instance a star such as Vega or Sirius) is situated at $\lambda_{max} = 2899$ Å, and therefore unobservable from the ground; the same applies to hotter stars. At the other extreme, a star having $T = 1 \times 10^3$ K has its maximum emissivity in the IR, and a cool body such as the Earth or Jupiter, in the far IR.

The wavelength corresponding to the frequency ν_{max} of maximum B_ν is given by:

$$Tc/\nu_{max} = 0.3544c_2 = 0.50996 \text{ cm·K}$$

and it is not c/λ_{max}. The flux at the peak of emissivity is:

$$\pi B(\lambda_{max}) = 1.2865 \times 10^{-4}T^5 \text{ erg·cm}^{-1}\cdot\text{cm}^{-2}\cdot\text{s}^{-1}\cdot\text{K}^{-5}$$

which rises extremely rapidly with temperature; expressing λ in μm, at $T = 10^4$ K the emittance is:

$$\pi B(\lambda_{max}) = 1.2865 \times 10^{12} \text{ erg·}\mu\text{m}^{-1}\cdot\text{cm}^{-2}\cdot\text{s}^{-1}$$

The gradient of the curve of emissivity is given by:

$$\Phi_\lambda(T) = \frac{c_2}{T}(1 - e^{c_2/\lambda T}) \tag{16.16}$$

which can be measured and used to evaluate the so-called color temperature of a star. This quantity can change according to the spectral range chosen for determining the gradient, because no real star radiates exactly as a black body.

It is also useful to express Planck's distributions in terms of photon emittance:

$$\pi N_\lambda(T) = \frac{2\pi h c^2}{\lambda^4} \frac{1}{e^{hc/k\lambda T} - 1} \quad (\text{ph}\cdot\text{s}^{-1}\cdot\text{cm}^{-2}\cdot\text{cm}^{-1})$$

$$\pi N_\nu(T) = \frac{2\pi \nu^2}{c^2} \frac{1}{e^{h\nu/kT} - 1} \quad (\text{ph}\cdot\text{s}^{-1}\cdot\text{cm}^{-2}\cdot\text{Hz}^{-1}) \tag{16.17}$$

Notice the exponents, 4 and not 5, and 3 and not 4, respectively, because each photon has energy $E = h\nu$.

The wavelength of maximum photon emissivity, and the corresponding photon flux are:

$$\lambda_m T = 0.25506 c_2 = 0.36698 \text{ cm}\cdot\text{K},$$

$$N(\lambda_m) = 2.1008 \times 10^{11} T^4 \text{ ph}\cdot\text{cm}^{-1}\cdot\text{cm}^{-2}\cdot\text{s}^{-1}\cdot\text{K}^{-4}$$

The Stefan–Boltzmann law becomes:

$$N = 1.52033 \times 10^{11} T^3 \text{ ph}\cdot\text{cm}^{-2}\cdot\text{s}^{-1}\cdot\text{K}^{-3}$$

A "universal black body curve" can be constructed, using as variable the product λT. The examination of this curve shows some general results:

- The energy radiated below λ_{max} is one quarter of the total.
- A whole 98% of the total is radiated between 0.5 and 8 λ_{max}, the rest being equally divided above and below those two limits.
- The decrease of emissivity is much steeper below λ_{max} than above it; below $0.17\lambda_{max}$ a fraction of only 10^{-10} is radiated.

Some useful numbers:

- At the normal laboratory temperature of 290 K (17°C), $\lambda_{max} = 10\ \mu\text{m}$, and the total emissivity is 401 W·m^{-2}, corresponding to approximately 1.2×10^{18} ph·cm^{-2}·s^{-1}·sr^{-1}, mostly emitted at $\lambda_{max} = 31.1\ \mu\text{m}$ (energy of about 4×10^{-2} eV).
- At a ten times higher temperature, corresponding to the temperature of cool stars, $\lambda_{max} \approx 1\ \mu\text{m}$, and the emissivity is several MW·m^{-2}; approximately 1% of the energy is emitted below 5000 Å or above 8 μm, and about 10% in the visible band.

- Raising the temperature by another factor of ten, namely reaching the surface temperature of a hot star, the emissivity rises by four orders of magnitude, the maximum moves to the far UV, and again, a very small percentage of energy is radiated in the visible band.

16.5 Color Indices and Two-Color Diagrams

Suppose we observe a star in two spectral bands, measuring $m(\lambda_1) = m_0(\lambda_1) - 2.5 \log_{10} i(\lambda_1)$ and $m(\lambda_2) = m_0(\lambda_2) - 2.5 \log_{10} i(\lambda_2)$, respectively. We define color index of the star the difference:

$$c_{12} = m(\lambda_1) - m(\lambda_2) = m_0(\lambda_1) - m_0(\lambda_2) - 2.5 \log_{10} \frac{i(\lambda_1)}{i(\lambda_2)} \quad (\lambda_1 < \lambda_2) \quad (16.18)$$

The name of this quantity comes from the visual sensation of color of a luminous source. We also have:

$$i(\lambda_2) = i_0 i(\lambda_1) \cdot 10^{0.4 c_{12}}$$

where the constant i_0 depends on the two wavelengths and the instrumental sensitivity, according to Equation 16.5.

If the star radiates as a black body, from Equation 16.12 we can easily derive the following expression:

$$c_{12} = m(\lambda_1) - m(\lambda_2) = c_0 + \frac{1.56}{T} \left(\frac{1}{\lambda_1} - \frac{1}{\lambda_2} \right) + f(T)$$

where $f(T)$ would be zero in the Wien approximation, and is always a small quantity for $T > 4000$ K. Therefore, for not too cool black bodies we have the relationship:

$$c_{12} = m(\lambda_1) - m(\lambda_2) \approx A_{12} + \frac{B_{12}}{T} \quad (16.19)$$

For instance:

$$m_{\text{blue}} - m_{\text{vis}} \approx -0.64 + \frac{7200}{T}$$

which is one of several useful formulae named after H.N. Russell.

However, real stars do not radiate as black bodies. Following Johnson and Morgan, it has been agreed that the zero point of the color indices is defined by a set of stars having surface temperatures similar to that of Vega, namely approximately 10,000 K. Those stars have the spectroscopic designation A0-V (Arabic zero, Roman 5), as explained in Chapter 17. For those standard stars, $c_{12} = 0$ by definition, no matter how the two wavelengths might have been selected:

$$U-B = B-V = V-R = R-I = \cdots = 0$$

(definition for a sample of A0-V stars)

It is plainly evident that color index equal to zero does not mean equal fluxes in the different spectral bands.

Let us take the black body radiation law and the sensitivity of the UBV photometric system. Formula (16.5) will allow calculation of the (U−B, B−V) color indices of the black body as a function of the temperature, as shown in Table 16.4.

The numbers in this table are only indicative values, they will change by a few hundredths of magnitude according to the particular realization of the UBV system (see for instance Matthews and Sandage, 1963). Two points are worth noting:

1. The color indices saturate with increasing temperature, as foreseen by Equation 16.19. Therefore, in this particular photometric system it is easier to estimate the temperature of a star cooler than 10^4 K than of a hotter one. The reason can be easily understood from Figure 16.8, which shows that in the blue-visual bands the black body curves for $T > 10^4$ K are almost parallel to each other, so that the gradients will be essentially equal.

2. The color indices of the 10^4 K black body are not zero, because real stars do not radiate exactly as black bodies.

TABLE 16.4

Color Indices of the Black Body in the UBV Photometric System

T	U−B	B−V	T	U−B	B−V
4,000	+0.37	+1.13	20,000	−1.01	−0.16
6,000	−0.25	+0.62	25,000	−1.06	−0.15
10,000	−0.69	+0.14	40,000	−1.14	−0.29
15,000	−0.91	−0.07	∞	−1.28	−0.44

Let us obtain the color indices of a large number of stars, randomly chosen over the celestial sphere, and plot their position in the (U−B, B−V) plane. The result is as in Figure 16.9, where all stars brighter than magnitude 6.5 have been used.

The great proportion of the stars are located on a main sequence running from the top left (the hottest stars) to the bottom right (the coolest stars). Stars also sparsely occupy the region to the right of this main sequence, towards a boundary defined by the black body locus. The precise interpretation of this two-color indices diagram (usually called in the abbreviated form two-color diagram) must wait until Chapter 17, although some preliminary notions are in order now:

- The A0 stars at (U−B, B−V) = (0,0) are actually the most distant ones from the black body locus. In Chapter 17, we shall see that the reason for this distance is the presence of hydrogen absorption lines (Balmer series), especially affecting the U filter.

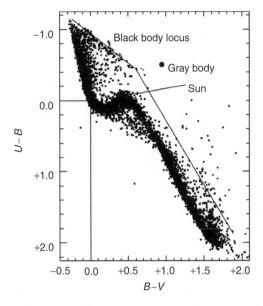

FIGURE 16.9
The (U–B, B–V) diagram for bright stars. In stellar photometry it is customary to draw the U–B axis positive downward. The black body locus, the gray body, and Sun, are also indicated.

- In the upper part of the diagram, the stars having colors very similar to those of the black body are the so-called white dwarfs, briefly mentioned in Chapter 13.
- Below the A0 stars, there is a short heavily filled region where stars of large radii and intermediate temperatures (giant stars) are located.
- Some points in the intermediate region between the main sequence and the black body line are simply unresolved binaries, with stars having different colors (say a hot plus a cool star seen as a single source). Unresolved normal galaxies, which in this context behave like a mixture of stars, are also found in the same region.
- There are points to the right of the black body line. Sometimes, the star simply varied in brightness during the observations; in other much more interesting cases, these peculiar color indices reveal the presence of nonthermal emission mechanisms (e.g., strong emission lines or continuous synchrotron mechanism). These cases occur in planetary nebulae or active galactic nuclei, where strong emission lines are superimposed on the continuous spectrum.
- The colors of some stars are not the true ones, they are affected by interstellar absorption, as explained in a later section. This effect is particularly evident in the upper part of the diagram, between the main sequence and the black body locus, because the stars in that region are usually very luminous, very distant, and close to the Milky Way plane, with a large quantity of intervening matter.

16.6 Calibration of the Apparent Magnitudes in Physical Units

Until now, the system of astronomical magnitudes has not been connected to any particular physical unit. This necessary calibration can be performed using the Sun, the Moon, or a set of convenient stars such as Vega. Regarding the Sun, the following text and Table 16.5 report some useful values:

- $f(\lambda_V) = 1.83 \times 10^5 \, \text{erg·cm}^{-2}\text{·s}^{-1}$ outside the atmosphere in the visual band.
- $f_{\text{bol}} = 1.36 \times 10^6 \, \text{erg·cm}^{-2}\text{·s}^{-1}$ outside the atmosphere (this quantity is also called solar constant; it is equivalent to $1.36 \, \text{kW·m}^{-2}$, or to $1.95 \, \text{cal·cm}^{-2}\text{·min}^{-1}$). The solar constant has only small oscillations in the present epoch, by no more than a few parts over a thousandth.

The bolometric correction is $C_V = -0.08 \, \text{mag}$. Using units of the photovisual system, the solar light flux outside the atmosphere is equivalent to $1.27 \times 10^5 \, \text{lux}$.

The direct utilization of the Sun to calibrate the normal stellar observations is clearly impossible, thus a number of stars with characteristics as close as possible to solar have been selected. The following stars are among the solar analogs: 16 Cyg B, VB64, HD 10,5590, and HR 2290.

From the calibration of Vega, we derive the following correspondences between magnitudes and fluxes (outside the terrestrial atmosphere):

$$U = 0 : f(3700 \, \text{A}) = 4.4 \times 10^{-9} \, \text{erg·cm}^{-2}\text{·s}^{-1}\text{A}^{-1}$$

$$B = 0 : f(4450 \, \text{A}) = 7.2 \times 10^{-9} \, \text{erg·cm}^{-2}\text{·s}^{-1}\text{A}^{-1}$$

$$V = 0 : f(5556 \, \text{A}) = 3.44 \times 10^{-9} \, \text{erg·cm}^{-2}\text{·s}^{-1}\text{A}^{-1} = 3.44 \times 10^{-8} \, \text{W·m}^{-2}\text{·}\mu^{-1}$$

For a generic star of magnitudes U, B, V the flux is given by:

$$f(\lambda_U) = 10^{-0.4(U+20.90)}, \quad f(\lambda_B) = 10^{-0.4(B+20.36)}, \quad f(\lambda_V) = 10^{-0.4(V+21.08)}$$

TABLE 16.5

Solar Color Indices

Color Index	Magnitude	Color Index	Magnitude
U–B	+0.195	V–I	+0.88
B–V	+0.650	J–H	+0.310
V–R	+0.540	H–K	+0.060
R–I	+0.340	K–L	+0.034
V–K	+1.486	L–M	−0.053

TABLE 16.6

Calibration Factors for Vega

Band	Flux Density $(W{\cdot}m^{-2}{\cdot}\mu^{-1})$	Photon Flux $(Ph{\cdot}m^{-2}{\cdot}s^{-1}{\cdot}\mu^{-1})$	Band	Flux Density $(W{\cdot}m^{-2}{\cdot}\mu^{-1})$	Photon Flux $(ph{\cdot}m^{-2}{\cdot}s^{-1}{\cdot}\mu^{-1})$
R = 0	7.8×10^{-8}	1.66×10^{11}	K = 0	4.0×10^{-10}	4.5×10^{9}
I = 0	8.3×10^{-9}	1.39×10^{11}	L = 0	8.1×10^{-11}	1.2×10^{9}
J = 0	3.3×10^{-9}	2.0×10^{10}	M = 0	2.1×10^{-11}	5.1×10^{8}
H = 0	3.9×10^{-10}	9.6×10^{9}	N = 0	1.0×10^{-12}	5.1×10^{7}

In photon fluxes, recalling that the energy of a photon is $E = 1.98610 \times 10^{-8}/\lambda(A)$ erg, we have:

$$\pi N_{ph}(U) \approx 800 \times 10^{-0.4U}, \quad \pi N_{ph}(B) \approx 1500 \times 10^{-0.4B},$$

$$\pi N_{ph}(V) \approx 1000 \times 10^{-0.4V} \ ph{\cdot}cm^{-2}{\cdot}s^{-1}A^{-1}$$

Other useful conversion factors from Vega (approximate within probably $\pm 5\%$) are given in Table 16.6.

16.7 Apparent Diameters and Absolute Magnitudes of the Stars

Let us assume that the star is a spherical body of radius R, emitting as a black body of temperature T, so that the luminous energy radiated over the whole sphere in the bandwidth $d\lambda$ is $dF_\lambda = 4\pi R^2 B_\lambda(T)d\lambda$. Furthermore, suppose that space is perfectly transparent, and that the object is at rest with respect to the observer located at distance d from the star. The luminous flux through the sphere of radius d will then be:

$$df_\lambda = 4\pi R^2 B_\lambda(T)d\lambda/4\pi d^2 = \left(\frac{R}{d}\right)^2 B_\lambda(T)d\lambda \qquad (16.20)$$

Therefore, the flux received from the star, in a transparent space, decreases according to the inverse square of the distance. Integrating over all wavelengths, the received bolometric flux is:

$$f_{bol} = f_0 \left(\frac{R}{d}\right)^2 T^4$$

The ratio (R/r) is proportional to the apparent diameter θ, and its square to the solid angle ω of the star as seen by the observer:

$$\frac{R}{d} = \tan\frac{\theta}{2} \approx \frac{\theta}{2}(rad), \quad \omega = \pi\left(\frac{R}{d}\right)^2 \approx \pi\left(\frac{\theta}{2}\right)^2 (sr)$$

so that:

$$df_\lambda = f_0 B_\lambda(T) \theta^2 \, d\lambda, \quad F_{bol} = F_0 T^4 \theta^2$$

The apparent magnitudes are then given by:

$$m_\lambda = m_0 - 2.5 \log B_\lambda(T) - 5 \log \theta,$$

$$m_{bol} = m_{0bol} - 2.5 \log T^4 - 5 \log \theta \tag{16.21}$$

From Equation 16.21, we see that the bolometric correction is independent from the distance and from the radius of the star:

$$C_V = m_{bol} - m_{5500} = m'_0 + 2.5 \log B_{5500}(T) - 10 \log T$$

The zero point of C_V was originally chosen to be zero for temperatures of 6500 K, so that it becomes negative for both higher and lower temperatures. For black bodies, the bolometric corrections are approximately: -5.4 at $T = 6 \times 10^4$ K, -0.4 at $T = 10^4$ K, -0.08 for the Sun, -1.0 at $T = 4 \times 10^3$ K, -4.0 at $T = 2.5 \times 10^3$ K. For real stars, the bolometric corrections were very uncertain for the hotter and the cooler stars, until the space age permitted observations in the far UV and far IR.

We explicitly mention that the quantity $f_0 B_\lambda(T) = f_\lambda/\theta^2$, namely a quantity proportional to the surface brightness of the star, is constant along the line of sight (provided the space is transparent) and, in particular, it equals the emissivity from the surface of the star.

Now, let us introduce the concept of absolute magnitude, by considering a particular distance $d = 10$ pc as standard distance. The absolute magnitude M is the apparent magnitude a star would have if located at the standard distance; from Equation 16.21:

$$M_\lambda = m_{0\lambda} - 2.5 \log \left(\frac{R}{10}\right)^2 - 2.5 \log B_\lambda(T),$$

$$M_\lambda - m_\lambda = -2.5 \log \left(\frac{d}{10}\right)^2 \tag{16.22}$$

and similar for the bolometric magnitude. The quantity y_λ:

$$y_\lambda = m_\lambda - M_\lambda = 5 \log \left(\frac{d}{10}\right) = -5 \log 10\pi \tag{16.23}$$

where d is in parsec and π is the annual parallax of the star in arcsec, is called distance modulus of the star. Therefore:

$$m_\lambda = M_\lambda + y_\lambda = M_\lambda + 5 \log d - 5 = M_\lambda - 5 \log \pi - 5,$$

$$M_\lambda = m_\lambda - y_\lambda = m_\lambda - 5 \log d + 5 = m_\lambda + 5 \log \pi + 5$$

From the definition given by Equation 16.23, it would seem that the distance modulus is independent of the wavelength. However, the interstellar (or intergalactic) space might not be fully transparent, so that the distance derived from inverting Equation 16.23 on the basis of some particular observational data, may well depend in a systematic way on the wavelength used. This is why we have explicitly kept the wavelength in the notation of the distance modulus y_λ.

A word of clarification about the notation: apparent magnitudes are indicated with small letters, m_λ, absolute magnitudes with capital letters, M_λ. However, the apparent magnitudes in Johnson's photometric system and its extensions (UBVRIYJHKLMNQ) are indicated with capital letters, B, V, etc.; for instance, an absolute magnitude in Johnson's V band will be indicated with M_V.

Regarding color indices, it is obvious that they do not depend on the distance (always in a transparent space), so that:

$$c_{12} = m(\lambda_1) - m(\lambda_2) = M(\lambda_1) - M(\lambda_2), \quad C_V = m_{bol} - m(\lambda) = M_{bol} - M(\lambda)$$

It is easy to extend the results found in Equation 16.20 and following, to cases where the star does not radiate as a black body, by substituting the appropriate spectral distribution I_λ in place of the Planck function; of course, the simple relation with temperature would be lost. Furthermore, the star is a gaseous sphere, so that its diameter can depend to some extent on the wavelength, through the optical depth of the atmosphere; therefore, even the apparent diameter will be a function of λ.

Let us consider in particular the Sun. Its distance is $1/206{,}265$ pc, so that its absolute magnitudes are:

$$M_\odot = m_\odot + 5 + 5 \log 20, 6265 = m_\odot + 31.57,$$

$$M_\odot(V) = -26.74 + 31.57 = +4.83, \quad M_\odot(bol) = -26.82 + 31.57 = +4.75$$

The average values of its radius, apparent diameter and apparent solid angle are, respectively:

$$R_\odot = 6.96 \times 10^{10} \text{ cm}, \quad \theta_\odot = 33' = 0.0095 \text{ rad}, \quad \omega_\odot = 6.81 \times 10^{-5} \text{ sr}$$

From these values and from bolometric flux observed outside the Earth's atmosphere, we derive the following quantities:

- Bolometric flux through the unit area of the solar surface:
  $F_\odot = 6.29 \times 10^{10}$ erg·cm^{-2}·s^{-1}
- Bolometric surface brightness:
  $I_\odot = F_\odot/\pi = 2.04 \times 10^{10}$ erg·cm^{-2}·s^{-1}·sr^{-1}
- Total luminosity: $L_\odot = 4\pi R_\odot^2 F_\odot = 3.83 \times 10^{33}$ erg·s^{-1}
- From the Stefan–Boltzmann law the effective temperature $T_\odot$, namely the temperature of a black body of same radius which would emit the same energy, $T_\odot = 5777$ K

From the previous considerations, we understand that these values will be only approximately correct for the real Sun. We shall not be surprised if the measured temperature depends on the wavelength or the position over the solar disk (higher at the center, lower at the limb, the so-called limb-darkening) because of the different optical depths.

Using the Sun as reference for radii and luminosities, we obtain for a generic star:

$$\log L = -3.147 + 2 \cdot \log R + 4 \cdot \log T_{eff},$$

$$M_{bol} = 4.75 - 2.5 \cdot \log \frac{L}{L_{\odot}} = 42.36 - 10 \cdot \log T_{eff} - 5 \cdot \log \frac{R}{R_{\odot}},$$

$$C_V = -42.54 + 10 \cdot \log T_{eff} + \frac{29,000}{T_{eff}},$$

$$\log T_{eff} = 4.221 - 0.1(V + C_V) - 0.5 \cdot \log \theta$$

(16.24)

where θ is expressed in milliarcsec. Equation 16.24 is the basis for the photometric determination of stellar diameters. For example, suppose we have measured the apparent magnitude V for a hot star, $V = 10$, having $T_{eff} = 2.7 \times 10^4$ K and $C_V = -2.9$. From Equation 16.24 we derive $\theta = 0.01$ mas. A star with $T_{eff} = 4 \times 10^3$ K and the same apparent magnitude would have a diameter approximately ten times higher. The difficulty of the direct measurement of the apparent diameter of stars is plainly evident if we recall the angular extent of the diffraction figure of the telescope and the considerations about the atmospheric seeing in Chapter 11.

Other useful formulae, usually associated with the name of H.N. Russell, and valid for sufficiently hot black bodies (when Wien's approximations is good), are:

$$\log F(4400) = -0.4 M_B + 9.11 - 2.5 \cdot \log \frac{R}{R_{\odot}},$$

$$\log F(5550) = -0.4 M_V + 8.85 - 2.5 \cdot \log \frac{R}{R_{\odot}}$$

(both in erg·cm^{-2}·s^{-1}A^{-1});

$$M_b = -0.72 + \frac{36,700}{T} - 5 \log \frac{R}{R_{\odot}}, \quad M_v = -0.08 + \frac{29,500}{T} - 5 \log \frac{R}{R_{\odot}}$$

$$M_b - M_v = m_b - m_v = -0.64 + \frac{7200}{T}$$

The monochromatic magnitudes in the blue and visual bands m_b, m_v used in the above relations are slightly different from the eterochromatic B, V magnitudes of Johnson's system.

16.8 The Hertzsprung–Russell Diagram

Around 1913, E. Hertzsprung and H.N. Russell independently discovered a most important relationship between the absolute magnitudes of normal stars and their temperatures (or equivalently, their color indices or their spectral types, as we shall see in Chapter 17). From the Stefan–Boltzmann law, we expect that stars can have any combination of radius and temperature, so that the distribution of stars in the plane ($\log L$, $\log T$) or in any of the equivalent planes (M_{bol}, $\log T$), (M_{bol}, B–V), (M_V, B–V) should be a uniform one. We shall call in the following the Hertzsprung–Russell (H-R) diagram any of those planes. Observation proves instead that this is not the case: stars are preferentially found in well-defined regions of the H-R diagram. Given the temperature, the radius assumes specific values only, for reasons that must be intrinsic to the equilibrium structure of the star. This behavior is represented in Figure 16.10. The left panel shows the diagram obtained by using data of the Hipparcos astrometric satellite; this sample of stars is therefore within approximately 50 parsec of the Sun. The right panel shows the theoretical H-R diagram which can be obtained from the Stefan–Boltzmann law; the diagonal lines are lines of equal radii in units of the solar radius.

These observational facts are at the very basis of all theories of stellar structure and evolution. Especially important has been, and still is, the examination of the H-R diagram for well-selected samples of stars, for instance, those belonging to galactic open clusters (e.g., the Hyades), to galactic globular clusters (e.g., M 3), or to small galaxies in the local group (e.g., Fornax). The selection of these stars is so important, not only because all stars have the same distance from the Sun and insterstellar reddening, but also because they presumably had a common origin in time and space, and

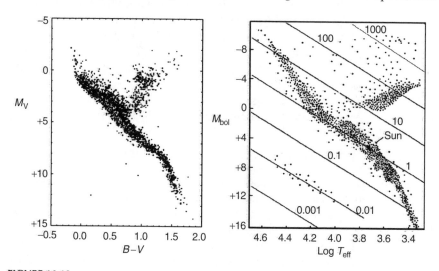

FIGURE 16.10

Left panel: the position of nearby stars in the (M_V, B–V) plane, from Hipparcos data. Right panel: observational data superimposed to the theoretical H-R diagram; the diagonal lines are lines of equal radii in units of the solar radius. Stars do not uniformly fill the plane, they preferentially populate a main sequence, two high radii sequences (giants and supergiants, at 100 and 1000 solar radii) and a low radius sequence (white dwarfs, at 0.01 solar radius).

an identical initial chemical composition. The very different H-R diagrams derived from these diverse samples led Baade, around 1955, to propose the concept of different stellar populations (Pop. I, represented by open galactic clusters, Pop. II, represented by globular clusters).

The density of points in the observed H-R diagrams is not representative of the true density of stars in space (namely of the so-called luminosity function, a statistical function which represents the distribution of stars among the luminosities). Faint stars greatly outnumber bright ones, so that the commonly held concept of the Sun being a normal star is somewhat misleading: no more than 5% of all stars are brighter than the Sun, and most are members of binary or multiple systems. The main sequence is observed extending to at least $M_V = +18$, where true stars give place to brown dwarfs (no nuclear reactions in their interiors), and further down to planets. Today planets around other stars are found with relative ease, thanks especially to very precise radial velocity measurements; they number around 150 at the time of writing and are constantly increasing.

These fascinating arguments cannot be discussed here but given the richness of this field of research, we provide some references in the Notes.

16.9   Interstellar Absorption

The assumption that the interstellar space is perfectly transparent is not correct, as was discovered by Trumpler and others around 1930. The Milky Way contains a quantity of diffuse matter in the form of interstellar clouds of gases and dust, which absorb and diffuse the radiation in a manner strongly dependent on the wavelength. Furthermore, the diffuse matter is preferentially distributed along the galactic plane, so that the absorption varies with the galactic coordinates of the line of sight.

Let $m(\lambda)$ be the observed magnitude; if the interstellar matter absorbs the amount A_λ (in magnitudes), then the true magnitude is $m_0(\lambda) = m(\lambda) - A(\lambda)$, and the absolute magnitude is:

$$M_0(\lambda) = m(\lambda) - A(\lambda) + 5 \cdot \log(d/10)$$

As remarked upon previously, until allowance is made for interstellar absorption, the measured distance modulus can depend on the wavelength.

In order to derive the function $A(\lambda)$, we make recourse to two sets of stars having the same spectral type as judged from the absorption lines (see Chapter 17): the first set of nearby and presumably unabsorbed stars, the second of much more distant stars. The classical works along these lines of reasoning were performed by Stebbins and Whitford around 1950, with observations extending from 3200 to 10,000 Å. They clearly demonstrated that the absorption is much more important in the blue than in the red, and therefore a heavily absorbed star will also show a color much redder than the true one (than expected from its spectral type). Therefore, interstellar absorption and interstellar reddening are used to indicate the same effect of absorption and scattering. This identification of absorption with reddening would not be correct if the interstellar matter also produced a gray absorption, which however has never been convincingly detected.

In the visible band, the interstellar reddening is very different from that caused by the terrestrial atmosphere. The latter depends essentially on λ^{-4} (Rayleigh scattering), as expected from scattering centers with dimensions much smaller than the wavelength (namely, molecules). The former has a dependence approximately as λ^{-1} (Mie scattering), as expected from particles having the same dimensions of the wavelength (dust). The explanation in terms of dust was firstly suggested by Lindblad around 1938 and then by Oort and van de Hulst. Modern observations by satellites have extended the knowledge of $A(\lambda)$ to the UV and to the IR, as detailed in the Notes.

Restricting our considerations to the UBV bands, let us introduce the so-called color excesses $E(U-B)$ and $E(B-V)$, namely the difference between the observed (and possibly reddened) color index and the unreddened one:

$$E(U-B) = (U-B) - (U-B)_0, \quad E(B-V) = (M-V) - (B-V)_0$$

Let R be the ratio between the absorption in the V band, $A(V)$, and $E(B-V)$. The observations provide the following values:

$$R = \frac{A(V)}{E(B-V)} = 3.2 \pm 0.1, \quad A(B) = \frac{4}{3}A(V), \tag{16.25}$$

$$E(U-B) \approx 0.72 \times E(B-V)$$

along any direction in the Milky Way (apart from some notable, but well-localized, exceptions, e.g., toward the Orion Nebula). Therefore, the parameter $Q = (U-B) - 0.72 \cdot (B-V)$ is essentially independent from the total absorption. In other words, Q is a reddening-free parameter function only of the spectral type (SpT, see Chapter 17) of the star. Unfortunately, the relationship $Q(SpT)$ is not a single valued one.

Now, let us consider an intrinsically hot (blue) and very luminous star, located in the upper left part of the two-color diagram (Figure 16.10; these stars have SpT indicated with the letters O and B). Imagine that an intervening dust cloud absorbs its light by $A(V)$ magnitude. Its color indices will be modified by the quantities $E(B-V) = A(V)/R, E(U-B) \approx 0.72 \cdot E(B-V)$, so that the star will move in the H-R diagram downwards to the right, along a line of slope 0.72, by the corresponding amounts, ending in the region between the main sequence and the black body locus. Notice that in the upper left region of the diagram, the slope of the main sequence is slightly steeper than the reddening line, otherwise we could not distinguish between a temperature and an absorption effect. This separation of effects is instead impossible (with the UBV system) for stars such as the Sun or cooler. Therefore, the UBV system is not the most suitable one to study the interstellar absorption. However, it provides useful reddening criteria for the bluest and most luminous stars, which sample the Milky Way plane to large distances.

Regarding the different amounts of extinction in the different galactic directions, the interstellar matter is distinctively concentrated at low galactic latitudes, where both the high luminosity and temperature O-B stars and the hydrogen clouds emitting the 21-cm (1420 MHz) spectral line of hydrogen are most abundant. Making an analogy with a plane-parallel atmosphere (thus ignoring the dependence on the galactic longitude l), suppose we observe a distant galaxy located beyond all absorbing matter at galactic latitude b. In our simple model, the galactic absorption is identical in the Northern and Southern galactic hemispheres, so that the modulus of b is employed in the following formulae. If τ_{90} is the total optical depth of the absorbing matter toward the galactic pole, the corresponding value in direction b will be:

$$\tau_{|b|} = \frac{\tau_{90}}{\sin|b|} = \tau_{90} \operatorname{cosec} |b| \tag{16.26}$$

so that in magnitudes:

$$m_{90} = m_0 + A_{90}, \quad m_b = m_0 + A_{90} \operatorname{cosec} |b| \tag{16.27}$$

The practical meaning of Equation 16.27 is as follows: suppose we count galaxies, well distributed over the celestial sphere, at increasingly fainter magnitudes. In a transparent Euclidean universe, their number increases with the volume, while their apparent luminosity f decreases with the square of the distance, so that the cumulative number of galaxies brighter than magnitude m increases with the power $3/2$, namely:

$$N(f) = N_0 f^{3/2}, \quad \log N(m) = \log N_0 + 0.6m \tag{16.28}$$

The galactic absorption will produce a decrease of this expected number equal to $A_{90} \operatorname{cosec}|b|$; therefore, from the counts we ought to be able to derive A_{90}, a procedure analogous to Bouguer's line for the air mass at the zenith. Hubble (1936) was indeed able to derive the classic values of absorption at the galactic poles, $A_{90}(V) = 0.18$ mag, $A_{90}(B) = 0.25$ mag. Hubble's original values have been revised several times, taking into account better data and also the fact that the expansion of the Universe decreases the slope of the counts below 0.6. This cosmological effect depends on the apparent magnitude, namely from the distance of each particular class of galaxy.

The total galactic extinction and reddening in a given direction have been calibrated using the brightness temperature T of the 21-cm spectral line, which is proportional to the total number of H atoms, N_H, in that direction. Knapp and Kerr (1974) found that for an extragalactic object:

$$N_H = 5.8 \times 10^{21} E(B-V) \text{ atoms·cm}^{-2}, \quad A(V) = 0.6 \times 10^{-21} N_H \text{ mag}$$

Therefore, accurate maps of $T(l, b)$ provide one of the best ways to correct the magnitudes and colors of extragalactic objects for interstellar reddening (see Notes).

As a final remark on interstellar matter, the scattering of radiation by interstellar particles will produce partially polarized light. The classical studies of interstellar polarization were performed by Hiltner around 1950, showing a fair regularity of the polarization according to the direction in the Milky Way. These observations provided the first evidence of a weak interstellar magnetic field (intensity of a few millionth of gauss), orienting in preferential directions the dust particles.

16.10 Extension to the Bodies of the Solar System

The previous considerations are not entirely suitable for Solar System objects which reflect the solar light; here, we shall not consider the far IR (thermal) radiation coming from the bodies nor the fluorescent light from comets. Furthermore, the AU, not the parsec, is the suitable distance unit, and the distance of the objects from the Sun and from the Earth is greatly variable with their ephemerides.

Let us indicate with r the distance from the Sun and with Δ the distance from the Earth, both in AU. The absolute magnitude $H(1,1)$ is defined as the observed one, when the body is at unit distance from Sun and Earth; the quantity $H(1,1)$ can be derived from the magnitude at the opposition:

$$H(1,1) = m(r, \Delta) - 5 \log r \cdot \Delta \qquad (16.29)$$

Or else, if the body is a fully illuminated perfectly reflecting circular disk of diameter $2\,s$, and $m_\odot$ is the apparent magnitude of the Sun ($m_\odot = -26.74$ in the visual band):

$$H(1,1) = m_\odot - 5 \log s(\mathrm{AU}) = 14.14 - 5 \log s(\mathrm{km}) \qquad (16.30)$$

For instance, the largest asteroid Ceres has a diameter of 913 km. According to Equation 16.30, it would have $H(1,1) = 0.84$. However, every body will reflect only a fraction of the light coming from the Sun, according to its surface characteristics (clouds, rocky or icy terrain, dust-covered land, etc.). If the reflectivity of Ceres was 10%, $H(1,1)$ would become $+3.34$ (indeed, Ceres has $H(1,1) = 3.32$). More precise considerations are therefore in order.

Let us take into account that the body, as seen from the Earth, is not entirely illuminated (the terminator is the line dividing the sunlit hemisphere from the shadowed regions). Let α be the so-called phase angle, namely the angle between the Sun and the Earth as seen from the body ($\alpha = 0°$ at opposition, $180°$ at conjunction), and $\varphi(\alpha)$ a phase function describing the fraction of the illuminated disk, normalized to unity at zero phase. For instance, if $\varphi(\alpha)$ is simply proportional to the area, then:

$$\varphi(\alpha) = \frac{1 + \cos \alpha}{2} \qquad (16.31)$$

All real bodies differ from this simple relationship. In particular, the observations prove that the Moon and most asteroids become decidedly brighter at zero phase angle than the value extrapolated from larger angles. This effect is partly due to the complex scattering law from the terrain, and partly to the influence of craters and mountains that shadow the illuminated area. Therefore, other functions are needed to describe the behavior of the reflection. The first of these functions is the (bolometric) geometric albedo p, that is the ratio between the total brightness of the body at $\alpha = 0°$ and that of a perfectly diffusing *disk* having the same diameter of the planet and at the same distance from the Sun. Introducing p, Equation 16.29 and Equation 16.30 become:

$$m(r, \Delta, 0) = m_\odot + 5 \log r \cdot \Delta - 5 \log s(\text{AU}) - 2.5 \log p,$$

$$\log 2s(\text{km}) = -0.2H(1,1) - 0.5 \log p + 3.12$$

(notice the coefficient 2.5 in front of $\log p$). At a generic phase angle:

$$m(r, \Delta, \alpha) = H(1,1) + 5 \log r \cdot \Delta - 2.5 \log p\varphi(\alpha)$$

For the Moon, $p = 0.12$, for Venus $p = 0.46$, for Jupiter and Io $p = 0.45$, for the carbonaceous-type asteroids (C-type) $p = 0.05$, for the silicate-rich asteroids (S-type) $p = 0.20$, and so on.

In addition to the geometric albedo, another useful quantity is the (bolometric) Bond albedo. Consider the integral of $\varphi(\alpha)$ over the entire solid angle:

$$q = \frac{1}{\pi} \int_{4\pi} \varphi(\alpha) d\omega = 2 \int_0^{\pi/2} \varphi(\alpha) \sin \alpha \, d\alpha$$

For a perfectly diffusing disk, $q = 1$, for a perfectly diffusing sphere, $q = 1.5$ (Lambert's law), for a metallic sphere, $q = 4$, and so on. The product $A_B = pq$ is called the Bond albedo. It expresses the ratio between the total light diffused from the sphere and the total light illuminating it. For the Moon, $A_B = 0.07$, for Ceres $A = 0.035$, for Venus and Jupiter $A = 0.7$, etc.

Furthermore, the reflectivities of the different bodies are wavelength dependent in different manners. For instance, the colors of the asteroids are usually redder than the solar; in addition, the reflected light can be partly polarized (see Figure 16.11).

Although the UBV photometric system has found wide application for asteroidal studies, more suitable band passes have been identified, for instance the eight-band ECAS photometry (Notes).

As a final remark, asteroids rotate, and thus the viewing geometry from Earth can appreciably change at different oppositions; furthermore, several objects show duplicity, with a further complication of the light curve.

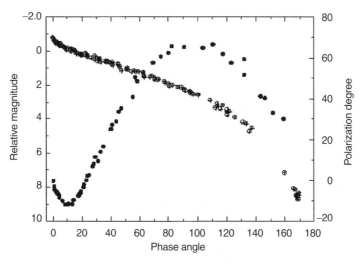

FIGURE 16.11
Photometry and polarimetry of an asteroid. The brightening at phase angle $\alpha = 0$ is seen, and hence the large variation of polarization with α.

16.11 Radiation Quantities

This section presents a more formal definition of the photometric quantities used until now. The specific intensity, or simply intensity, I is the flux of radiation at a given point in a given direction across the unit surface normal to that direction per unit time and per unit solid angle. In general, the intensity is a function of two angular variables (θ, φ), with $0 \leq \theta \leq \pi$, $0 \leq \varphi \leq 2\pi$. For simplicity, we shall assume azimuthal symmetry around φ. Integrating over the solid angle $d\omega = \sin\theta\, d\theta\, d\varphi$, we obtain the flux of radiation through the unit surface, or flux density F:

$$\pi F = \int_{4\pi} I \cos\theta\, d\omega = 2\pi \int_0^\pi I(\theta) \cos\theta \sin\theta\, d\theta$$

$$= 2\pi \int_{-1}^1 I(\theta) \cos\theta\, d(\cos\theta) \qquad (16.32)$$

where θ is the angle between the ray and the outward normal to the surface (notice, several authors do not have the factor π entering in the definition of flux, so they would indicate with $\mathcal{F}$ what is here πF; some authors even call astrophysical flux the quantity πF). The flux of radiation emitted from a unit surface (more properly, the emittance) is obtained by integration over the outward hemisphere:

$$\pi F = \int_{2\pi} I \cos\theta \, d\omega = 2\pi \int_0^{\pi/2} I \cos\theta \sin\theta \, d\theta \qquad (16.33)$$

which, for the isotropic radiation of the black body, is $\pi F = \pi I = \pi B$, $I = F/\pi$ (as already stated, π not 2π, because of the projection effect, as can be seen by resolving the integral at constant I).

The radiation density is:

$$u = \frac{1}{c} \int_{4\pi} I d\omega = \frac{4\pi}{c} J \qquad (16.34)$$

where J indicates the mean intensity. The radiation quantities per unit frequency, or unit wavelength, are written respectively as I_ν, I_λ, πF_ν, πF_λ, u_ν, u_λ, J_ν, J_λ. According to Equation 16.1, the following relations apply:

$$I = \int_0^\infty I_\nu d\nu = \int_0^\infty I_\lambda d\lambda, \quad I_\lambda = \frac{c}{\lambda^2} I_\nu = \frac{\nu^2}{c} I_\nu, \quad \lambda I_\lambda = \nu I_\nu$$

The observer receives the radiation emitted, in all directions, by each surface element of the hemisphere facing the observer. Owing to the great distance of all stars (including the Sun), each surface element at angular distance θ from the center is seen by the observer at the same angle θ from its normal (see Figure 16.12); if R is the radius of the star, the area connecting all elements having the same θ is therefore an annulus of radius $\rho = R \sin\theta$ and thickness $Rd\theta$, namely $dA = 2\pi R \sin\theta \cdot Rd\,\theta$.

Summing over the visible star, the total amount of energy radiated in a solid angle $d\omega$ is:

$$E_\lambda = 2\pi R^2 d\omega \int_0^{\pi/2} I_\lambda(\theta) \cos\theta \sin\theta \, d\theta$$

Let us take, as solid angle $d\omega$, a unit area $\Delta\sigma = 1$ at the observer, distant d from the star. Therefore $d\omega = 1/d^2$. If the space is perfectly transparent and

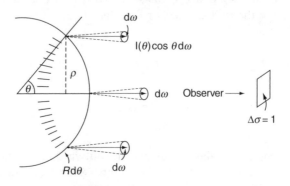

FIGURE 16.12

The stellar radiating hemisphere. The observer at infinity receives the flux coming from all directions θ between 0 and $\pi/2$; the dashed region indicates the area with θ between $(\theta, \theta + d\theta)$.

the emission isotropic, all the energy E_λ will flow through $\Delta\sigma$, so that the observed flux f_λ will be:

$$f_\lambda = I_\lambda \frac{\pi R^2}{d^2} = \omega I_\lambda \qquad (16.35)$$

In other words, the observed spectrum gives the energy distribution of the outwards flux, and the star behaves like a disk of surface πR^2 subtending a solid angle ω (for instance, the Sun viewed from the Earth has $R/d = 214.94$, the subtended solid angle is $\omega = 6.8 \times 10^{-5}$sr). Equation 16.35 is valid even if the emissivity is nonisotropic, provided I_λ is meant as an average intensity.

These last considerations, in particular Equation 16.35, have shown a remarkable degree of symmetry between source and observer, namely the quantity $\sigma_\lambda = f_\lambda/\omega$, that we can call surface brightness, which stays constant and equal to I_λ, independent of the distance. The surface brightness of the Sun is the same at Mercury, the Earth, and Pluto. This result, of distance-independent surface brightness, also applies to diffuse nebulae, or even to distant galaxies, provided the interstellar absorption and the metrics of the expanding Universe for objects at cosmologic distance, are properly taken into account.

Notes

- The paper by N. Pogson was published in Monthly Notices of the *Royal Astronomical Society*, vol. XVII, 1856, with the title *Magnitudes of the 36 Minor Planets for Each Day of Each Month of the Year 1857*. The previous and subsequent history can be found in a paper by Hearnshaw, J.B., 1992, Origins of the stellar magnitude scale, *Sky and Telescope*, p. 494, November.

- The mechanism of vision of the human eye can be found in: http://webvision.med.utah.edu/
  For the averted vision see: http://www.roboticobservatory.com/jeff/observing/averted.html and: http://hyperphysics.phy-astr.gsu.edu/hbase/vision/rodcone.html

- How faint is the faintest star the human eye can see at the focus of a telescope, was discussed by Bowen, I.S., 1947, Limiting visual magnitude, *Publications of the Astronomical Society of the Pacific*, **59**, p. 253. The formula given there is: $m_{lim} = 7.5 + 2.5 \log D + 2.5 \log M$, where D is the diameter of the telescope in centimeters, and M the magnification of the eyepiece. For $D = 150$ cm and $M = 500$, the formula gives $m_{lim} = 19.7$. However, the seeing conditions and the brightness of the night sky can lead to differences of more than one magnitude with respect to the formula.

- The paper describing the photometer used for the UBV system is: Johnson, H.L., Morgan, W.W., 1951, On the color-magnitude diagram of the Pleiades, *Astrophysics Journal*, **114**, p. 522. See also the paper: Johnson, H.L., Morgan, W.W., 1953, Fundamental stellar photometry for standards of spectral type on the revised system of the Yerkes spectral atlas, *Astrophysics Journal*, **117**, p. 313. Since then, the production of some filters has been discontinued and others changed name (e.g., Schott GG13 became GG385, GG11 became GG495).

- For the extension to the near IR: Johnson, H.L., 1966, Astronomical measurements in the infrared, *Annual Review of Astronomy and Astrophysics*, **4**, p. 193.

- For the (U−B, B−V) colors of the black body and of the synchrotron radiator see: Matthews A.T., Sandage A.R., 1963, Optical identification of 3C 48, 3C 196 and 3C 286 with stellar objects, *Astrophysical Journal*, **138**, p. 30.

- For a general discussion of light detectors see: Rieke, G.H., 2002, *Detection of Light: From the Ultraviolet to the Submillimeter*, Cambridge University Press, Cambridge. The Single Photon Avalanche Photodiodes (SPADs) are described in the paper by S. Cova, M. Ghioni, A. Lotito, I. Rech, F. Zappa, 2004, Evolution and prospects for single-photon avalanche diodes and quenching circuits, *Journal of Modern Optics*, **51**, pp. 1267–1288.

- The extinction coefficients for Mauna Kea are taken from the Gemini website: http://www.gemini.edu/sciops/ObsProcess/obsConstraints/ocTransparency.html#MK%20optical%20extinction%20curve

- For the air-mass, night sky, artificial lights, zodiacal light:
  Kasten, F., Young, A.T., 1989, Revised optical air mass tables and approximation formula, *Applied Optics*, **28**, pp. 4735–4738.
  Young A.T., 1976, Atmospheric extinction, *Methods in Experimental Physiology, vol. 12A, Optical and Infrared*, M.L. Meeks, Ed., Academic Press, New York, pp. 123–180, Chap. 3.1.
  Garstang, R.H., 1989, Night sky brightness and observatories and sites, *Publications of the Astronomical Society of the Pacific*, **101**, p. 306.
  Garstang, H.R., 1991, Dust and light pollution, *Publications of the Astronomical Society of the Pacific*, **103**, p. 1109 and the site of the International Dark Sky association: http://www.darksky.org/index.html

- For the zodiacal light measurement from space see: Torr, M.R., Torr, G.D., Stencel, R., 1979, Zodiacal light surface brightness measurements by Atmospheric Explorer-C, *Icarus*, **40**, p. 49.

- For the calibration of Vega and of solar analogs:
  Hayes, D.S., Latham, D.W., 1975, A rediscussion of the atmospheric extinction and the absolute spectral distribution of Vega, *The Astrophysics Journal*, **197**, p. 593.
  Campins, H., Rieke, G.H., Lebofsky, M.J., 1985, Absolute calibration of photometry at 1 through 5 microns, *Astronomical Journal*, **90**, pp. 896–899.
  Useful references for a wealth of photometric systems are given by: Mermilliod, J.-C., Hauck, B., Mermilliod, M., *The General Catalogue of Photometric Data*, University of Lausanne, Switzerland, http://obswww.unige.ch/gcpd/gcpd.html and: Munari, U., Fiorucci, M., Moro, D., 2002, *The Asiago Data Base on Photometric Systems*, http://ulisse.pd.astro.it/Astro/ADPS/

- The classical work on stellar populations is contained in the volume: O'Connell, D.J.K., ed., 1957, *Stellar Populations*, Vatican Observatory Press, Vatican.

- A review on H-R diagrams is given by Chiosi, C., Bertelli, G., Bressan, A., 1992, New developments in understanding the H-R Diagram, *Annual Review of Astronomy and Astrophysics*, **30**, p. 235.

- For the interstellar absorption law:
  Whitford, A.E., 1951, An extension of the interstellar absorption curve, *Astrophysics Journal*, **107**, p. 102.
  Cardelli, J.A., Clayton, G.C., Mathis, J.S., 1989, The relationship between infrared, optical and ultraviolet extinction, *Astrophysics Journal*, **345**, p. 245.

- For the interstellar polarization:
  Hiltner, W.A. 1952, Photometric, polarization and spectrographic observations of O and B stars, *Astrophysics Journal*, **106**, p. 231.
  Coyne, G.V., ed., 1987, *Polarized Radiation of Interstellar Origin*, Vatican Observatory Press, Vatican.

- For the relationship between HI temperature and interstellar absorption:
  Knapp, G.R., Kerr, F.J., 1974, The galactic dust-to-gas ratio from observations of 81 globular clusters, *Astronomy and Astrophysics*, **35**, p. 361.
  Burstein, D., Heiles, C., 1984, Reddening estimates for galaxies, *Astrophysics Journal*, **54**, Suppl., p. 33.

- The classical work of E. Hubble on the expansion of the Universe and the galactic absorption was published in 1936 with the title: *The Realm of the Nebulae* (reprinted by Dover Inc.).

- For asteroidal photometry:
  The ECAS eight-color system is described by Zellner, B., Tholen, D.J., Tedesco, E.F., 1985, The eight-color asteroid survey — Results for 589 minor planets, *Icarus*, **61**, pp. 355–416.

The infrared satellite IRAS made great contributions to the determination of the albedos and diameters of the asteroids, see Tedesco, E.F., Veeder, G.J., Fowler, J.W., Chillemi, J.R., 1992, *The IRAS Minor Planet Survey*, PL-TR-92-2049.

For a large volume of data and methods we recommend the volumes:

Asteroids I, Asteroids II and *Asteroids III*, University of Arizona Press.

See also: Bus, S.J., Binzel, R.P., 2002, Phase II of the small main-belt asteroid spectroscopic survey. A feature-based taxonomy, *Icarus*, **158**, p. 146.

Figure 16.11 has been taken from: http://www.univer.kharkov. ua/astron/asteroids/

Exercises

1. Calculate the (U–B, B–V) color indices of a gray body, namely of a body having $F_U = F_B = F_V$. By definition (and using for simplicity monochromatic magnitudes):

$$U-B = c_{UB}^0 - 2.5\log\frac{F_U}{F_B}, \quad B-V = c_{BV}^0 - 2.5\log\frac{F_B}{F_U}$$

Using Vega's calibration: $F_U/F_B = 4.4/7.2$, $F_B/F_V = 7.2/3.7$, $c_{UB}^0 = -0.53$, $c_{UB}^0 = +0.72$, which are therefore the color indices of a spectrum having a flat (gray) emission spectrum.

2. Calculate the photon flux collected in the visual band by a CCD receiver at the Cassegrain focus of a 2-m telescope from a star of apparent magnitude V at the zenith.

The effective area of a telescope with a primary mirror of radius r and area occultation ε^2 by the secondary is: $A_{eff} = \pi r^2(1 - \varepsilon^2)$. Usually ε^2 is of the order of 10%, so that the effective area of the 2-m telescope is approximately 28,000 cm². The overall reflectivity (two mirrors) in the visual band is around $0.9^2 \approx 0.8$; the CCD quantum efficiency for a good device is around 0.8; the sky transparency at the zenith is around 0.8; the filter transmissivity is about 0.9.

If the bandpass of the filter is 800 Å, we finally obtain the detection rate:

$$N_{ph} \approx 1080 \times 28,000 \times 0.8 \times 0.8 \times 0.9 \times 800 \times 10^{-0.4\,V}$$

$$\approx 1.39 \times 10^{10} \times 10^{-0.4\,V}\,\text{counts/s}$$

Integrating for Δt seconds, the total number of detected photons is: $N_{ph}\cdot\Delta t \approx 1.39 \times 10^{10} \times 10^{-0.4V}\Delta t$. For an ideal (noise-free) detector, the signal-to-noise (S/N) ratio is limited by Poisson's statistics, according to which the noise is simply the square root of the signal:

$$\frac{S}{N} = \frac{N_{ph} \cdot \Delta t}{\sqrt{N_{ph} \cdot \Delta t}} = \sqrt{N_{ph}} \cdot \sqrt{\Delta t}$$

The value of S/N must be at least 5 for a sure detection, and larger than 50 for a good photometry. Notice that in this ideal case, the value of S/N increases with the square root of the area of the telescope and of the exposure time. In practice, however, the detector is not noise free, and the atmospheric sky brightness contributes an additional signal, which is an added noise and which varies from site to site. In a dark observatory, a typical value of the sky brightness is 1 star of $V = 21.0$ per square arcsec.

The background brightness is present even in observations from space, because there is always a number of unresolved stars and galaxies inside each pixel. To give some numbers, at low galactic latitudes the sky background is equivalent to 250 stars of apparent magnitude $V = 10$ per square degree, at high galactic latitudes to 35 such stars. In different units, this galactic background is equivalent to one star of $V = 21.8/\text{arcsec}^2$ and of $V = 23.9/\text{arcsec}^2$, or to $N_{ph} \approx 2.0 \times 10^{-2}$ $m^{-2} s^{-1} A^{-1}$ arcsec^{-2} and to $N_{ph} \approx 2.8 \times 10^{-3}$ $m^{-2} s^{-1} A^{-1}$ arcsec^{-2}, respectively. Furthermore, at low ecliptic latitude there is a background of solar light scattered by the interplanetary dust known as zodiacal light (and more specifically, gegenshein at 180° from the Sun).

3. Calculate the magnitude and color indices of an unresolved binary star.

Suppose the two stars have the following characteristics:

star A, $V_A = 4.40$, $(U-B)_A = +0.05$, $(B-V)_A = +0.58$

star B, $V_B = 5.90$, $(U-B)_B = +0.47$, $(B-V)_B = +0.89$

Then, from the general relation $f_{A+B} = f_A(1 + 10^{-0.4(m_B - m_A)})$, we have:

$V_{A+B} = 4.40 - 0.24 = 4.16$, $B_{A+B} = 4.98 - 0.19 = 4.79$,

$U_{A+B} = 5.03 - 0.13 = 4.90$, $(U-B)_{A+B} = +0.11$, $(B-V)_{A+B} = +0.63$

This calculation can be extended to an arbitrary number of unresolved stars, e.g., a distant globular cluster or a distant galaxy.

4. Given that the interstellar absorption is larger in U than in B and in V, discuss why $E(U-B)$ is only 72% of $E(B-V)$.

5. Calculate the equilibrium temperature of a planet of radius R at distance d from the Sun. The total energy radiated from the Sun is

$$E_\odot = \sigma \cdot T_\odot^4 \cdot 4\pi R_\odot^2$$

the flux at d (AU) is

$$f = \frac{\sigma \cdot T_\odot^4 \, 4\pi R_\odot^2}{4\pi \cdot d^2} = \sigma T_\odot^4 (R_\odot/d)^2$$

and the cross section of the planet is πR^2. If A is the Bond albedo, the fraction $(1 - A)f\pi R^2$ will be absorbed to heat the planet. Furthermore, we assume that the planet rotates fast enough to distribute the heat over its entire body, so that the temperature is the same over the entire surface. Therefore, we have:

$$\sigma \cdot T^4 = \frac{\sigma \cdot T_\odot^4}{4\pi \cdot d^2} \cdot \frac{4\pi \cdot R_\odot^2}{4\pi \cdot R^2} \cdot \pi \cdot R^2 (1 - A),$$

$$T = T_\odot \left[\left(\frac{R_\odot}{d} \right)^2 \frac{1 - A}{4} \right]^{\frac{1}{4}} = 4.1 \times 10^3 (1 - A)^{1/4} \sqrt{\frac{R_\odot}{d}}$$

In conclusion, the equilibrium temperature of a planet does not depend on its radius, but only (weakly) on its albedo. For instance, for the Earth: $\sqrt{R_\odot/d} \approx 0.068$, $A = 0.39$, $(1 - A)^{1/4} = 0.88$, $T = 248$ K. With the same reasoning we would find for Jupiter, $T = 111$ K.

For both the Earth and Jupiter, these values are lower than the measured ones, hinting at more complex physical situations. Internal heat sources are required in addition to the solar one (certainly, the explanation of the Earth's internal heat, namely a hot liquid core and radioactive decay of elements in the crust, is very different from that of Jupiter, namely phase transition in the metallic H_2 core). All other gaseous planets have considerable internal sources.

17

Elements of Astronomical Spectroscopy

Spectroscopic examination is one of the most powerful ways of studying the physical, chemical, and kinematical analysis of the heavenly bodies, from planets to stars, to nebulae, to distant extragalactic objects. From a stellar spectrum, we can obtain the temperature, pressure, and chemical composition, and derive, by means of suitable calibration, fundamental quantities such as total luminosity, mass, and radius of the star. We can also measure the radial velocity, rotational status, possible duplicity, radiation mechanisms, and the presence of magnetic fields.

In this chapter, we provide a few notions of the observational techniques, used from the near UV to the near IR, then we consider some theoretical arguments to help with the interpretation, and discuss the classification schemes of spectra of normal stars. Only mention of the spectra of other bodies (peculiar stars, planets, minor bodies, active galaxies, etc.) will be made.

Among many important events, we recall the first observations of Newton on the dispersive power of prisms, and the discovery made by Fraunhofer in 1814 of dark bands in the solar spectrum. In 1860, Foucault, Kirchoff, and Bunsen identified such dark lines with those produced by known chemical elements. In 1868, Angstrom produced a very accurate solar spectrum. In 1882, Rowland produced diffraction gratings of excellent quality, obtaining a spectral resolution and precision in wavelength much superior to that given by prisms.

This chapter assumes basic knowledge of atomic and molecular spectroscopy. Atomic lines are produced by radiative transitions among the quantized internal energy levels; each line has a frequency corresponding to the difference between the energies of the two levels. For molecular lines and bands, the situation is more complex: in addition to electronic energy levels, the molecule has quantized degrees of freedom associated with its vibrations and rotations. As a matter of specific notations, astronomers indicate the ionization stages of atoms with a Roman numeral +1; therefore neutral helium is indicated with He I, singly ionized helium with He II, two-times ionized carbon with C III, and so on. The notation H II would then indicate the proton. Astronomical spectra often display lines of very high ionization stages, corresponding to temperatures impossible to reach in the laboratory, for instance Fe XV or Ni XVI in the solar corona, whose temperature exceeds millions of degrees. For molecules, the notation with "$+$" is used, for example ionized water H_2O^+ seen in comets. Forbidden lines (namely those arising from metastable states via magnetic dipole or electric quadrupole transitions) are indicated by the wavelength of that transition followed by the symbol of the element within square brackets, for example λ 5577 [O I], λ 5007 [O III]. Again, these transitions are possible in astrophysical conditions but often impossible to reproduce in the laboratory. The forbidden lines were, indeed, first observed in celestial sources, and only afterwards assigned to a particular element (see Notes).

17.1 Spectroscopic Techniques

In the visible, near UV and near IR, spectra are generally obtained by means of spectrographs having as basic dispersive elements either prisms or gratings (or a combination of the two, the so-called grisms). Very schematically, the light focused on the focal plane of the telescope is intercepted by an aperture (usually a rectangular one, therefore a slit) whose function is to limit the solid angle of the accepted beam, and sometimes to provide a convenient reflective surface on which the star and the surrounding field can be seen. As discussed in Chapter 11, the Earth's atmosphere produces a dispersion of the stellar light in the vertical plane; therefore, it is advisable to align the slit of the spectrograph in that plane, to avoid systematic chromatic errors. After the slit, the light goes through a collimator, namely a lens system or a mirror, which produces a parallel beam, which in its turn illuminates the dispersive element (or elements). Finally, another complex lens system (camera) produces on the detector a series of quasimonochromatic images of the illuminated portion of the slit (see an example in Figure 17.1).

In some cases, the dispersive element is placed directly in parallel light before the entrance pupil of the telescope; this is, for instance the case of the objective prisms used with wide-field astrographs or Schmidt telescopes.

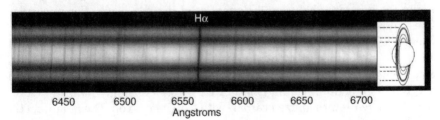

FIGURE 17.1

In this spectrum of Saturn covering the Hα region, the slit went through the disk and the rings, as indicated on the right side; the inclination of the spectral lines is due to the rotation of the disk and of the rings. (CCD spectrum obtained with the 1.2m telescope at Asiago Observatory.)

In this manner, all stars in the field produce simultaneously a low resolution spectrum. In other cases, the focal plane of the telescope is equipped with a number of adjacent slits or of movable optical fibers, in order to obtain simultaneous spectra of dozens of stars. Finally, the dispersion of light can be realized by entirely different means, for example, by Fourier spectroscopy in the near IR.

In contemporary astronomical practice, the spectrum is examined and displayed with the wavelength increasing to the right. In addition to the covered spectral range, two characteristics of the spectrum are usually given, namely the reciprocal dispersion D (in A/mm, often improperly called dispersion) and the spectral resolution $R = \lambda/\Delta\lambda$, which are determined by the slit width, the dispersive element, the optical configuration, and the detector pixel size. Both D and R can be wavelength dependent if the dispersive device is a prism. In astronomical applications, two contrasting elements often need some kind of compromise: it would be desirable to obtain the maximum theoretical resolution by keeping the slit width at its minimum; however, the seeing conditions often require widening the slit to collect as much light as possible. The dimensions of the slit on the telescope focal plane (the height, namely the coordinate perpendicular to the spectral dispersion, and the width) are usually given in arcseconds. The dimensions of the slit on the detector are determined by the ratio of the focal length of the collimator to that of the camera (briefly, the collimator/camera ratio), and, if the case, by the anamorphic magnification of the diffraction grating.

In the past, when the detector was photographic emulsion, the height of the spectrum used to be enlarged by moving the star up and down along the slit during the exposure time. This technique facilitated the examination of the spectral features against the grain of the emulsion, but it considerably increased the telescope time, and it could not be applied to extended objects such as comets, planets, nebulae, and galaxies, otherwise all the spatial information along the slit would have been lost. Modern detectors have largely overcome this problem. In all cases, the spectral lines are the monochromatic images of the illuminated slit in the different wavelengths.

Figure 17.2 shows the spectrum of the blue (hot) star α And. The spectrum of every normal star has two components, a continuum in emission and atomic lines in absorption. In cooler stars, in addition to atomic lines molecular absorption bands are also seen.

There are important cases of bodies with emission lines (peculiar stars, diffuse, and planetary nebulae, active galaxies, etc.) and/or emission bands, for example comets (see Figure 17.3 showing three prismatic spectra of Halley's comet in its passage of 1986). In other cases, only the continuum is visible, for instance in hot white dwarfs, while at the other extreme the continuum is so faint that only emission lines are readily visible, for example in some planetary nebulae or active galactic nuclei (see Figure 17.4). A word of clarification: on the photographic plates, the light darkens the grains of the emulsion, so that the bright continuum appears dark and the absorption lines appear clear on the originals (negative); the situation reverses on

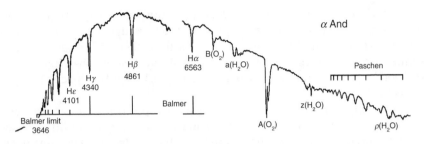

FIGURE 17.2
Low resolution uncalibrated spectrum of the blue star α And, with prominent hydrogen lines in absorption. Left: from the Balmer discontinuity in the near UV to the second Balmer line Hβ. Right: from the first Balmer line Hα to the lines of the Paschen series in the near IR. The strongest telluric features (B, a, A, z, ρ) are indicated. (Spectrum by the author.)

positive prints such as in Figure 17.3. On solid-state devices, the absorptions appear as dips of the continuum (the absorptions remove energy from the continuum) and the emissions are bright (they add energy to the continuum).

Utilizing linear detectors such as photoelectric cells or solid-state devices (e.g., CCDs), one can measure with great precision the luminous flux through the slit, in the different spectral bands. The measurement of a calibrated comparison star, made under the same observation conditions (sky transparency, airmass, etc.) will provide the spectrum of the star in physical units, namely its spectrophotometry, both for the continuum component and for the discrete lines and bands. Spectrophotometric

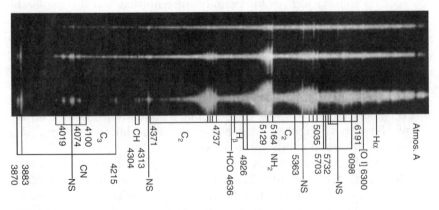

FIGURE 17.3
Several molecular bands of CN, C_3, C_2 (Swan's band), NH_2, and atomic lines are visible in emission in the spectrum of Halley's periodic comet, here observed at three dates in 1985 and 1986. The continuum is solar light scattered by cometary dust particles. Notice that the molecular band intensities decrease to the blue, with sharp red edges. As in the case of Saturn, the spectra have spatial resolution perpendicular to the dispersion. The dispersion of the spectra was produced by prisms, so that the blue region is much more dispersed than the red one. (Spectra by the author.)

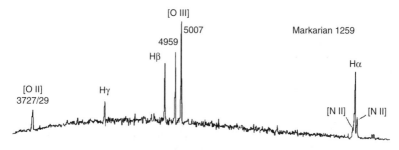

FIGURE 17.4
Grating spectrum of the emission line galaxy Markarian 1259. The spectrum is dominated by the strong emission lines of the H I Balmer series and of the forbidden transitions of O II, O III, and N II. The dispersion is essentially constant from the blue to the red. (Spectrum by the author.)

measurements are of fundamental value in verifying and calibrating the theoretical models of radiation from all celestial bodies.

The radiation from stars is usually unpolarized, but there are noticeable exceptions, such as peculiar stars, comets, and active galactic nuclei. Even the Sun, in high spatial resolution spectra covering small portions of the disk, exhibits polarization. This polarization may be intrinsic to the radiation mechanism (e.g., synchrotron light), due to slanted reflection from a solid surface (e.g., the asteroids), or induced by the intervening medium (e.g., the interstellar dust). Spectropolarimetric studies are therefore a powerful means for studying the physical conditions of the celestial bodies. Some references are given in the Notes.

17.2 The Analysis of the Spectral Lines

The first step in the analysis of spectral lines are the measurement of their wavelengths and the identification of the responsible element and of its ionization stage, in order to permit the assignment of each line to a given atomic transition (this first step is not always possible, even today several lines of the solar spectrum defy identification). Then, two most useful quantities can be obtained for each line, namely the equivalent area and the equivalent width. Suppose we have a well-calibrated spectrum, with reciprocal dispersion high enough to discern strong and weak absorption lines. We can measure with precision the line profile F_λ, namely the intensity along the curve extending from λ_1 to λ_2, λ_0 being the wavelength at the center (see Figure 17.5). The continuum has been arbitrarily normalized to 1.

The intensity is specified by the so-called equivalent area $A_{\lambda 0}$ defined as:

$$A_{\lambda 0} = \int_{\lambda_1}^{\lambda_2} (1 - F_\lambda)\mathrm{d}\lambda \tag{17.1}$$

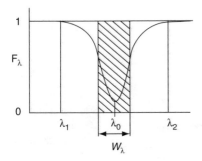

FIGURE 17.5
The line profile, equivalent area, and equivalent width. On the line profile, we distinguish between the center (or core) and the wings.

a quantity which is a measure of the energy subtracted by the line from the continuum. The equivalent width W_λ of the line is the width in angstrom of a rectangle having the same total integrated energy as the true line. Notice that the equivalent width alone does not discriminate between a shallow wide line and a narrow deep one. Its value can critically depend on the choice of the continuum, and sometimes it can be fairly uncertain (see Figure 17.6); great caution is therefore needed in the quantitative interpretation of the spectra.

The equivalent width depends on the number of atoms N in the atmosphere capable of absorbing that transition. Therefore, it depends on the temperature and density (or pressure) inside the gas, and on the chemical composition of the star (number of atoms in the present context means the total number of those atoms in the column of unit cross-section through the stellar atmosphere). Furthermore, the strength of a given line will depend on the probability of that transition, summarized by a so-called f factor (or oscillator strength, described in a later section). The relationship between the line intensity and the number of atoms, or better the product $N \cdot f$, is far from linear. In addition, it depends on the broadening mechanisms affecting the line.

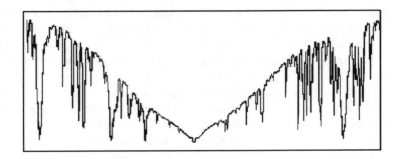

FIGURE 17.6
The K line of Ca II in the solar atmosphere shows how difficult the determination of the continuum level can be; this high-resolution spectrum shows the very broad wings of the K line affected by several weaker absorption lines of other elements.

Suppose for instance that the only broadening mechanism is the natural width of the line, namely the width dictated by Heisenberg's uncertainty principle. The absorptivity of the gas will be very high at the center of the line, and it will drop to almost zero at a very small distance from it. Therefore, even a small number of atoms will produce a strong absorption at the very center of the line, and almost no wing: the line will be deep and very narrow. By adding increasing numbers of atoms, the center of the line will tend to saturate, and the wings will rapidly grow. Therefore the equivalent width regime will pass from a linear dependence on $N{\cdot}f$ at low numbers, to a much slower one, essentially $W = \sqrt{N{\cdot}f}$, at higher column densities. In most stellar types, the broadening due to collisions between the atoms is much more important than that due to the natural width, however, the resulting profile is the same as natural broadening (except that for H and He, as explained later). The H and K lines of Ca II in the solar atmosphere provide a good example of collisionally broadened line profile (Figure 17.6).

Another important cause of broadening is thermal agitation, namely the superposition of the individual Doppler shifts. In this case, at low densities the line is broad and not very deep. By adding successively more atoms, the linear regime continues for a longer interval than in the case of collisional broadening, until eventually the intensity saturates; no matter how many more atoms are added, W_λ will stay constant.

Superimposing natural, collisional, and Doppler broadening, we obtain the overall behavior of W_λ with $N{\cdot}f$, namely the so-called curve of growth. The previous oversimplified discussion has shown that this curve will be composed of three parts (see Figure 17.7):

1. At low $N{\cdot}f$, the intensity will be proportional to $N{\cdot}f$.
2. For intermediate values of $N{\cdot}f$, the center of the line is deep, but the wings are relatively unimportant; the intensity will remain almost constant for a range of $N{\cdot}f$.
3. For very large values of $N{\cdot}f$, the intensity will increase with $\sqrt{N{\cdot}f}$.

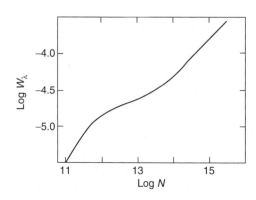

FIGURE 17.7
An idealized behavior of the curve of growth.

The crossover between the three regimes will depend on the relative importance of the three mechanisms; Doppler broadening is sensitive only to temperature, but collisions are also sensitive to density (or pressure). In reality, other broadening mechanisms must be considered, for instance the Stark effect, which is so important for H and He lines that the previous considerations on the curve of growth are incapable of explaining the observed profiles. Similar difficult problems are caused by the Zeeman effect in sunspots or magnetic stars. On the other hand, stellar rotation causes a characteristic broadening of the profiles, but does not affect the curve of growth. Therefore, the precise theory has considerable complexity.

We have already said that several factors contribute to the width of the line. Heisenberg's uncertainty principle prevents a line from being strictly monochromatic, and moreover, the hyperfine structure adds to the natural width, but these factors usually do not contribute more than a few mÅ in the visible. Macroscopic effects are much more important, such as the thermal Doppler broadening. In a gas in thermal equilibrium with temperature T, each particle of mass m would acquire a kinetic energy given by:

$$\frac{1}{2}mv^2 = \frac{3}{2}kT \tag{17.2}$$

where v is the velocity vector modulus. Maxwell's function (see also Equation 12.27) gives the particles distribution as a function of v:

$$dN(v) = \frac{N}{\sqrt{\pi}}e^{-(v^2/2kT)}\frac{dv}{v} = 4\pi N\left(\frac{m}{2\pi kT}\right)^{3/2}v^2 e^{-mv^2/2kT}dv \tag{17.3}$$

(Maxwell's distribution is not always appropriate to astrophysical conditions, for instance highly collimated jets can be described as gases with three different temperatures along the three principal directions). By virtue of the Doppler effect, each particle of mass m will absorb at larger or smaller wavelengths according to its velocity, so that the line profile will be described as a Gaussian function with a FWHM given by:

$$\text{FWHM} = 7.16 \times 10^{-7}\lambda\sqrt{\frac{T}{m}} \tag{17.4}$$

Therefore, the lines of the lighter elements, in particular hydrogen, will be the widest.

Mutual collisions among the particles are another cause of broadening because they alter the energy levels of atoms and molecules, and the wavelengths of the corresponding transitions. For instance, the great majority of particles in the solar atmosphere are H I atoms, whose density is proportional to the gaseous pressure P_g. In the hottest stars, instead, the fastest particles are free electrons, so that the collisional broadening is proportional to the electron pressure P_e.

In the classic theory of radiation emitted by oscillating charges, the resulting alteration of wavelength can be interpreted as damping of the oscillator, in this case collisional damping. The line profile can be approximated by a Lorentz function:

$$\frac{F(\lambda)}{F(\lambda_0)} = \frac{1}{1 + \left(\dfrac{\lambda - \lambda_0}{\text{FWHM}}\right)^2} \tag{17.5}$$

While the Doppler broadening is more efficient at the center of the line, the collisional damping contributes more to the wings.

A third effect, the Stark broadening, is due to the microscopic electric fields caused by free charges passing near the radiating atom or ion. These short-lived fields of random direction and amplitude are sufficient to perturb the energy levels (especially the outer ones) of the ion. The effect is particularly important for hydrogen, and contributes to the different aspect of the lines between dwarfs, giants, and supergiants. In dwarfs, the density is higher, so that the Stark effect is stronger. In giants and supergiants, the density is decidedly lower, and the broadening is essentially Doppler. The broadening of the H I lines increases as n^2 (where n is the principal quantum number), so the higher lines of the Balmer series become progressively broader until they finally merge with one another. The last visible line on a high-resolution spectrum is related to the electron density N_e (cm^{-3}) by the Inglis-Teller formula:

$$\log N_e = 22.0 - 7.0 n_{\text{last}}$$

The lines of helium and other heavier elements exhibit a more complex behavior, showing the so-called quadratic Stark effect. Detailed calculations have been given, for instance by Vidal et al. 1973 (see Notes).

In the Sun, high spatial resolution allows the detection of macroscopic electric fields (well ordered over lengths of several hundred kilometers although of short temporal duration), in particular, in the proximity of dark spots when a magnetic polarity inversion produces a flare. The spectral lines then separate into several components by virtue of this macroscopic Stark effect, and the different components are differently polarized.

A macroscopic magnetic field produces the Zeeman effect, well observable in the Sun especially in the proximity of dark spots (see Figure 17.8). The lines of particular metals are easily seen resolved in different components having different polarization.

Finally, the broadening can be due to ordered movements of the atmosphere (currents), to turbulence, or more simply to the rotation of the entire star around an axis with arbitrary inclination i to the line of sight. The generally unknown inclination allows the measurement of $v_{\text{rot}} \cdot \sin i$. Several blue stars have rotational velocities exceeding 300 km/sec, while stars cooler than F0, for instance the Sun, are much slower rotators. As a consequence of this slow rotation, while the mass of the solar system is dominated by the mass of the Sun, its angular momentum is dominated by the planetary revolutions.

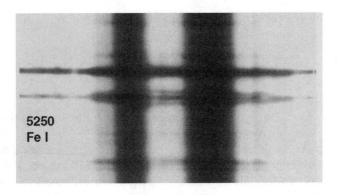

FIGURE 17.8
The Zeeman effect in a large solar dark spot. The Fe I line at 5250A is split in two components separated by 0.12 A, corresponding to a magnetic field of 3600 gauss. The field is almost as strong in the penumbra (center) as in the umbra (the two vertical dark streaks). In a generic star, the varying direction of the magnetic field along the line of sight usually prevents the resolution of the different components, and the Zeeman effect is seen as a broadening of the line. The magnetic field can be very strong on peculiar stars and on white dwarfs, reaching intensities of several hundred megagauss.

17.3   Detailed Balance and the Boltzmann Equation

In a closed system, we may require the fulfillment of the detailed balance condition: each and every transition is balanced by the opposite process, e.g., the emission of a photon of energy $h\nu$ by a jump from state B to state A, is balanced by the absorption of a photon of the same energy in a transition from A to B. The medium is then in a condition of thermodynamic equilibrium, at least locally (condition referred to as LTE): the population of states depends only on the temperature, while the density of the medium affects the processes rates only. Therefore, from Kirchoff's first law, at each place the ratio between volume emissivity and linear absorption coefficient, namely the source function, is a function of temperature only:

$$S_\nu = j_\nu/k_\nu \qquad (17.6)$$

In the following, $S_\nu(T)$ will be identified with Planck's radiation function, although there are astrophysical situations where this identification is not even approximately legitimate.

Consider the detailed balance between two states, n (upper) and m (lower), separated by an energy $\Delta E_{nm} = h\nu$. Downward transitions (emissions) can take place either by spontaneous or stimulated emission. A radiation density u_ν produces stimulated emission at a rate $N_n \cdot u_\nu \cdot B_{nm}$, where N_n is the number of atoms per cubic centimeter in the nth level (the photons produced by stimulated emissions are coherent with the incident electromagnetic field, this is the basic process of masers and lasers).

Spontaneous emissions occur at a rate $N_n \cdot A_{nm}$ even in the absence of the radiation field. Upward transitions (absorptions), are stimulated by the incident radiation field, and occur at a rate $N_m \cdot u_\nu \cdot B_{mn}$. The coefficients A_{nm}, B_{mn} are determined from the detailed balance condition:

$$N_n(A_{nm} + u_\nu B_{nm}) = N_m \cdot B_{mn} \cdot u_\nu$$

In most astronomical applications, stimulated emission is negligible, so that:

$$N_n \cdot A_{nm} \approx N_m \cdot B_{mn} \cdot u_\nu$$

In a gas composed of a given element (either neutral or ionized), in thermodynamic equilibrium, the relative population of two energy levels n, m separated by energy ΔE_{mn} is given by the Boltzmann law:

$$\frac{N_n}{N_m} = \frac{g_n}{g_m} e^{-\Delta E_{mn}/kT} = \frac{g_n}{g_m} e^{-h\nu/kT} \tag{17.7}$$

The terms g_n and g_m are the statistical weights of the two levels, namely the number of indistinguishable states having the same energy in each level, or else the number of electrons that can occupy that level without violating the Pauli exclusion principle. Einstein showed that if the populations are given by Equation 17.7, then the radiation density u_ν must be:

$$u_\nu = \frac{4\pi}{c} B_\nu = \frac{8\pi h\nu^3}{c^3} \frac{1}{e^{h\nu/kT} - 1}$$

where B_ν is the Planck function, and furthermore, that the transition probabilities are related by:

$$A_{nm} = \frac{8\pi h\nu^3}{c^3} B_{nm} = \frac{8\pi h\nu^3}{c^3} \frac{g_m}{g_n} B_{mn}, \quad g_m B_{mn} = g_n B_{nm} = \frac{8\pi^3}{3h^2} S_{mn} \tag{17.8}$$

where $S_{mn} = S_{nm}$ is the so-called line-strength (electric dipole for permitted lines, or magnetic dipole or electric quadrupole for forbidden lines). In the astronomical literature, the quantity $g \cdot f$ (named weighted oscillator strength) is usually found instead of S_{nm}:

$$gf = g_m f_{mn} = -g_n f_{nm} = \frac{8\pi^2 m_e \nu}{3he^2} S_{mn}, \quad A_{nm} = \text{const} \cdot \frac{gf}{g_n \lambda^2}$$

where m_e and e are, respectively, mass and electric charge of the electron; the value of the numerical constant depends on the adopted unit system.

Let us consider again the Boltzmann formula 17.7 applied to the case of H I. Each energy level i is referred to the ground level ($n = 1$):

$$\frac{N_i}{N_1} = \frac{g_i}{g_1} e^{-E_i/kT}, \quad i = 2, 3, \ldots$$

where E_i is the excitation potential of the ith level above the ground state. By summation over all i's we obtain the total population of that atom:

$$N = \sum N_i = \frac{N_1}{g_1}\left[\frac{g_2}{g_1}e^{-E_2/kT} + \frac{g_3}{g_1}e^{-E_3/kT} + \cdots\right] = \frac{N_1}{g_1}U(T) \quad (17.9)$$

where $U(T)$ is called the partition function. Owing to the facts that E_i tends to a limit and that $g_i \propto i^2$, this series seems to be unbound. However, this apparent mathematical divergence is physically removed by the collisions with the nearby atoms (Bohr's radii increase with i^2). The correction for the highest terms, which depends on pressure and temperature, could be applied by artificially lowering the ionization potential by a small amount.

With appropriate methods, the correct partition function $U(T)$ can be calculated for H I and for all astrophysically important atoms and ions (see from instance Gray, 1976; note that several authors give the $\log_{10}U$).

Table 17.1 gives approximate values of $U(T)$ for a few elements, for two temperature values. In most cases $U \approx g_1$ independent of T, but there are exceptions, such as Na I, Ca, and Fe.

The Boltzmann formula can be written in a more practical way. By expressing ΔE_{mn} in electronvolts (eV), letting $\Delta E_{mn}(eV) = \chi_{mn}$, using powers of 10, and introducing the usual notation in the astronomical literature $\Theta = 5040/T$, the law can be written as:

$$\frac{N_n}{N_m} = \frac{g_n}{g_m}10^{-\Delta E_{mn}(eV)\cdot 5040/T}, \quad \log\frac{N_n}{N_m} = \log\frac{g_n}{g_m} - \chi_{mn}\cdot\Theta$$

17.4 The Saha Equation

The ratio between the partition functions of ion plus electron and of neutral atom, in other words the ionization balance in thermal equilibrium, is given

TABLE 17.1

Partition Functions of Selected Ions

Element	g_1	Neutral U		g_1	First Ionization U	
		$T = 5,000$	$T = 10,000$		$T = 5,000$	$T = 10,000$
H	2	1.995	1.995	1	1.000	1.000
He	1	1.000	1.000	2	1.995	1.995
C	9	9.913	10.000	6	6.026	6.026
N	4	4.074	4.571	9	8.913	9.332
O	9	8.710	9.333	4	3.981	4.074
Ne	1	1.000	1.000	6	5.370	5.623
Na	2	2.042	3.981	1	1.00	1.00
Ca	1	1.174	3.548	2	2.188	3.467
Fe	25	26.915	54.954	30	42.658	63.096

by the Saha equation:

$$\frac{N_{i+1}}{N_i}N_e = \frac{2u_{i+1}}{u_i}\frac{(2\pi m)^{3/2}(kT)^{3/2}}{h^3}e^{-E_i/kT} = 4.83\times10^{15}T^{3/2}\frac{g_{i+1}}{g_i}e^{-E_i/kT} \quad (17.10)$$

where E_i is the ionization potential of the ground state of the ith ion.

The term $u_e = 2\dfrac{(2\pi mkT)^{3/2}}{h^3}$ represents the partition function of the free electron, namely its density of states in phase space; the factor 2 accounts for the two possible spin states of the electron.

The Saha equation does not have the same degree of general validity as does the Boltzmann equation, because ionization can also be caused by largely non-LTE phenomena, e.g., by collision with a stream of particles having a preferred direction. Therefore, in specific cases the Saha equation does not represent the true ionization degree of the plasma.

Table 17.2 shows the ionization potentials χ_i (expressed in eV) for the same elements of Table 17.1.

Furthermore, let us call U the partition function of the ion, and use the variable $\Theta = 5040/T$. In logarithmic form, Saha's equation becomes:

$$\log_{10}\left(\frac{N_{i+1}}{N_i}N_e\right) = -\chi_i\cdot\Theta - \frac{3}{2}\log_{10}\Theta + 20.9366 + \log_{10}\frac{2U_{i+1}}{U_i} \quad (17.11)$$

Making use of the perfect gas law applied to the electronic component $P_e = N_e kT$, the equation can be written also as:

$$\log_{10}\left(\frac{N_{i+1}}{N_i}P_e\right) = -\chi_i\cdot\Theta + \frac{5}{2}\log_{10}T - 0.4772 + \log_{10}\frac{2U_{i+1}}{U_i} \quad (17.12)$$

The pressure is expressed in dyn cm^{-2} (1 dyn cm^{-2} = 10^{-6} bar = 0.1 Pa).

An example of the Saha equation is given in Figure 17.9, which refers to conditions of temperature and pressure appropriate to the solar photosphere and corona. In the photosphere, all calcium would be neutral until the temperature reaches 3000 K, and then a fraction begins to be singly ionized.

TABLE 17.2

Ionization Potentials (eV) of Selected Elements

Element	I	II	III	IV	V	VI	VII	VIII	IX
H	13.60								
He	24.59	54.42							
C	11.26	24.38	47.89	64.49	392.08	489.98			
N	14.53	29.60	47.45	77.47	97.89	552.06	667.03		
O	13.62	35.12	54.93	77.41	113.90	138.12	739.32	871.30	
Ne	21.56	40.96	63.45	97.11	126.21	157.93	207.26	239.09	1195.8
Ca	6.11	11.87	50.91	67.15	84.43	108.78	127.70	147.40	188.70
Fe	7.87	16.16	30.65	54.80	75.50	100.00	128.30	151.12	235.00

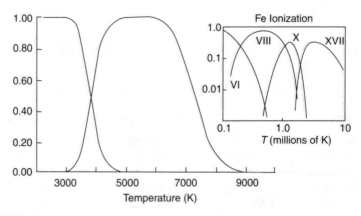

FIGURE 17.9
Ionization equilibrium between neutral and ionized calcium at solar photospheric conditions of
temperature and pressure. The inset shows the situation of iron in the solar corona, where the
temperature exceeds 1×10^6 K.

At 4000 K, Ca I and Ca II are equally abundant; at 5000 K, all calcium
becomes singly ionized; at 7000 K, Ca II and Ca III are in the same
proportion, and so on. In the corona, the temperature rises above millions of
K, the pressure is extremely low, and iron and other heavy elements become
multiply ionized.

The Saha equation shows that at constant temperature the degree of
ionization will depend on the electron pressure P_e. Let us therefore consider
two stars having equal masses and effective temperatures, but very different
radii, for instance a dwarf and a giant. The ionization will be lower in the
dwarf. Conversely, the dwarf will need a higher temperature to reach the
same ionization as the giant. This fact introduced a severe complication in
the earlier interpretation of the stellar spectra, which was clarified by the
work of Russell, Fowler, and Milne.

Table 17.3 shows the electron pressure P_e (in barye) in the atmospheres of
dwarf, giant, and supergiant stars.

From the above discussion regarding the Boltzmann and Saha equations,
we have concluded that the intensities of the absorption lines depend on two
fundamental variables, namely temperature and electron pressure (all other
variables, such as chemical composition, magnetic fields, rotation, etc., are

TABLE 17.3

Typical Stellar Electron Pressures P_e

		Electron Pressure P_e (barye)		
T (K)	Θ	Dwarfs	Giants	Supergiants
10,080	0.5	320	100	33
5,040	1.0	4	0.8	0.2
3,360	1.5	0.57	0.02	0.004

ignored in this context). This fact implies that the lines of several elements, having different ionization potentials, must be simultaneously taken into account for proper classification of a stellar spectrum.

Instead of electron pressure, we can use luminosity as a second variable. The stellar mass is indeed a fairly stable parameter for stars of the same temperature; therefore two stars having the same temperature but different radii will have very different densities and surface pressures, and different mechanisms of interaction between atoms and molecules. The consequence of this is that the spectral lines and bands of a giant star have an appearance different from that of a star of the same temperature but smaller radius, a difference that can be easily noticed even on spectra of moderate spectral resolution. For instance, in the hotter stars the interatomic Stark effect produces Balmer lines that are narrower in giants than in dwarfs; in the cooler stars, the blue band of CN is stronger in giants than in dwarfs. The accurate examination of the stellar spectra must show the influence of the luminosity (or equivalently of the radius, surface gravity, or pressure) on the aspect of the lines and bands, in addition to that of the temperature, and must permit the determination of both variables.

Spectroscopy therefore provides a powerful means to determine the luminosity, and hence the distance, of a star too far away for the application of the trigonometric parallax method. The foundations of this method of spectroscopic parallaxes go back to the works of two pioneers of astrophysics, namely Kohlschutter and Adams in 1914. Their work was then expanded upon by Adams and Joy, around 1935.

To complete these considerations, let us examine the total gas pressure, which is the sum of the ion and electron partial pressures. We assume, as before, that the equation of state of a stellar atmosphere can be described by that of a perfect gas (an assumption which has to be taken with great caution in a planetary atmosphere). The total gas pressure is due to ions and electrons:

$$P_{gas} = (N_i + N_e)kT \tag{17.13}$$

where N_i is summed over all ionization stages (including neutral). In a stellar atmosphere, all matter is essentially composed of hydrogen and helium, with traces of other elements collectively indicated by Z; although their density is low, Z can contribute an appreciable number of free electrons, so that:

$$N_i \approx N(\text{H I}) + N(\text{H II}) + N(\text{He I}) + N(\text{He II}) + N(\text{Z I}) + N(\text{Z II}) + \text{traces}$$

$$N_e \approx N(\text{H II}) + N(\text{He II}) + N(\text{Z II}) + \text{traces}$$

Two limiting regimes can be considered in this respect:

- A hot star ($T > 10^4$ K). Hydrogen is totally ionized, so that $N_e \approx N_i$, $P_e/P_{gas} \approx 1/2$, independent of the precise chemical composition of the atmosphere.

- A cool star $(T < 5 \times 10^3 \text{ K})$. Only metals are ionized, so that $N_e \approx N(\text{Z II})$ and the ratio $P_e/P_{\text{gas}} \approx N(\text{Z II})/[N(\text{H I}) + N(\text{He I}) + N(\text{Z I}) + N(\text{Z II})]$ is much smaller than 1, and sensitive to the chemical composition (although in precise considerations the contribution of molecules should be included).

Therefore, the Sun is in an intermediate situation. The degree of ionization of an average stellar atmosphere (an average between a metal rich and a metal poor composition) is given in Table 17.4, relating gas pressure, electron pressure, and temperature. As expected, at constant electron pressure P_e the gas pressure P_g varies by many orders of magnitude with T.

The general result that the ionization increases with the decrease of density, applies also to the interstellar medium (ISM). Lines of strongly ionized elements (e.g., C IV) are observed in the ISM even if the temperature of the gas is extremely low; the reason for this is that, once an element is ionized by the absorption of a UV photon the probability of recombination is essentially nil because of the very low density.

The Saha equation can be extended to molecules. Consider for simplicity a diatomic molecule AB, composed by two atoms A and B. The dissociation rate is governed by a similar equation:

$$N_A N_B / N_{AB} = \frac{U_A U_B}{Q_{AB}} \left(\frac{2\pi m_{AB} kT}{h^2} \right)^{3/2} e^{-D/kT} \tag{17.14}$$

where $m_{AB} = m_A m_B / (m_A + m_B)$ is the reduced mass of the molecule, U_A and U_B are the partition functions of the two atoms, D is the dissociation energy, and $Q_{AB} = Q_{\text{rot}} \cdot Q_{\text{vib}} \cdot Q_{\text{el}}$ is the molecular partition function. Data for some astrophysically relevant molecules are provided in Table 17.5 and Table 17.6. Numerically, expressing m in amu, D in eV, N in cm^{-3}:

$$\log \frac{N_A N_B}{N_{AB}} = 20.2735 + \frac{3}{2} \log m_{AB} + \frac{3}{2} \log T - D/\Theta + \log \frac{U_A U_B}{Q_{AB}} \tag{17.15}$$

Notice for instance how much easier is the dissociation of CO_2 compared with CO. Equation 17.14 can be considered a particular case of the general law of mass action found by Guldberg and Wage.

TABLE 17.4

Ionization Degree of an Average Stellar Atmosphere

	$\log_{10}P_g$			
$\text{Log}_{10} P_e$	$\Theta = 0.2$ $(T = 25{,}200)$	$\Theta = 0.6$ $(T = 8{,}400)$	$\Theta = 1.0$ $(T = 5{,}040)$	$\Theta = 1.4$ $(T = 3{,}600)$
-2	-1.8	-1.67	$+0.78$	$+2.4$
0	$+0.29$	$+0.35$	3.9	5.3
2	2.30	2.98	6.7	8.5
4	4.31	6.84	10.0	12.4

TABLE 17.5

Diatomic Molecules

Name	Dissociation Potential (eV)	Ionization Potential (ev)	Equilibrium Distance (Å)
H_2	4.5	15.4	0.74
C_2	6.3	12.2	1.24
CH	3.5	10.6	1.12
CO	11.1	14.0	1.13
CN	7.8	14.2	1.17
O_2	5.1	12.1	1.21
OH	4.4	12.9	0.97

17.5 Criteria of Spectral Classification of Stars

In order to devise a coherent scheme of spectral classification of stellar spectra, namely to assign stars into spectral types SpT, the generally adopted criteria are the behavior of the continuum with the wavelength (in other words the color of the star as a measure of its surface temperature), and the relative intensities of absorption lines or bands. Each scheme is strongly dependent on the spectral resolution or on the spectral response of the available equipment. As it occurred for laboratory spectroscopy, many classification schemes were devised well in advance of the physical understanding of the stellar atmospheres and interiors.

In each classification, stars having well-known properties are used as a template, against which the spectrum of each particular star can be examined. Then, the empirical classification is calibrated in terms of the fundamental physical quantities temperature T and luminosity L (by virtue of the Stefan-Boltzmann law, luminosity and radius play essentially the same role). Other variables can be included in the scheme, e.g., the chemical composition μ, the magnetic field **H**, the rotation, but in order to do so finer observations are needed. The book by Jaschek and Jaschek (1990) provides additional details.

TABLE 17.6

Polyatomic Molecules

Name	Dissociation Potential (eV)	Ionization Potential (ev)	Equilibrium Distance (Å)
H_2O	5.1	12.6	3.5
N_2O	1.7	12.9	4.0
CO_2	5.3	13.8	3.8
NH_3	4.3	10.2	3.0
CH_4	4.4	13.0	3.5

Therefore, each particular classification scheme is at least bidimensional: SpT = SpT(T,L) = SpT(T,R). Let us examine the influence of the temperature T, which ranges for normal stars from approximately 50,000 to 1000 K, by ordering the stars as a function of their color, from the bluest to the reddest. We easily see that the appearance of the absorption lines and bands changes appreciably. For example, consider the behavior of the Balmer series of hydrogen going from the hotter to the cooler stars: the intensity of the lines increases with the decrease of T from 50,000 K to reach a maximum at 10,000 K, and decreases again to almost disappear around 3500 K. This behavior therefore is a temperature, not a chemical composition effect.

Indeed, all stars have essentially the same composition. Hydrogen and helium are by far the dominant elements; if we put conventionally equal to 1 the number of oxygen atoms per unit volume, then the standard mix μ is:

$$\mu = 1600\,H + 160\,He + 1\,O + 0.5\,N + 0.3\,C$$
$$+ 0.2\,Ne + 0.1\,Fe + 0.06\,Mg + \dots$$

Compositional differences from star to star are essentially due to the heavier elements, to the point that all elements other than H and He are often called metals, even if their chemical properties are certainly not metallic. The composition can be given by number of atoms, but also by mass (see Table 17.7). In theoretical applications, the percentages by mass are indicated by X (hydrogen), Y (helium), Z (all the rest), with $X + Y + Z = 1$, as in Table 17.8.

It is plainly evident that the standard stellar composition is very different from that of the Earth, Mercury, the Moon, or Mars. Indeed, considering that the average density of the Sun is so much lower than that of the Earth, namely 1.4 against 5.4, we conclude that the Sun's composition must be dominated by the lighter elements and the Earth's by the heavier ones. Jupiter, Saturn, and the other giant planets have densities comparable to the solar one. On the other hand, stars dominate the mass of the visible Universe, so that hydrogen and helium are the main universal constituents. The strong depletion of light elements (Li, Be, B) in the stellar atmospheres and interiors is of great significance, both for the theories of stellar evolution and for the understanding of the initial chemical composition of the Universe.

TABLE 17.7

Schematic Standard Mix of Chemical Elements, by Number and by Mass

Element	Number	Mass	Electrons
H	100	100	100
He	9.8	39	20
C, N, O, Ne	0.15	2.2	1.1
All the rest	0.01	0.4	0.21

TABLE 17.8

Theoretical Notation of the Standard Mix by Mass

X (mass fraction of H) = 0.71
Y (mass fraction of He) = 0.27
Z (mass fraction of "metals") = 0.02

We now describe two widely used classification schemes, devised at Harvard and Yerkes, respectively.

17.6   The Harvard and the MK Classification Schemes

Following the pioneering attempts by Father A. Secchi, S.J., at the Specola Vaticana in 1870, W.H. Pickering and coworkers produced the Harvard classification, at the beginning of the 20th century, by examining more than 250,000 photographic spectra obtained with objective prisms from both hemispheres. The stars were brighter than approximately the ninth magnitude. The spectra, covering the blue range from 3300 to 4900 Å, were ordered in a discrete sequence of seven classes having decreasing temperatures. The classes were designated with the following capital letters: O, B, A, F, G, K, M (this succession of letters reflects the many steps used to arrive at the final ordering). The planetary nebulae were indicated with P and the novae with Q, both objects being characterized by strong emission lines. In 1922, the IAU recommended some modifications to the original notation, which are still in use today. For instance, the letter "e" indicates the presence of emission lines, "k" the presence of interstellar lines, "p" a peculiarity. The Harvard catalog, known as HD (Harvard Durchmunsterung, but later also Henry Draper Catalog), was published starting in 1918. The HD has been thoroughly revised by Houck (see Notes).

The HD classification was followed by that produced at Yerkes Observatory by Morgan, Keenan, and Kellman, known as MKK (often simply as MK) classification. As in the HD system, the Yerkes spectra were obtained with prismatic dispersion. The typical value of R was approximately 100 A/mm at Hγ. The dominant variable T was estimated inside seven intervals, designated with the same capital letters O, B, A, F, G, K, M used in the HD system. Each interval was subdivided into ten finer subclasses with Arabic numbers from 0 to 9, always in the sense of decreasing temperature:

MKK SpT: O4, ..., O9; B0, B1, ..., B9; A0, A2, ..., A9;

G0, G2, ..., G9; K0, K1, ..., K9; M0, M1, ..., M9

Not all subclasses were actually used; for instance A1 or G1 were lacking, in the sense that the template did not contain any star of that subtype. For other subtypes, a fractional classification was needed, e.g., O9.5 or B0.5.

With respect to Harvard's scheme, Yerkes classification takes into account, the absolute luminosity, indicated with the five Roman numerals I, II, III, IV, V. These five luminosity classes are: I, supergiants (subdivided into Ia, Ib, and Iab); II: bright giants; III: giants; IV: subgiants; V: dwarfs (also called main sequence, or zero age sequence in theoretical works). Unfortunately, the designation "dwarf" is misleading, and often confused with white dwarfs, which are actually not part of the MKK scheme of luminosities. Indeed, luminosity classes VI and VII, used by several authors to indicate subdwarfs and white dwarfs, do not belong to the original Yerkes scheme.

Therefore, examples of complete SpT designations in the MKK classification are: B2-Ia, A0-V, F5-III, and so on. The Sun belongs to the MKK SpT G2-V, while in the original Harvard classification it was a G0 star.

Examples of photographic and CCD spectra of stars classified in the MKK system can be found on the websites indicated in the Notes. Here we give some of the main criteria of classification.

Type O (old Harvard types Pd, Oe, Oe5): blue color, $50,000 \leq T \leq 25,000$ K; strong absorption lines of the Pickering series of He II, whose maximum intensity is reached at O5; a significant percentage of the stars hotter than O4 also display emission lines. From O4 to O9 the relative intensities of the absorptions of He II, He I, H, Si IV are used to fix the SpT. He I increases from O4 to O9. At O7, the lines of N III reach their maximum, and C III $\lambda4647$ begins to appear.

Typical stars: O5 ζ Pup; O6, λ Cep; O7, S Mon; O8, λ Ori; O9, 10 Lac.

Type B: blue–white color, $25,000 \leq T \leq 12,000$ K. He II disappears, He I reaches its maximum intensity, H I gradually increases and is a good criterium of luminosity (as was shown by Abetti, Lindblad, and Schalen). The ratios Si III/Si IV and Si II/He I are good indicators of SpT; for instance the ratio $\lambda4552$ Si III/$\lambda4089$ Si IV is <1 at B0 and >1 at B1. As luminosity criteria, the ratio $\lambda4089$Si IV/$\lambda4009$ He I can be used in addition to the Balmer lines. Among the B stars, the appearance of Balmer lines in emission, often inside the broader absorptions, is not infrequent; those stars are indicated as Be.

Typical stars: ε Ori B0 Ia; δ Sco B0 V; γ Ori B2 III; η UMa B3 V; β Ori B8 Iab; α Leo B7 V.

Type A: white color, $12,000 \leq T \leq 8000$ K. The He I lines disappear and the Balmer lines, usually quite broad, reach their maximum. Figure 17.10 shows examples of transition from B to A for main sequence stars. The K-line of Ca II at $\lambda3933$ is clearly visible, while the H-line of Ca II at $\lambda3970$ is masked on low- and medium-resolution spectra by the Hε line of the Balmer series. The Balmer series can be resolved up to the 11th line. The increasing intensity of the metal lines can be used as a temperature indicator; for instance, $\lambda4030/34$ of Mn II starts to be seen at A2. The Balmer lines are

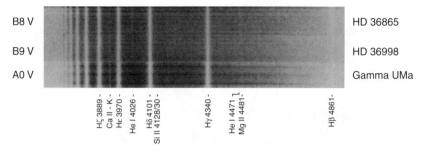

FIGURE 17.10

Stars from B8 V to A0 V. Notice the progressive decrease of the He I lines, and the increase of the Ca II, Si II, and Mg II lines. The spectra have been taken with a grating spectrograph and blue emulsion, so that the plate sensitivity has a strong decrease already at Hβ. The spectra have been widened by trailing along the slit. (Adapted from the Abt, H.A., Meinel, A.B., Morgan, W.W., Tapscott, J., 1968, *An Atlas of Low Dispersion Grating Stellar Spectra*, Kitt Peak National Observatory, Tucson, Arizona.)

indicators of luminosity as they are for the B stars, and so are the Fe II lines, whose intensity increases with L.

Typical stars: α Dra A0 III; γ Gem A0 IV; α Lyr A0 V; α Cyg A2 Ia; β Tri A3 III; γ Lyr A7 III.

Type F: white–yellow color, $8000 \leq T \leq 6000$ K. The spectrum is similar to A-type, although the Balmer lines are weaker and the H, K of Ca II and other metal lines are stronger. The molecular band at λ4310 (the so-called G-band, a blend of CH and metal lines) is clearly visible. Several lines of ionized metals are good indicators of luminosity, e.g., Sr λ4077 and λ4215, Ti II λ4161 and λ4399, Fe II λ4233, which are stronger on the giants. There is a noticeable luminosity effect also in the violet continuum, which is bluer in the dwarfs than in the giants; the effect increases on going to cooler stars, and becomes very evident in the K type.

Typical stars: α Lep F0 Ib; γ Vir F0 V; β Cas F2 IV; γ Cyg F5 Ib; β Vir F5 V.

Type G: yellow color, $6000 \leq T \leq 5000$ K. Spectrum similar to the solar one. Strong metal lines; in particular H, K of Ca II are stronger than the Balmer lines. The ratio Fe II λ4325/Hγ is a good indicator of T. The CN violet bands at λ4216 and λ4144 are good indicators of luminosity, being much stronger in giants than in dwarfs. A similar but weaker effect is shown by the G-band.

Typical stars: α Aqr G0 Ib; α Sag G0 II; η Boo G0 IV; Sun (see Figure 17.11), and the solar analog 16 Cyg G2 V; μ Her G5 IV; ζ Cyg G8 II; δ Boo G8 III; β Aql G8 IV; ζ Boo A G8 V.

The Moon, Mercury, Mars, Pluto, the asteroids (and the dust component of the cometary comae and the zodiacal light) reflect the solar light in the visible region of the spectrum, with only slight modifications due to the reflectance from the solid materials of their surfaces. Figure 17.12 shows as examples the reflectance spectra from the blue to the near IR of two so-called near Earth asteroids. The spectra have been normalized to the solar spectrum by using the intermediate of solar analog stars. Notice how different the two spectra are, an indication of different surface properties of the two bodies.

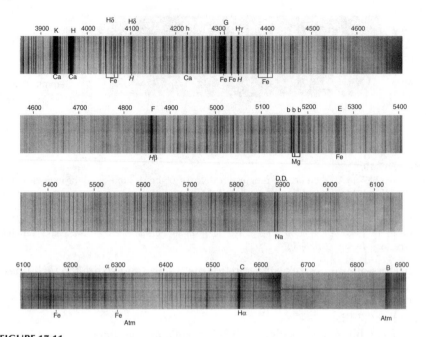

FIGURE 17.11
The photographic solar spectrum from the H and K lines of Ca II (top left) to the telluric B-band of O_2 (bottom right). The absorption lines are black. Atm indicates the α and B telluric features. (Adapted from a spectrum taken at Mount Wilson Observatory.)

Figure 17.13 shows the spectrum (taken with the TNG) of one of the most distant solar system bodies, namely of the trans-Neptunian object (TNO) 2400 DW. The object has semimajor axis $a = 39.5$ AU, eccentricity $e = 0.218$, inclination $i = 20°.6$, period $P = 248$ year (notice how similar these elements are to those of Pluto). Its estimated diameter is 1500 km. At the distance of the 2004 DW, the Sun is an almost point-like source. The spectrum shows a

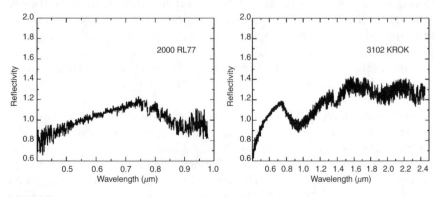

FIGURE 17.12
Reflectance spectra of two near Earth asteroids. (Adapted from Lazzarin, M., Marchi, S., Barucci, M.A., di Martino, M., Barbieri, C., 2004, *Icarus*, **169**, pp. 373–384, see Notes.)

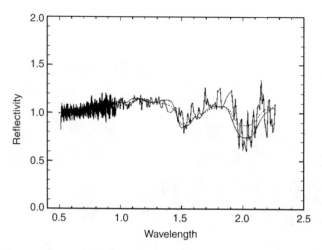

FIGURE 17.13

Spectrum of the TNO 2004 DW, obtained with the TNG. (Adapted from Fornasier, S., Dotto, E., Barucci, A.M., Barbieri, C., 2004, *Astronomy and Astrophysics*, **422**, pp. L43–46, see Notes.)

neutral and nearly featureless visible continuum, and water ices absorption bands at 1.5 and 2.0 μm. Venus, Jupiter, and the other giant planets have a very different situation, as their atmospheres produce important absorption bands. We see examples in Figure 17.14. The spectrum of Jupiter and Saturn show bands of ammonia ices (NH_3), the spectrum of Uranus and Neptune is dominated by the bands of methane (CH_4) ices. These methane bands, removing the reflected red spectrum, provide a distinctive green color, easily seen by eye with a modest telescope, to Uranus and Neptune.

Type K: yellow–orange color, $5000 \leq T \leq 4000$ K. The blue continuum becomes weaker and weaker, the spectrum resembles that of the solar spots. The H, K lines of Ca II reach their maximum at K0, while the intensity of the Ca I lines at $\lambda 4227$ and $\lambda 4454$ increases along the sequence. The T criteria are the same used for the G-type, and the CN band at $\lambda 4216$ can still be used for L.

Typical stars: ε Cyg K0 III; η Cep K0 IV; γ Cep K1 IV; ε Peg K2 Ib; α Ari K2 III; ε Eri K2 V; η Per K3 Ib; γ Aql K3 II; δ And K3 III; ζ Cyg K5 II; α Tau K5 III; 61 CygA K5 V.

Type M: reddish color, $4000 \leq T \leq 2500$ K. The spectrum is dominated by the appearance of molecular bands, typically those of TiO, ScO, and VO. The temperature is estimated by the strength of these bands or of the Ca I line $\lambda 4227$. The Balmer series can still be used for the determination of luminosity.

Typical stars: β And M0 III; μ Cep M2 Ia; α Ori M2 Ib; α Her M5 II.

The blue-visible spectral region is very rich in absorption lines and bands, and therefore it provides good classification criteria. The other

Wavelength (A)

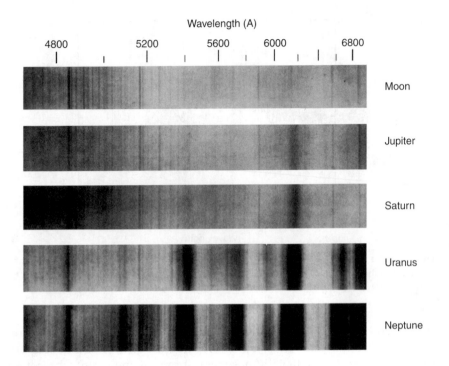

FIGURE 17.14
Low resolution photographic prismatic spectra of the Moon (which simply reflects the Sun's light) and of the gaseous planets with strong molecular absorptions of NH_3 and CH_4.

spectral regions accessible from ground and from Space, namely the UV, the near IR from 7000 Å to 2.5 μm and the far IR, also provide very useful information. The study of the UV spectrum was greatly extended by the long-life satellite International Ultraviolet Explorer (IUE), which operated for 18 years from 1986 until 1994, observing more than 10,000 stars. The Hubble Space Telescope has returneds an exceptional amount of high quality data for a decade. Other orbiting telescopes, such as IRAS, ISO, and Spitzer, have greatly extended the knowledge of IR spectra.

17.7 Low Temperature Stars

In the Harvard classification scheme, additional spectral types were used at low temperatures, indicated by the letters S, N, R. The main difference among them was not temperature itself, but different intensities of different molecular bands. While in M stars the most prominent bands are those of TiO, ScO, and VO, the S-type spectra are dominated by the oxides of zirconium, yttrium, and barium, with evidence also of technetium, a very important element for chronology because its lifetime is of only 2×10^6 years. The S stars often display emission lines, they all are giant, with irregular variability over with long periods.

The N, R stars were later put together in type C of carbon stars, subdivided in C–R (warmer) and C–N (cooler). The spectra of these stars show weak metal oxides, but prominent CN bands at $\lambda\lambda$ 3590, 3883, 4216, 4606. Moreover, the C_2 bands (called also Swan's bands, and visible in emission in cometary spectra) are present, with heads at $\lambda\lambda$ 4737, 5165, 5635 (notice that the intensity of the CN and C_2 bands decline to the blue side, those of TiO and ZrO to the red, with a sharp blue edge). The differences between M and C must be attributed to a different chemical composition, more precisely, to the relative abundances of C and O. Furthermore, the isotopic composition of C stars is peculiar: while on the Earth $^{13}C/^{12}C \approx 1/90$, in C stars $^{13}C/^{12}C \approx 1/4$.

The known population of dwarfs with spectral types later than M has seen an impressive increase in recent times, thanks to the availability of detectors with good sensitivity in the near IR. These very cool and very red dwarfs, intrinsically very faint at optical wavelengths, fall into two new categories: L dwarfs, whose near-IR spectra feature carbon exclusively in the form of CO; and even cooler T dwarfs, whose near-IR spectra show carbon predominantly in methane (CH_4). The L and T spectral types thus make the first substantial addition to the original Harvard and MKK schemes, and complete a continuous spectral sequence from the hottest O stars ($T_{\text{eff}} \approx 50,000$ K) to the coolest known brown dwarfs ($T_{\text{eff}} \approx 750$ K). The Notes provide references to recent work.

In general, the majority of cool stars are variable; a notable example is Mira Ceti (o Cet), the prototype of the Mira Ceti variables, classified as M9e, because for a great part of its light cycle the Balmer series and the lines of Fe II appear in emission. The presence of emission lines in cool stars is also indicative of chromospheric activity, which can be observed not only at optical wavelengths, but also in the x-ray domain (see Notes).

17.8 Peculiarities

Several stars defy a simple classification, requiring additional parameters to describe their spectra properly. For instance, stars with strong lines of Ba, Sr, and rare earths are found among the G and K giants. At higher temperatures, between F5 and B5, there are the so-called peculiar As (Ap). These stars have a strong and rapidly varying magnetic field, with an associated variability of the spectral lines of metals such as Mn, Eu, Cr, Sr, Si. On the other hand, in the same temperature range, Am (A metallic) stars are found, with no strong magnetic field, with underabundance of Ca and great overabundance of heavy elements and rare earths.

A good percentage of the hotter stars show emission lines of H I, Fe II, Si II, Ti II inside the corresponding absorptions. The explanation is the presence of a thin, rotating disk of matter around the star. The appearance of the spectrum also depends on the inclination of such a disk to the line of sight. The hot star P Cyg displays a remarkable feature, namely strong narrow emissions accompanied on the blue side by wide absorptions. The stars with these characteristics, obviously named P-Cyg stars, must be surrounded by a large thick expanding envelope. P Cyg itself has shown large variations in brightness over the past centuries; today it is around the fifth magnitude.

Among the hotter stars, we find the so-called Wolf-Rayet (W stars, the old Harvard types Oa, Ob, Oc), which have $T > 50,000$ K. W stars are very rare in space, but extremely luminous ($M \approx -9$), so that they can be observed even in external galaxies. W stars are also found at the center of planetary nebulae. They are presumably binaries, very old and very massive. Their spectra are characterized by broad emission lines of He I and He II; C II, C III, and C IV; N II, N III, and N IV; O II, O III, O IV, O V, and OVI. It has been found advisable to divide the W type into two subclasses, namely the WN and the WC. The WN display strong lines of N, the WC have strong lines of C and O. The lines of N on one hand, and those of C and O on the other, seem to be mutually exclusive.

White dwarfs are indicated by the prefix D in front of the spectral type, e.g., DB, DA. The spectrum of the DB shows essentially only He lines; that of the DA only the Balmer lines. The difference is too large to be attributable only to a temperature effect, a chemical diversity must also play an important role. Furthermore, the masses are significantly different, those of the DBs being around 0.3, those of the DAs around 0.7 solar masses. There are white dwarfs with essentially continuous, featureless spectra (designates as DCs), and others where the magnetic field is so large, several megagauss, that the spectral lines are exceedingly large due to the Zeeman effect (see Notes).

17.9   Relationship between the MK Classification and Photometric Parameters

We have already seen in Chapter 16 the relationship between the temperature and the color indices of a black body, and the position of a sample of bright stars in the (U–B, B–V) two-color diagram. We can now be more precise, with reference to the MK spectral types. Table 17.9, adapted from the original calibration by Keenan (1963, in the collection Basic Astronomical Data), presents the (U–B, B–V) color indices as a function of the SpT. Table 17.10, adapted from the same source, provides the absolute visual magnitudes and effective temperatures. Keenan's original calibrations have been revised several times, taking advantage of the availability of UV and IR data from orbiting telescopes. One notices that, starting at F2, a dwarf

TABLE 17.9

Values of Color Indices (U–B, B–V) as Function of SpT for the Different Luminosity Classes

SpT	U–B	B–V	SpT	V U–B	V B–V	III U–B	III B–V	Ib U–B	Ib B–V
O5–V	−1.19	−0.33	F0	+0.03	+0.30			+0.15	+0.17
O9–V	−1.12	−0.31	F2	+0.00	+0.35			+0.18	+0.23
B0–V	−1.08	−0.30	F8	+0.02	+0.52			+0.41	+0.56
B2–V	−0.84	−0.24	G2	+0.12	+0.63			+0.63	+0.87
B5–V	−0.58	−0.17	G5	+0.20	+0.68	+0.56	+0.86	+0.83	+1.02
B8–V	−0.34	−0.11	K0	+0.45	+0.81	+0.84	+1.00	+1.17	+1.25
A0–V	0.00	0.00	K5	+1.08	+1.15	+1.81	+1.50	+1.80	+1.60
A2–V	+0.05	+0.05	M2	+1.18	+1.49	+1.89	+1.60	+1.95	+1.71
A5–V	+0.10	+0.15	M5	+1.24	+1.64	+1.58	+1.63	+1.60	+1.80

TABLE 17.10

Absolute Visual Magnitudes M_V and Effective Temperatures T_{eff} of the Different Spectral Types and Luminosity Classes

TSp	V M_v	V T_{eff}	I M_v	I T_{eff}	TSp	V M_v	V T_{eff}	III M_v	III T_{eff}	I M_v	I T_{eff}
O5	−5.7	4,2000			F2	+3.5	7,000			−6.5	7,000
O9	−4.5	34,000	−6.5	32,000	F8	+4.0	6,300			−6.5	5,750
B0	−4.1	30,000			G0	+4.4	5,900			−6.4	5,400
B2	−2.5	21,000	−6.4	17,600	G2	+4.7	5,800			−6.3	5,000
B5	−1.2	15,000	−6.2	13,600	G5	+5.2	5,560	+0.9	5,000	−6.2	4,900
B8	−0.3	11,500	−6.2	11,000	K0	+5.9	5,150	+0.7	4,700	−6.0	4,500
A0	+0.6	9,800	−6.3	10,000	K5	+7.4	4,400	−0.3	4,100	−5.8	4,000
A2	+1.4	9,000	−6.5	9,400	M0	+8.8	3,850	−0.4	3,700	−5.5	3,600
A5	+2.0	8,200	−6.6	8,600	M2	+10.0	3,550	−0.6	3,500	−5.5	3,400
F0	+2.6	7,300	−6.6	7,500	M5	+12.3	3,200	−0.3	3,400	−5.5	2,900

is decidedly hotter than a giant or a supergiant; this is due to the utilization of the spectral lines instead of the continuum as the main criterion for evaluating temperature, as discussed before.

Notes

- For an introduction on atomic spectroscopy see: Martin, W.C., Wiese, W.L., *Atomic Spectroscopy*, http://physics.nist.gov/Pubs/AtSpec/AtSpec.PDF. A nonastronomical but didactically instructive site on visual quantum mechanics is: http://phys.educ.ksu.edu/vqm/index.html

- The literature on astronomical spectroscopy is extremely rich. Here we give only a few of the many available sources.

  Regarding the forbidden lines, Bowen (Bowen, I.S., 1927, The origin of the chief nebular lines, *Publications of the Astronomical Society of the Pacific*, **39**(231), p. 295) was able to identify the so called nebulium lines, so prominent in planetary and diffuse nebulae, as forbidden transitions of O II, O III, and N II. The forbidden lines in the solar corona were identified by Edlén, B., 1942, *Zeitschrift fr Astrophysik*, **22**, p. 30. See also Garstang, R.H., 1995, Radiative hyperfine transitions, *The Astrophysical Journal*, **447**, p. 962.

  A fundamental catalog of reference spectral lines was published by Moore, C.E., 1945, *A Multiplet Table of Astrophysical Interest*, Princeton University Observatory, Princeton, NJ, and in several publications of the National Bureau of Standard.

  Atlases of lines emitted by commercial lamps useful for astronomical purposes are found in all major observatories, for instance in:

  D'Odorico, S., Ghigo, M., Ponz, D., 1987, *An Atlas of the Thorium–Argon Spectrum for the ESO Echelle Spectrograph in the $\lambda\lambda 3400–9000\,\text{Å}$ region*, ESO Scientific Report no. 6.

  A very useful catalog of emission lines from celestial sources, including those due to the night sky, was given by:

  Meinel, A.B., Aveni, A.F., Stockton, M.W., 1968, Catalog of emission lines in astrophysical objects, Tucson.

- The original Harvard (HD) classification is found in:

  Cannon, A.J., Pickering, E.C., *Harvard Observatory Annals*, **91–100**, (1924 on), and its revision in:

  Houck, N., 1975–1988, *An Atlas of Objective Prism Spectra*, University of Michigan, Michigan.

- For the MKK spectral classification:

  Morgan, W.W., Keenan, P.C., Kellman, E., 1943, *An Atlas of Stellar Spectra*, University of Chicago Press, Chicago, IL.

  Morgan, W.W., Keenan, P.C., 1973, Spectral classification, *Annual Review of Astronomy and Astrophysics*, **11**, p. 29.

  Abt, H.A., Meinel, A.B., Morgan, W.W., Tapscott, J., 1968, *An Atlas of Low Dispersion Grating Stellar Spectra*, Kitt Peak National Observatory, Tucson, Arizona.

  Keenan, P.C., McNeil, R.C., 1976, *An Atlas of Spectra of the Cooler Stars*, Ohio State University Press, Ohio.

  Garrison, R.F. ed., 1984, *The MKK Process and Spectral Classification*, David Dunlap Observatory, Toronto.

  Jacoby, G.H., Hunter, D.A., Christian, C.A., 1984, A library of stellar spectra, *Astrophysical Journal Supplement Series*, **56**, p. 257.

- A digital reproduction of the MK and Gray Atlases, together with other interesting material on stellar spectroscopy, can be

found at: http://nedwww.ipac.caltech.edu/level5/ASS/_Atlas/
frames.html

http://nedwww.ipac.caltech.edu/level5/spec_stars.html

- Spectrophotometry of stars can be found for instance in:
  Breger, M., 1976, A catalog of spectrophotometric scans of stars,
  Astrophysical Journal Supplement Series, **32**.
  Hayes, D.S., Oke, J.B., Schild, R.E., 1979, A comparison of the
  Heidelberg and Nbs-Palomar spectrophotometric calibrations,
  Astrophysical Journal, **162**, p. 361.

  The ELODIE library (Ph. Prugniel & C. Soubiran) includes
  1962 spectra of 1388 stars obtained with the ELODIE spectrograph at the Observatoire de Haute–Provence 193 cm telescope
  in the wavelength range 400 to 680 nm. It provides a large
  coverage of atmospheric parameters: T_{eff} from 3000 K to
  60,000 K, log g from -0.3 to 5.9 and [Fe/H] from -3.2 to
  $+1.4$. A detailed description is available at the ELODIE.3
  website:
  http://www.obs.u-bordeaux1.fr/public/astro/CSO/
  elodie_library.html

- The IUE and ISO websites are:
  http://www.vilspa.esa.es/iue/iue.html
  http://www.iso.vilspa.esa.es/

- For IRAS, HST, and Spitzer see:
  Olnon, F.M., Raimond, E., Neugebauer, G., van Duinen, R.J.,
  Habing, H.J., Aumann, H.H., Beintema, D.A., Boggess, N.,
  Borgman, J., Clegg, P.E., Gillett, F.C., Hauser, M.G., Houck, J.R.,
  Jennings, R.E., de Jong, T., Low, F.J., Marsden, P.L., Pottasch, S.R.,
  Soifer, B.T., Walker, R.G., Emerson, J.P., Rowan-Robinson, M.,
  Wesselius, P.R., Baud, B., Beichman, C.A., Gautier, T.N., Harris, S.,
  Miley, G.K., Young, E., 1986, IRAS catalogues and atlases — Atlas
  of low-resolution spectra, *Astronomy and Astrophysics Supplement
  Series*, **65**(4), pp. 607–1065, http://irsa.ipac.caltech.edu/IRASdocs/
  iras.html
  http://www.stsci.edu/institute/
  http://www.spitzer.caltech.edu/

- For a discussion of the hottest stars and associated stellar wind, see
  for instance:
  Nota, A., Pasquali, A., Drissen, L., Leitherer, C., Robert, C., Moffat,
  A.F.J., Schmutz, W., 1996, O stars in transition. I. Optical
  spectroscopy of Ofpe/WN9 and related stars, *Astrophysical Journal
  Supplement*, **102**, p. 383 and paper II, 1997, *The Astrophysical Journal*,
  478, p. 340.

- For a determination of the temperatures of dwarfs see:
  Gray, D.F., Johanson, H.L., 1991, Precise measurements of stellar

temperatures using line-depth ratios, *Publications of the Astronomical Society of the Pacific*, **103**, p. 439.

- For the recent extension to the L and T types see for instance:
  Basri, G., Mohanty, S., Allard, F., Hauschildt, P.H., Delfosse, X., Martín, E.L., Forveille, T., Goldman, B., 2000, An effective temperature scale for late-M and L dwarfs, from resonance absorption lines of Cs I and Rb I, *The Astrophysical Journal*, **538**, pp. 363–385.
  Geballe, T.R., Knapp, G.R., Leggett, S.K., Fan, X., Golimowski, D.A., Anderson, S., Brinkmann, J., Csabai, I., Gunn, J.E., Hawley, S.L., and 19 coauthors, 2002, Toward spectral classification of L and T dwarfs: infrared and optical spectroscopy and analysis, *The Astrophysical Journal*, **564**, p. 466.

- The works of Struve and Elvey (ca. 1930) were central to evaluating the influence of rotation. See:
  Gray, F., 1999, *Stellar Rotation and Precise Radial Velocities*, IAU Colloquium170, J.B. Hearnshaw, C.D. Scarfe, eds.

- For the emissivity in x-rays:
  Micela, G., Sciortino, S., Serio, S., Vaiana, G.S., Bookbinder, J., Glub, L., Hardnen, F.R., Rosner, R., 1985, Einstein x-ray survey of the Pleiades — The dependence of x-ray emission on stellar age, *Astrophysical Journal*, **292**, p. 172.

- For the Stark broadening of hydrogen lines:
  Vidal, C.R., Cooper, J., Smith, E.W., 1973, Hydrogen stark-broadening tables, *Astrophysical Journal Supplement Series*, **25**, p. 37.

- For spectropolarimetry, see for instance:
  Coyne, G.V. Ed., 1987, *Polarized Radiation of Interstellar Origin*, Vatican Observatory Press.
  di Serego, A.S., Cimatti, A., Fosbury, R.A.E., Perez-Fournon, I., 1996, Spectropolarimetry of 3C 265, a misaligned radio galaxy, *Monthly Notices of the Royal Astronomical Society*, **279**, p. L57.

- Examples of spectra of asteroids and near Earth objects have been taken from:
  Fornasier, S., Dotto, E., Barucci, A.M., Barbieri, C., 2004, Water ice on the surface of the large TNO 2004 DW, *Astronomy and Astrophysics*, **422**, pp. L43–46.
  Lazzarin, M., Marchi, S., Barucci, M.A., di Martino, M., Barbieri, C., 2004, Visible and near-infrared spectroscopic investigation of near-Earth objects at ESO: first results, *Icarus*, **169**, pp. 373–384.

- Spectra of comets and planets can be found for example in:
  http://www.fiz.uni-lj.si/astro/cometeso2.html, http://speclab.cr.usgs.gov/planetary.spectra/planetary-sp.html

Bibliography

Textbooks

Bennet, J., Donahue, M., Schneider, N., Voit, M., 1999, *The Cosmic Perspective*, Addison-Wesley Longman Inc., Reading, MA, see http://www.astrospot.com.

Karttunen, H., Kröger, P., Oja, H., Poutanen, M., Donner, K.J., 1987, *Fundamental Astronomy*, Springer, New York.

Proterohe, W.M., Capriotti, G., Newsom, G.H., 1989, *Exploring the Universe*, Merrill, Columbus.

Roth, G.D., 1994, *Compendium of Practical Astronomy*, Vol. 3, Springer, New York.

Roy, A.E., Clarke, D., 2003, *Astronomy, Principles and Practice*, Institute of Physics Publishing, Bristol.

Spherical Astronomy, Astrometry, Celestial Mechanics, Gravitation

Allen, R.H., 1963, *Star Names, Their Lore and Meaning*, Dover, New York.

Baker, J.L., Gollub, J.P., 1996, *Chaotic Dynamics, An Introduction*, 2nd ed., Cambridge University Press, Cambridge.

Bakich, M.E., 1995, *The Cambridge Guide to Constellations*, Cambridge University Press, Cambridge.

Bate, R.R., Mueller, D.D., White, J.E., 1971, *Fundamentals of Astrodynamics*, Dover, New York.

Bertotti, B. Ed., 1974, *Experimental Gravitation*, School of Physics E. Fermi, Academic Press, New York.

Bertotti, B., Farinella, P., 1999, *Physics of the Earth and of the Solar System*, Kluwer Academic Publishers, Dordrecht.

Bertotti, B., Farinella, P., Vokroulický, D., 2003, *Physics of the Solar System*, Kluwer, Dordrecht.

Blanco, V.M., McCuskey, S.W., 1961, *Basic Physics of the Solar System*, Addison Wesley, Reading, MA.

Bohm, D., 1965, *The Special Theory of Relativity*, Benjamin, New York.

Brouwer, D., Clemence, G.M., 1961, *Methods of Celestial Mechanics*, Academic Press, New York.

Ciufolini, I., Wheeler, J.A., 1995, *Gravitation and Inertia*, Princeton University Press, Princeton.

Couteu, P., 1982, *Observing Visual Double Stars*, MIT Press, Cambridge.

Danby, J.M.A., 1988, *Fundamentals of Celestial Mechanics*, Willmann–Bell, Richmond.

Danjon, A., 1980, *Astronomie Generale*, Librairie Blanchard, Paris.

Dick, S., McCarthy, D., Luzum, M., 2000, *Polar Motion: Historical and Scientific Problems, IAU Colloquium 178, ASP Conference Series Vol. 208.*

Eddington, A., 1920, *Space, Time and Gravitation*, Cambridge University Press, Cambridge.

Eichhorn, H., 1974, *Astronomy of Star Positions*, Frederick Ungar, New York.

Einstein, A., 1905, Zur Elektrodynamik bewegter Koerper, *Ann. d. Phys.*, **17**, p. 891.

Einstein, A., 1915, Zur allgemeinen Relativitätstheorie, *Sitzungsberichte der Königlich Preußischen Akademie der Wissenschaften (Berlin), Seite*, pp. 778–786.

Green, R., 1985, *Spherical Astronomy*, Cambridge University Press, Cambridge.

Kowalesky, J., 1995, *Modern Astrometry*, Springer, Berlin.

Landau, L., Lifschitz, E., 1982, *Mechanics*, 4th Ed. MIR, Moscow.

Laskar, J., Joutel, F., Robutel, P., 1993, Stabilization of the earth's obliquity by the moon, *Nature*, **361**(6413), pp. 615–617.

Luyten, W.J., 1963, *Basic Astronomical Data*, K.A. Strand, Ed., University of Chicago Press, Chicago.

Møller, C., 1962, *The Theory of Relativity*, Oxford at Clarendon Press, London.

Mueller, I.I., 1969, *Spherical and Practical Astronomy as Applied to Geodesy*, Frederick Ungar, New York.

Murray, C.A., 1983, *Vectorial Astrometry*, Adam Hilger, Bristol.

Murray, C.D., Dermott, S.F., 1999, *Solar System Dynamics*, Cambridge University Press, Cambridge.

Newcomb, S., 1906, *A Compendium of Spherical Astronomy*, Dover, New York.

Nieto, A., 1972, *The Titius–Bode Law of Planetary Distances: Its History and Theory*, Pergamon Press, New York.

Roy, A.E., 1979, *Orbital Motion*, Adam Hilger, Bristol.

Schwarszchild, K., 1916, Berlin Berichte, p. 189.

Smart, W.M., 1965, *Text Book on Spherical Astronomy*, Cambridge University Press, Cambridge.

Smart, W.M., 1953, *Celestial Mechanics*, Longmans and Green, London.

Sobel, D., 1996, *Longitudes*, Penguin Books, Baltimore.

Thomson, W.T., 1986, *Introduction to Space Dynamics*, Dover, New York.

Will, C.M., 1993, *Theory and Experiment in Gravitational Physics*, Cambridge University Press, Cambridge.

Walter, H.G., Sovers, O.J., 2000, *Astrometry of Fundamental Catalogues*, Springer, Berlin.

Woolard, E.W., Clemence, G.M., 1966, *Spherical Astronomy*, Academic Press, New York.

Astrophysics

Aller, L.H., 1991, *Atoms, Stars and Nebulae*, Cambridge University Press, Cambridge.

Atreya, S.A., 1986, *Atmospheres and Ionospheres of the Outer Planets and Their Satellites*, Springer, Berlin.

Binney, J., Merrifield, M., 1998, *Galactic Astronomy, Princeton Series in Astrophysics*, Princeton University Press, Princeton.

Böhm-Vitense, E., 1989, *Stellar Astrophysics*, Vol. 3, Cambridge University Press, Cambridge.

Chandrasekhar, S., 1960, *Radiative Transfer*, Dover, New York.

Cowley, C.R., 1995, *An Introduction to Cosmochemistry*, Cambridge University Press, Cambridge.

Herzberg, G., 1944, *Atomic Spectra and Atomic Structure*, Dover, New York.

Herzberg, G., 1950, *Spectra of Diatomic Molecules*, Van Nostrand, Princeton, NJ.

Gray, D.F., 1976, *The Observation and Analysis of Stellar Photospheres*, Wiley/Cambridge University Press, New York.

Jaschek, C., Jaschek, M., 1990, *The Classification of Stars*, Cambridge University Press, Cambridge.

Kaler, J.B., 1997, *Stars and Their Spectra: An Introduction to the Spectral Sequence*, Cambridge University Press, Cambridge.

Kraus, J.D., 1966, *Radio Astronomy*, McGraw-Hill, New York.

Mendillo, M., Nagy, A., Waite, J.H., Eds., 2002, Atmospheres in the Solar System: Comparative Aeronomy, *American Geophysical Union*, Washington.

Mihalas, D., Binney, J., 1981, *Galactic Astronomy: Structure and Kinematics*, W.H. Freeman, San Francisco.

Minnaert, M.G.J., 1993, *Light and Color in the Outdoors*, Springer, New York.

Osterbrock, D.E., 1974, *Astrophysics of Gaseous Nebulae*, W.H. Freeman, San Francisco.

Pauling, L., Goudsmit, S., 1930, *The Structure of Line Spectra*, McGraw-Hill, New York.

Schatzman, E., Praderie, F., 1990, Les Etoiles, *InterEditions et Editions du CNRS*, Paris.

Shu, F.H., 1991, *The Physics of Astrophysics*, University Science Books, Mill Valley.

Straughan, B., Walter, S., 1976, *Spectroscopy*, Chapman and Hall, London.

Thorne, A.P., 1974, *Spectrophysics*, Chapman and Hall, London, (Halsted Press, New York).

Trumpler, R.J., Weaver, H., 1953, *Statistical Astronomy*, Dover, New York.

Unsold, A., 1978, *The New Cosmos*, Springer, Berlin.

van de Hulst, H., 1957, *Light Scattering by Small Particles*, Dover, New York.

White, H.E., 1944, *Atomic Spectra*, Dover, New York.

Zirin, H., 1988, *Astrophysics of the Sun*, Cambridge University Press, Cambridge.

Astronomical Almanacs, Coordinates, Ephemerides and Their Computations

Supplement to the Astronomical Almanac, 1984, U.S. and H.M. Nautical Almanac Offices, Washington and London.

Astronomical Almanac, 1984, U.S. and H.M. Nautical Almanac Offices, Washington and London.

Astronomical Almanac for Computers, 1988, U.S. Nautical Almanac Office, U.S. Government Printing Office, Washington.

Explanatory Supplement to the Astronomical Ephemeris, 1961, H.M. Nautical Almanac Office, HM Stationary Office, London, (corrected imprint 1977).

Seidelmann, K.P., Ed., 1992, *Explanatory Supplement to the Astronomical Almanac, Nautical Almanac Offices*, University Science Books, Mill Valley.

Apparent Places of Fundamental Stars, Astronomisches Rechen-Institut, Heidelberg.

Multiyear Interactive Computer Almanac, 1989 and 2005, US Naval Observatory, Willmann–Bell, Richmond (from 1800 to 2050).

de Vaucouleurs, G., Peters, W.L., 1984, The dependence on distance and redshift of the velocity vectors of the sun, the galaxy and the local group with respect to different extragalactic frames of reference, *The Astrophysics Journal,* **287**, pp. 1–16.

Duffett-Smith, P., 1992, *Practical Astronomy with your Personal Computer,* Cambridge University Press, Cambridge.

Duffett-Smith, P., 1992, *Practical Astronomy with Your Calculator,* Cambridge University Press, Cambridge.

Franz, M., 2002, *Heliospheric Coordinate Systems,* http://www.space-plasma.qmul.ac.uk/heliocoords/

Hapgood, M., 1992, Space physics coordinate transformations: A user guide, *Planetary and Space Science,* **40**, pp. 711–717, Corrigendum, Space physics coordinate transformations: A user guide, *Planetary and Space Science,* **45**(8), p. 1047.

Hapgood, M., 1995, Space physics coordinate transformations: The role of precession, *Annals of Geophysics,* **13**(7), pp. 713–716.

Hapgood, M., 2002, http://sspg1.bnsc.rl.ac.uk/Share/Coordinates/ct_home.htm

IAU Division I: Fundamental Astronomy, http://syrte.obspm.fr/iaudiv1/

IAU RESOLUTIONS: http://syrte.obspm.fr/IAU_resolutions/Resol-UAI.htm

Meeus, J., 1991, *Astronomical Algorithms,* Willmann–Bell, Richmond.

Simon, J., Bretagnon, P., Chapront, J., Chapront-Touzé, M., Francou, G., Laskar, J., 1994, Numerical expressions for precession formulas and mean elements for the moon and the planets, *Astronomy and Astrophysics,* **282**(2), pp. 663–683.

Standish, E., 1990, The observational basis for JPL's DE 200, the planetary ephemerides for the Astronomical Almanac, *Astronomy and Astrophysics,* **233**, pp. 252–271.

Standish, E., 1998a, *JPL, Planetary and Lunar Ephemerides, DE405/LE405.* Interoffice memorandum IOM312.F, JPL, Los Angeles.

Standish, E., 1998b, Time scales in the JPL and CfA ephemerides, *Astronomy and Astrophysics,* **336**(1), pp. 381–384.

Taff, L.G., 1985, *Computational Spherical Astronomy,* Wiley, New York, and 1991, Krieger Publishing Company, Malabar.

Van Flandern, T.C., Pulkkinen, K.F., 1979, Low precision formulae for planetary positions, *Astrophysics Journal,* **41**(Suppl.), pp. 391–411.

Catalogs, Positions, Finding Charts

1. Few classical stellar catalogs:

Aitken, R.G., 1935, *The Binary Stars,* Dover, New York.

Fricke, W., Schwan, H., Lederle, T., 1988, *Fifth Fundamental Catalogue FK5,* Braun, Karlsruhe.

Gliese, W., 1969, *Catalogue of Nearby Stars,* Braun, Karlsruhe.

Gliese, W., Jahreiss, H., 1979, Nearby star data published 1969–1978, *Astronomy and Astrophysics,* **38**(Suppl.), p. 423.

Hoffleit, D., Jascheck, C., 1984, *The Bright Stars Catalogue,* Yale University Observatory, New Haven.

Luyten, W.J., 1963, Proper motion surveys, *Basic Astronomical Data,* K.A. Strand, ed., University of Chicago Press, Chicago.

PPM Star Catalogue (positions and proper motions of 181731 stars North of −2.5 deg), Astronomisches Rechen–Institut, Heidelberg.

SAO Star Catalogue, 1963, Cambridge, MA.

The Hipparcos and Tycho Catalogues, 1997, ESA SP-1200, 17 Volumes — with 14 contributions by Fabricius, C., Høg, E., Makarov, V.V., see: http://astro.estec.esa.nl/SA-general/Projects/Hipparcos/hipparcos.html

The Tycho Reference Catalog (*Astronomy and Astrophysics*, 1998) has been replaced completely by the Tycho-2 Catalog: An astrometric and photometric reference catalogue of the 2.5 million brightest stars on the entire sky (*Astronomy and Astrophysics* 355.2, P L19-L22, 2000). See: http://www.astro.ku.dk/%7Eerik/Tycho-2/ (from Høg, E. homepage, last updated 11 February 2002).

The Hubble Space Telescope Institutes (http://www.stci.edu) provides the following Catalogs and Surveys: Guide Star Catalog; Digitized Sky Survey; Guide Star Photometric Catalog; Cataclysmic Variables; Spectra of Nearby Galaxies.

The US Naval Observatory (USNO, http://ad.usno.navy.mil/; it can be accessed also through The Canadian Astronomy Data Centre http://cadcwww.hia.nrc.ca/usno/) provides:

- The on-line USNO-A V1.0 Catalog, of 488,006,860 sources whose positions can be used for astrometric references, compiled by David Monet and collaborators.
- The USNO-A2.0 Catalog of 526,280,881 stars, based on a re-reduction of the Precision Measuring Machine (PMM) scans that were the basis for the USNO-A1.0 catalog. The major difference between A2.0 and A1.0 is that A1.0 used the Guide Star Catalog (Lasker et al., 1990, The guide star catalog. I-Astronomical foundations and image processing, *Astronomical Journal*, **99**, pp. 2019–2058) as its reference frame whereas A2.0 uses the ICRF as realized by the USNO ACT catalog (Urban et al., 1997, *Bulletin of the American Astrological Society*, **29**, p. 1306).
- UCAC2, now available also from (Vizier, Strasbourg) as entry I/289, and from Project Pluto (http://www.projectpluto.com/ucac2.htm).

2. The Strasbourg astronomical Data Center (CDS, http://cdsweb.u-strasbg.fr) collects and distributes astronomical data catalogs, related to observations of stars and galaxies, and other galactic and extragalactic objects. Catalogs about the solar system bodies and atomic data are also included.

The catalogs and tables managed by CDS can be summarized as follows:

5620 Catalogs available from CDS, of which 4581 are available on-line (as full ASCII or FITS files), and 4264 are also available through the VizieR browser.

3. Digital Images and archives of data.

Through the web sites of the HST and of ESO one can access the digitized versions of the Sky Surveys (DSS) covering the entire celestial sphere.

Another useful web site is The Canadian Astronomy Data Centre http://cadcwww.hia.nrc.ca, which contains a wealth of archives (even from nonoptical telescopes), and the ESO Skycat tool, that combines visualization of images and access to catalogs and archive data for astronomy.

The NASA Extragalactic Database NED (http://nedwww.ipac.caltech.edu/) gives coordinates, magnitudes, redshifts, cross references, finding charts, etc., for 7,600,000 extragalactic objects, as well as *thesis abstracts* of doctoral dissertations on extragalactic topics.

4. Catalogs and finding charts are available also through: http://skyview.gsfc. nasa.gov/

SkyView is a Virtual Observatory on the Net generating images of any part of the sky at wavelengths in all regimes from Radio to Gamma-Ray.

5. For solar system objects:

http://asteroid.lowell.edu provides asteroid observing services, including finding charts with respect to USNO stars.

The IUA Commission 4 (Ephemerides) is presently located in http://danof.obspm.fr/ iaucom4 and provides a List of the Ephemerides centers with their products.

The JPL site http://ssd.jpl.nasa.gov/iau-comm4/ was the previous site of IAU Commission 4; it still gives information about Planetary and Lunar Ephemerides.

The tool http://ssd.jpl.nasa.gov/horizons.html: The JPL HORIZONS *On-Line* Solar System Data and Ephemeris Computation Service provides access to key solar system data and flexible production of highly accurate ephemerides for solar system objects (163000 + asteroids and comets, 128 natural satellites, 9 planets, the Sun, L1, L2, select spacecraft, and system barycenters).

Constants, Formulae, Tables

Abramowitz, M., Stegun, I.A., 1965, *Handbook of Mathematical Functions*, Dover, New York.

Allen, C.W., 2000, *Astrophysical Quantities*, 4th ed., A.N. Cox, Editor In Chief, AIP, New York, (the fourth edition has been thoroughly revised and enlarged, available also on CD-ROM).

Anderson, H.L., (Editor In Chief) 1989, *A Physicist's Desk Reference*, AIP, New York.

Lang, K.R., 1974, *Astrophysical Formulae*, Springer, Berlin.

Lang, K.R., 1991, *Astrophysical Data: Planets and Stars*, Springer, Berlin.

Stern, D.P., 2003, A Math Refresher, http://www-spof.gsfc.nasa.gov/stargaze/ smath.htm

Zombeck, M.V., 1990, *Handbook of Space Astronomy and Astrophysics*, 2nd ed., Cambridge University Press, Cambridge, can be accessed on line.

Earth Rotation, Tides, Clocks

1. Classic books, which provide a starting point for all further study:

Munk, W.H., MacDonald, G.J.F., 1960, *The Rotation of the Earth: A Geophysical Discussion*, Cambridge University Press, Cambridge.

Lambeck, K., 1980, *The Earth's Variable Rotation: Geophysical Causes and Consequences*, Cambridge University Press, Cambridge.

2. Few recent citations of tidal effects and variations in Earth rotation:

Brosche, P., Seiler, U., Suendermann, J., Wuensch, J., 1989, Periodic changes in the Earth's rotation due to oceanic tides, *Astronomy and Astrophysics*, **220**, pp. 318–320.

Chao, B.F., Ray, R.D., 1997, Oceanic tidal angular momentum and Earth's rotational variations, *Progress in Oceanography*, **40**, pp. 399–422.

Eubanks, T.M., 1993, Variation in the orientation of the Earth, *Contributions of Space Geodesy to Geodynamics: Earth Dynamics*, American Geophysical Union, Washington, pp. 1–54.

Gross, R.S., 1993, The effect of ocean tides on the Earth's rotation as predicted by the results of an ocean tide model, *Geophysical Research Letters*, **20**, pp. 293–296.

Herring, T.A., Dong, D., 1994, Measurement of diurnal and semidiurnal rotational variations and tidal parameters of the Earth, *Journal of Geophysical Research*, **99**, pp. 18051–18071.

Ray, R.D., Steinberg, D.J., Chao, B.F., Cartwright, D.E., 1994, Diurnal and semidiurnal variations in the Earth's rotation rate induced by oceanic tides, *Science*, **264**, pp. 830–832.

3. Future Clocks: http://clockdev.usno.navy.mil/ion.html

4. Earth Rotation Services:

IVS International VLBI Service for Geodesy and Astrometry: http://ivscc.gsfc.nasa.gov/. IAU Commission 19 was originally established as "Latitude Variation", and later renamed "Rotation of the Earth".

The Earth Orientation Department of the U.S. Naval Observatory determines and predicts the time-varying alignment of the Earth's terrestrial reference frame with respect to the celestial reference frame. The Earth Orientation Parameters (EOP) determined are the coordinates of the Earth's pole (polar motion), the rotation angle about the pole (Universal Time or UT1), precession, and nutation. Observational results are published regularly in the *IERS Bulletin A*.

http://www.iers.org/

http://maia.usno.navy.mil/

http://hpiers.obspm.fr/

EOP Product Center provides Earth orientation parameters for:
- Terrestrial and space navigation/positioning
- Artificial satellite orbit determination
- Astronomical and geophysical research

ICRS Product Center maintains the International Celestial Reference System (ICRS) for:

- International Astronomical Union needs
- Earth orientation determination
- Astronomical and geophysical research

Tides: http://bowie.gsfc.nasa.gov/ggfc/tides/intro.html

Collections, Encyclopedias, Web Sites

The student will find most useful sources of information in Collections and Encyclopedias, few of which are here nominated:

Acta of the IAU (Highlights, Symposia and Colloquia):
http://www.iau.org/IAU/Activities/publications/
Annual Review of Astronomy and Astrophysics (Ann Rev. Inc.).
Annual Review of Earth and Planetary Sciences (Ann. Rev. Inc.).

Basic Astronomical Data, University of Chicago Press (a collection of 12 old but still useful volumes).

Conference Series Volumes of the Astronomical Society of the Pacific.

New Astronomy Reviews (formerly known as Vistas in Astronomy), Elsevier B.V.

Encyclopedia of Astronomy and Astrophysics Online (P. Murdin, Editor in Chief) http://eaa.iop.org/index.cfm?action = home.

Shirley, J.H., Fairbridge, R.W., Eds., 1997, _Encyclopedia of Planetary Sciences_, Kluwer, Dordrecht, with a CD-ROM of planetary images.

Mark, H., Ed., 2003, _Encyclopedia of Space Science and Technology_, Wiley InterScience, New York, http://www3.interscience.wiley.com/cgi-bin/mrwhome/104554787/HOME

Weissman, P.R., McFadden, L.A., Johnson, T.V., 1999, _Encyclopedia of the Solar System_, Academic Press, New York.

Landolt-Bornstein Group 6, _Astronomy and Astrophysics_ (1965 with supplements and extensions), Springer, Berlin, can be accessed on line http://www.springeronline.com/sgw/cda/frontpage/0,11855,4-10113-0-0-0,00.html

Maran, S.P., Ed., 1992, _The Astronomy and Astrophysics Encyclopedia_, Van Nostrand Reinhold, New York, foreword by C. Sagan.

Beatty, J.K., Petersen, C.C., Chaikin, A., 1999, _The New Solar System_, Sky Publishing Co./Cambridge University Press, Cambridge.

The _NASA Astrophysics Data System_ (ADS, The Digital Library for Physics, Astrophysics, and Instrumentation) provides a powerful search engine for papers published in all major journals: http://adswww.harvard.edu/

Many web sites of Astronomical Departments and Institutions contain extremely useful and updated papers, lecture notes and resources.

Two examples:

1. The Observatories of the Carnegie Institution of Washington, (http://www.ociw.edu/) under "Astrophysics Series".

2. The CalTech (http://www.astro.caltech.edu/resources/software.html#ASTRO) provides several useful software and tools.

A collection of pointers to astronomically relevant Internet resources (including professional and amateurs organizations) can be found in AstroWeb: http://cdsweb.u-strasbg.fr/astroweb/

Most web sites of the major astronomical institutions provide convenient links to astronomical resources, for instance through the ESO library:

http://www.eso.org/gen-fac/libraries/datacenters.html

The ESO site (http://www.eso.org/science/scisoft/) provides also _Scisoft_, a collection of astronomical software utilities, mostly public domain tools developed outside ESO (for Linux only). Major data-analysis packages (e.g., IRAF/STSDAS, ESO-MIDAS and IDL) are included, as well as many smaller utilities.

Timely information on variable stars, supernovae, comets, asteroids, etc., can be found in:

IAU Central Bureau for Astronomical Telegrams, Circulars and Minor Planets Electronic Circulars:

http://cfa-www.harvard.edu/iau/cbat.html

http://cfa-www.harvard.edu/iauc/RecentIAUCs.html

http://cfa-www.harvard.edu/mpec/RecentMPECs.html

For Gamma Ray Bursters (GRB): http://gcn.gsfc.nasa.gov/gcn3_circulars.html

Preprints are usually found in: http://arXiv.org, http://xxx.lanl.gov/archive/astro-ph

Index